William Henry Bragg

William Lawrence Bragg

Howard Walter Florey

Frank Macfarlane Burnet

John Carew Eccles

Aleksandr Mikhailovich Prokhorov

Bernard Katz

Patrick White

John Warcup Cornforth

Peter Charles Doherty

ABIE

Australia's Nobel Laureates is an imprint of *ABIE*,
Australia's annual journal of business, technology and innovation.

Published through ETN Communications Pty Ltd.
ISBN 0-9580207-2-8
Ph: (02) 9410 3660 Fax: (02) 9410 3662.
14 Bancroft Avenue, Roseville NSW 2069, Australia.

Editorial and commercial enquiries welcome: info@etncom.com

Printed and bound by Everbest Printing Co Ltd Hong Kong.

Australia's Nobel Laureates

ADVENTURES IN INNOVATION

ABIE

from the Prime Minister

One of Australia's most valuable assets is the great capacity of our people to innovate and adapt to change.

Our history and achievements provide proof that Australians are resourceful and flexible. In today's increasingly globalised world, these qualities are even more important.

Australia's Nobel laureates illustrate the power of curiosity, a willingness to take on big challenges and the benefits of initiative and perseverance.

Nine out of the 10 Nobel Prizes awarded to laureates from Australia have been in the fields of science and medicine. They are true Australian heroes who deserve wide recognition for their magnificent endeavours.

It is important to raise the profile of our successful innovators and pioneers in the scientific community.

The publication *Australia's Nobel Laureates* celebrates their work.

The inclusion of contemporary profiles of innovation demonstrates how companies and institutions are adapting and innovating to meet new challenges.

Innovative techniques have applications for all aspects of the Australian economy, including finance, education, human resources, mining, manufacturing, design and construction, science, medicine and agriculture.

Investment in innovation is crucial to secure our economic future. Backing Australia's Ability – the Federal Government's $3 billion commitment to innovation over five years – is built on pursuing these endeavours.

This degree of financial commitment heightens our capacity to turn great ideas into real jobs and wealth for Australia.

Australia's Nobel Laureates will assist in the nationwide recognition of the economic and social benefits flowing from scientific advances and innovation.

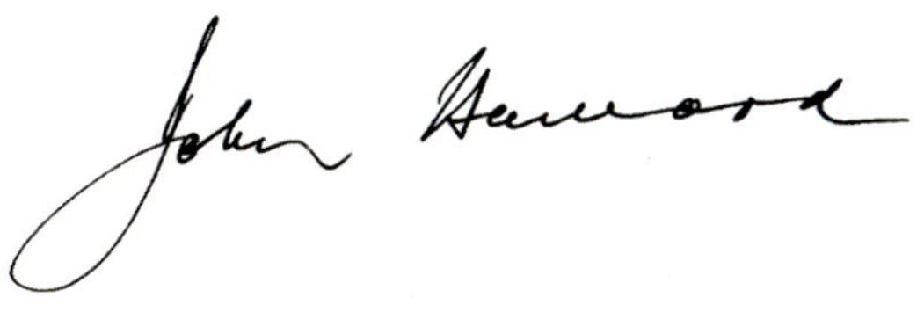

John Howard
PRIME MINISTER

Editorial

EDITORIAL

EDITOR-IN-CHIEF
John Keeney

MANAGING EDITOR
Valerie Khoo

ASSOCIATE EDITOR
Gayle Bryant

PUBLISHER
Ashley Russell

ASSOCIATE PUBLISHER
Heather Boothroyd

DESIGN
Scribbles Art & Design

PRODUCTION EDITORS
Katrina Lobley
Luisa Volpato
Tim Ireland

WRITERS
Kristin Austin-Bird
Kevin Bailey
Robin Batterham
Bob Beale
Karen Boalch
Jane Breusch
Gayle Bryant
Kevin Childs
Brad Collis
Ian Cooke
Julian Cribb
Lucio Di Bartolomeo
Tracey Evans
Geoff Garrett
Shirley Glaister
Kathy Graham
Julie Hare
Helen Hewitt
Mark Hollands
Cathy Johnson
Alan Jones
Mark Kestigian
Valerie Khoo
Katrina Lobley
Roger Lough
Rosemary Lynch
Jan McGuinness
Tim Murphy
Gustav Nossal
Edward O'Daly
Alan Pettigrew
Phil Robertson
David Salt
Anthony Schmidt
Nic Svenson
Lisa Tait
Colin Taylor
Abbie Thomas
Tim Treadgold
Stephan Wellink
Judith Whitworth
David Woolbank
Paul Wright

from the editor-in-chief

Australia has the highest per capita achievement of Nobel awards in the world – there have been nine Nobel laureates in science and one in literature. Incredibly, a book has never been published on these fascinating achievers. Our Nobel laureates deserve widespread recognition alongside the more numerous sporting champions and artists prominent in the public mind. A broader appreciation of them, and all that flows from this, can improve not only our economy but also our culture.

As publishers of the *Australian Business and Investment Explorer* (*ABIE*), the journal for "every thinking Australian", we voiced our commitment to the role of science, innovation and technology as vital partners to business. During the production of *ABIE*'s Volume Two, we conducted an informal survey of more than 200 senior executives on their knowledge of Australia's Nobel awards. The result? An awareness level of less than 3 per cent. This figure has meaning.

We were inspired to present the third edition of *ABIE* in a different form and style.

Part one offers biographies of our Nobel laureates, which examine both their lives and their work. Edited for a general audience, we sought to present them whole to demonstrate, among other things, their role as fellow Australians. To demystify their contributions and lives, so as to enlighten our readers. And yes, unashamedly, to make them more "popular". As they should be.

There is value in perceiving them this way. While very talented, it was commitment and hard work – often in times of great difficulty – which drove their success.

No school child should feel that the achievement of a Nobel award is beyond his or her grasp. Our most recent laureate, Peter Doherty, started his career modestly, with ambitions to be a rural veterinarian. Curiosity, fascination and a love of endeavour fuelled his steady progress. There are other wonderfully human stories here which were played out in different countries across the whole of the 20th century – but all are inextricably tied to Australia.

All of the Nobel laureates were great innovators. In part two, we project this theme across a wider canvas, examining more than 65 areas of innovation occurring today in this country – in science, but also in education, architecture, project management, finance, manufacturing, medicine, electronics, accounting, IT, mining and government policy. This is, to our knowledge, the largest collection of innovation profiles ever assembled.

With the business audience in mind, we added articles from corporate leaders and other innovation specialists – those who have made new thinking come alive within organisations, those who have used management to achieve it, and others who have formally and rigorously studied how to make innovation spark and grow, thereby enriching all our institutions – and our lives.

John Keeney
Editor-in-chief

Project Leaders

3M Australia

Australia Post

Australian Business Limited

Australian Industry InnovationXchange Network

National Australia Bank
Bligh Voller Nield
Bovis Lend Lease
General Property Trust
Lend Lease

Canon Australia Pty Ltd

Dupont (Australia) Ltd

Grains Research and Development Corporation

IDP Education Australia

Johnson & Johnson Family of Companies

National Health and Medical Research Council

The Pharmacy Guild of Australia

PricewaterhouseCoopers

University of Technology, Sydney

sponsors

Lead Sponsors

ADI Limited

AMCOR Australasia

ANSTO (Australian Nuclear Science and Technology Organisation)

Australian Centre for International Agricultural Research

Bayer CropScience

Bishop Technology Group Limited

Cochlear Limited

The Como Melbourne

Commonwealth Bank

CSIRO

Defence Science and Technology Organisation (DSTO)

Electrolux Home Products Pty Ltd

Ergon Energy Pty Ltd

Financial Planning Association of Australia

Gartner

Invetech Pty Ltd

Macquarie Bank Limited

Mayne Health Diagnostic Imaging

National ICT Australia

Orica Limited

ResMed Ltd

RMIT University

Spotless Group Limited

Thales Underwater Systems

The Walter and Eliza Hall Institute of Medical Research

WMC Resources Ltd

Sponsors

Aon

ARRB Transport Research Ltd

ATP Innovations Pty Ltd

Australian Academy of Science

Australian Bureau of Statistics

Australian Genome Research Facility Ltd

Australian Institute for Commercialisation

Australian Institute of Marine Science

Australian National University

Australian Proteome Analysis Facility Ltd

Australian Research Council

Australian Trade Commission (Austrade)

BioMelbourne Network

BlueScope Steel

Boral Limited

Brisbane Airport Corporation Limited

Curtin University of Technology, Office of Research & Development

Department of the Premier & Cabinet, Office of Science and Innovation, Government of Western Australia

Department of Primary Industries, Government of Victoria

Drake International Pty Ltd

Edison Mission Energy Holdings Pty Ltd

EMC Corporation

Griffith Hack

Horticulture Australia Limited

Hydro Tasmania

Feelike Pty Ltd

The Institute of Chartered Accountants in Australia

The Institute of Patent and Trade Mark Attorneys of Australia

IP Australia

John Curtin School of Medical Research

The Macfarlane Burnet Institute for Medical Research and Public Health (Burnet Institute)

Orbital Engine Corporation

Pacifica Group Technologies Pty Ltd

RABiT (Research Centre for Advanced By-Wire Technologies)

The Queensland Institute of Medical Research

Science, Technology and Innovation Directorate, Department of Further Education, Employment, Science and Technology, Government of South Australia

SC Johnson & Sons Pty Ltd

Standards Australia

Starpharma Pooled Development Limited

Steritech Pty Ltd

Swinburne University of Technology

Telstra Corporation Limited

Titan Resources NL

University of Ballarat

The University of Queensland

University of Southern Queensland

The University of Sydney

University of Western Sydney

Advisory board

We thank our advisory board for their ideas, contributions, insights and commitment to the promotion of Australian innovation.

PATRON:

Dr Geoff Garrett
Chief Executive Officer – CSIRO

ADVISORY BOARD:

Mr David Miles
Chairman – National Innovation Awareness Strategy

Dr Fred Bradner
Technical Director – 3M Australia

Professor Sue Serjeantson
Executive Secretary – Academy of Science

Mr Rowan Gilmore
Chief Executive Officer – Australian Institute for Comercialisation

Mr Rob Durie
Executive Director – Australian Information Industry Association

Mr Dennis Trewin
Chief Statistician – Australian Bureau of Statistics

Mr Grant Kearney
Chief Executive – Australian Industry, InnovationXchange

Professor Ian W Chubb
Vice-Chancellor and President – Australian National University

Dr Colin Grant
Deputy Executive Director – Bureau of Rural Sciences

Professor Phil Robertson
Director – Canon Information Systems Research Australia

Mr Bob Lim
Executive Director – CEDA

Dr John Parker
Chief Technology Officer – Cochlear

Ms Janine Kirk
Executive Director – Committee for Melbourne

Dr Ian Chessell
Chief Defence Scientist – Defence Science & Technology Organisation

Dr Alan Jones
Director Innovation Analysis
– Department of Industry Tourism & Resources

Mr Steve Redman
Managing Director – EMC

Dr Michael Elliott
Medical Director – Glaxo Smith Kline

Dr Steven Duvall
Director, Strategic Investment – INTEL

Mr Alan McElrea
Managing Director – MINCOM Ltd

Ms Andrea Galloway
Group Managing Director – Spherion Technology

Dr Hugh Bradlow
Managing Director and Chief Technology Officer
– Telstra Research Laboratories

Professor Deryck Schreuder
Vice-Chancellor – University of Western Australia

I must offer a dedication to four people with a long and passionate interest in our Nobel laureates.

To Dr Barry Jones who first mentioned their importance to me over 20 years ago and fostered knowledge of them tirelessly.

Also to Dr Stephan Wellink who has long been writing a book on them and has provided special inspiration and assistance to this publication.

And to A.D'A & D.D'A.Webling who published a fine monograph on the subject in 2000.

JOHN KEENEY
EDITOR-IN-CHIEF

thank you

ETN Communications would like to thank the following individuals. It is through their support and initiative that this edition of ABIE has been possible.

Brad Allen
Tony Andreeski
Sylvia Atiken
Victor Avrutis
Dr Mark Baker
Louise Banks
Brian Barrett
Paul Bidgood
Dr Mark Bradley
Dr Fred Bradner
Helen Brain
Heather Briggs
David Brookes
Louise Brooks
Gary Butler
Nick Campbell
Jim Carden
Dr Julian Clarke
Ric Clark
Colleen Clur
Shaun Coffey
Maree Conway
Professor Kerry Cox
Paul Cull
Ashley Deacon
Lucio Di Bartolomeo
Andrew Dingjan
Professor Graham Durant
Lance Emerson
Lars Erikson
Dr Peter Farrell
Paul Field
Dr Sue Forrest
Professor Neil Furlong
Alex Galante
Warrick Gilbert
Michael Givoni
Dr Barney Glover
Tom Godfrey
Professor Michael Good
Bruce Grey
James Grose

thank you

Kate Gunn
Peter Hamsworth
Nigel Harrison
Stephen Harrison
Sandi Harwood
Karen Hellwig
Maurice Hermann
Troy Hey
Jeff Higgins
Dr Bruce Hobbs
Pip Hoyle
Lindy Hyam
Paul Hyslop
Martine Jaeger
Ken Johnson
Professor Lesley Johnson
Louise Kanis
Michael Kavanagh
Grant Kearney
Ron Keeley
Cleve Killiby
Rosemary Kirkby
Gary Kichenside
Harry Kollatz Jnr
Hugh Lamberton
Jill Lester
Sue Lewington
Hugh Liney
Anne Long
Professor John Lovett
Jane MacAulay
Colin Macauley
Kevin MacDonald
Anna Manzoney
Chris Masuark
Keith Maxted
Andrew Miedler
Theresa Millard
Audrey Muire
Tim Murphy
Stewart Murrihy
Jag Nangla
Oona Nielssen
Lee-anne Norris
Ian Olton
Madan Patel
Craig Pearce
Professor Neil Phillips
Mark Plaisted
Andrew Potter
Hutch Ranck
Dr George Riffkin
Umberto Righetti
Martin Ritchie
Harry Rodden
Kylie Rubans
Bill Ryan
Karilyn Ryan
Dr Ron Sandland
Peter Schultz
Beth Sibly
Professor Sue Serjeantson
Fiona Skivington
Kay Smith
Colin Taylor
Kylie Taylor
Ray Taylor
Sandra Taylor-Bowman
Deniz Tek
Euan Thomas
Clare Verberne
Stephen Walter
Steve Wesselingh
George West
Julie White
Trina White
Professor Judith Whitworth
Hans Wijgh
Peter Wilson
Lindsay Wood
Stephen Woodhill
Lindsay Woodland
David Woolbank
Robert Wulff
Rob Younger
Bias Electronics Sydney
Raco Town Cars Melbourne

Give me a place to stand –

and I will move the Earth.

ARCHIMEDES

Contents

LEADERS SPEAK reveals insights on innovation from some of Australia's leading business and scientific minds. These are found throughout the book.

inventas vitam juvat
excoluisse per artes.

INSCRIPTION ON THE MEDALS WON
BY AUSTRALIA'S NOBEL LAUREATES

Inventions enhance life,
which is beautified through art.
(translation)

ABIE

Australia's Nobel laureates

By Stephan Wellink

WILLIAM HENRY BRAGG
Nobel Prize in Physics 1915

WILLIAM LAWRENCE BRAGG
Nobel Prize in Physics 1915

HOWARD WALTER FLOREY
Nobel Prize in Physiology or Medicine 1945

Nobel Australians

The tyranny of distance has resulted in Australia becoming a nation of innovators. The achievements of the Australia's Nobel laureates are a reflection of the country's inventiveness and drive.

The year 1901 is significant for both Australia and the Nobel Prize. This was the year when Australia achieved nationhood and also the year when the first Nobel Prizes were awarded. Both Australia and the Nobel Prize have grown up together, jointly celebrating their centennial year.

The Nobel Prizes are renown throughout the world and bring the recipients great kudos and fame. Alfred Bernhard Nobel's (1833-1896) legacy rested with the fact that he was able to combine his background in science with the vision of an industrialist and the drive of an entrepreneur.

Nobel wrote literary works and was a keen observer of social matters and issues surrounding world peace. His legacy, the Nobel Prizes, reflect the interests he held during his life.

Nobel nominated five areas for recognition: physics, chemistry, peace, literature and physiology or medical works. Another award, the Bank of Sweden Prize in Economic Sciences in Memory of Alfred Nobel, was established in 1968 and first awarded in 1969.

Celebrity Status

Each year, the announcement of the winners of the Nobel Prizes causes great interest and excitement. It is the chance for scientists, writers, economists and peace advocates to gain recognition for excellence in their spheres of intellectual pursuit. For a small window of time in October and December each year, the recipients acquire the celebrity of a sports champion or film star.

Australia is recognised as an inventive and creative nation and these attributes can be traced back to 1768 when Captain James Cook was given command of the HMAS Endeavour. His orders were to take scientists on an expedition from England that led to the discovery of Australia via Tahiti and New Zealand in 1770. It can be said that Australia was founded as a result of a scientific expedition.

The tyranny of distance has made it difficult for Australians to obtain, collaborate and share their knowledge outside of Australia's shores. The introduction of air travel and the information technology revolution has greatly assisted Australians in contributing to the stock of human knowledge from home.

PhDs were first awarded in Australia as recently as 1948 when Sir Rupert Myers became the first Australian to receive one. In the same year, Erica Wolff, an immigrant from Germany, was the first woman to receive a PhD in this country. Both graduated from the University of Melbourne. Until this time, Australians had to

The Walter and Eliza Hall Institute

FRANK MACFARLANE BURNET
Nobel Prize in Physiology or Medicine 1960

Australian Academy of Science

JOHN CAREW ECCLES
Nobel Prize in Physiology or Medicine 1963

The Nobel Foundation

ALEKSANDR MIKHAILOVICH PROKHOROV
Nobel Prize in Physics 1964

The Nobel Foundation

BERNARD KATZ
Nobel Prize in Physiology or Medicine 1970

The Nobel Foundation

PATRICK WHITE
Nobel Prize in Literature 1973

National Institute for Medical Research, London

JOHN WARCUP CORNFORTH
Nobel Prize in Chemistry 1975

University of Melbourne Archives

PETER CHARLES DOHERTY
Nobel Prize in Physiology or Medicine 1996

travel overseas for their doctorates and to do collaborative research.

In spite of this disadvantage, Australia has built an excellent research infrastructure and has since trained many eminent researchers who have made and will continue to make significant impacts internationally.

No other award in the world brings the prestige that a Nobel award delivers. The Prize is seen as the watershed for all awards that honour scientists and significant achievers in literature, peace and economics.

Why should we laud the Australian recipients of the Prize and who are they? To answer this question there is a need to understand the impact of the Prize itself. No other award in the world brings the prestige that a Nobel award delivers. The Prize is seen as the watershed for all awards that honour scientists and significant achievers in literature, peace and economics.

The honourees are those who have completed a piece of work that is considered by the Nobel committee to be a breakthrough and a significant contribution of new knowledge for the benefit of society.

Australia has produced Nobel laureates in Chemistry (John Cornforth), Medicine or Physiology (John Eccles, Bernard Katz, Howard Florey, Frank Macfarlane Burnet and Peter Doherty), Physics (Lawrence and William Bragg, Aleksandr Prokhorov) and Literature (Patrick White).

Each Australian laureate is unique and the achievements they have made transcend politics, prejudice, physical impairment and refugee status. With the exception of Patrick White, all the Australian laureates have been scientists.

Like many great people, they have been comfortable in themselves and have not sought public recognition.

Patrick White also possessed this quality. David Marr, author and biographer of Patrick White, relates the story of how the Swedish Ambassador to Australia tried to contact White to advise him of his Nobel Prize win for Literature.

White, a very private man, had an unlisted telephone number and even when his friends and colleagues were contacted they would not break White's rule about giving out his number. According to Marr's book: "Chaos followed."

Australia's Nobel Laureates attempts to raise awareness of the contribution these great Australians have made to society and acknowledge the credit they have brought to Australia through their achievements.

These people flourished in an era when our country was growing up and finding its character. The country relied very much on allowing immigrants to enter and contribute to the welfare of the nation. Australia was a very different place last century. Katz and Prokhorov were both refugees.

Katz (Medicine or Physiology 1970) found a home here after being persecuted as a Jew in Germany. He served in the Royal Australian Air Force in World War II. Prokhorov (Physics 1964) was born in Atherton, Queensland. His parents fled Russia at about the time of the Tzarist revolution and returned in 1923.

Nobel links

There have been other Nobel laureates with links to Australia. An American, Daniel Gajdusek (Physiology or Medicine 1976), was an investigator at the Institute of Medical Research in Melbourne in the 1950s when he began the work that would lead to his being awarded the Nobel Prize.

John Harsanyi was awarded the Nobel Prize for Economics in 1994, shared with John Nash Jnr and Reinhard Selten, for their pioneering analyses of equilibria in the theory of non-cooperative games.

Althought residing in the US at the time, he had completed a masters degree at the University of Sydney (1953) and had also worked in a research position at the Australian National University.

The 2003 Nobel Prize for Literature was awarded to South African-born John Maxwell Coetzee who resides in Australia where he is attached to the University of Adelaide.

In times past, Australians have been compelled to pursue their careers abroad in order to be recognised as successes at home. This need to compete, combined with the quest for knowledge and creative pursuit, has meant Australians punch well above their weight in cultural, sporting and scientific activities.

Much has been written about the exploits of our sporting heroes and movie stars. This book attempts to go some way towards raising the awareness of a different type of hero, to celebrate their achievements and provide inspiration.

Professor Istvan Hargittai states in his book *The Road to Stockholm* that, from very early on, the Nobel Prizes have been surrounded by an "aura" and that science in particular needs icons.

The Australians who are featured in the following pages have all been high achievers who were rewarded by the ultimate recognition of excellence: the Nobel Prize.

Stephan Wellink is director of Research and Commercialisation, University of Technology, Sydney.

The Nobel Prize medal

Each medal features a different image depicting what the award is for. The medals for **Physics and Chemistry** feature an image that respresents Nature in the form of a goddess resembling Isis emerging from the clouds, holding a cornucopia in her arms. The veil which covers her face is held up by the Genius of Science. It is the medal of The Royal Swedish Academy of Sciences.

The Genius of Medicine holds an open book in her lap on the medal for **Physiology or Medicine**. She is collecting the water pouring out from a rock in order to quench a sick girl's thirst. This is the medal of the Nobel Assembly at the Karolinska Institute.

The medal for **Literature** depicts a young man sitting under a laurel tree writing down the song of the Muse. It is the medal of the Swedish Academy.

Each of these three medals feature the Latin inscription "*inventas vitam juvat excoluisse per artes*", meaning "Inventions enhance life which is beautified through art".

The image on the **Peace** medal depicts a group of three men forming a fraternal bond. Its Latin inscription is "*pro pace et fraternitate gentium*" which means: "For the peace and brotherhood of men". It is the medal of the Norwegian Nobel Committee.

The medal for **Economics** shows the North Star emblem of the Royal Swedish Academy of Sciences, dating from 1815, with the words "*Kungliga Vetenskaps Akademien*" (The Royal Swedish Academy of Sciences) around the edge.

Alfred Nobel's face also features on one side of each medal. The image on the medals for Physics and Chemistry, Physiology or Medicine and Literature is the same, featuring a portrait of Alfred Nobel and the years of his birth and death in Latin. As seen here, the design of Alfred Nobel's face on the medals for Peace and Economics is slightly different.

PEACE

ECONOMICS

References: 1. Australian Academy of Science Newsletter July-September 1997, Number 37 2. The Nobel Prize. 100 years of honouring achievement, Stanford Report, October 3, 2001. 3. The Nobel Prize- the first 100 years, Agneta Wallin Levinovitz and Nils Ringertz (editors) the Nobel Foundation Stockholm Sweden, Imperial College Press, London, 2001. 4. Newsgallery, 5. Alan Symons, Nobel Laureates 1901-2000, Polo Publishing, London, 2000. 6. Patrick White Letters, David Marr (Editor), page 421, Random House Australia, 1994. 7. Walter Gratzer, Eurekas and euphorias. The Oxford book of scientific anecdotes, Oxford University Press, New York, 2002. 8. The Angus and Robertson Concise Australian Encyclopedia, Angus and Robertson Publishers, 1984. 9. Nobel e-Museum, http://www.nobel.se/index.html 10. Gajdusek, D. Carleton, Unconventional Viruses and the origin and disappearance of Kuru, Nobel Lecture, December 13, 1976.

What is the Nobel Prize?

Each year, the world eagerly awaits the announcement of who will be awarded a Nobel Prize. But what is it and when did it first come into existence?

The Nobel Prize is the first international award given yearly since 1901 for achievements in physics, chemistry, medicine, literature and peace. The Prizes were instigated by Swedish scientist Alfred Nobel. When Nobel signed his last will and testament on November 27 1895, he laid the foundation for what would become the Nobel Prizes. He stated in his will, "The whole of my remaining estate shall be dealt with in the following way: the capital, invested in safe securities by my executors, shall constitute a fund, the interest on which shall be annually distributed in the form of prizes to those who, during the preceding year, shall have conferred the greatest benefit on mankind."

Nobel added: "It is my express wish that in awarding the prizes no consideration whatever shall be given to the nationality of the candidates, but that the most worthy shall receive the prize, whether he be Scandinavian or not." The words ensured that the Nobel Prizes would be truly international awards.

A private institution known as the Nobel Foundation was established in 1900 based on the will of Alfred Nobel. It manages the assets made available through the will for the awarding of the Nobel Prizes. In 1968, the Sveriges Riksbank (Bank of Sweden) instituted the Prize in Economic Sciences in Memory of Alfred Nobel.

Eligibility

To be considered eligible for an award, it is necessary to be nominated in writing by a person competent to make such a nomination. According to the Statutes of the Nobel Foundation, information about the nominations is not to be disclosed, publicly or privately, for a period of 50 years. The restriction not only concerns the nominees and nominators, but also investigations and opinions in the awarding of a prize.

PHYSIOLOGY OR MEDICINE

PHYSICS AND CHEMISTRY

LITERATURE

The Nobel Foundation

The actual prize

The prize consists of a medal, a personal diploma and a prize amount.

The announcement of the Nobel Prize winners for the year is made on the same day that the Prize-awarding institutions choose from among the names recommended by the Nobel committees. This takes place in the month of October. Immediately after the vote, a press conference is held by the Prize-awarding institutions involved. These institutions are: The Royal Swedish Academy of Sciences (Prize in Physics, Chemistry and Economics), Karolinska Institute (Prize in Physiology or Medicine), The Swedish Academy (Prize in Literature) and The Norwegian Nobel Institute (Peace Prize).

Prize ceremonies

The Prizes are all awarded on December 10 each year, the anniversary of Alfred Nobel's death. The Prize award ceremonies and banquets take place in Stockholm (for the Nobel Prizes in Physics, Chemistry, Physiology or Medicine, Literature and The Bank of Sweden Prize in Economic Sciences in Memory of Alfred Nobel), while the Nobel Peace Prize ceremony takes place in Oslo, Norway on the same date. At the Prize Award ceremony in Stockholm, His Majesty the King hands each laureate a diploma and a medal. The Peace Prize is presented by the Chairman of the Norwegian Nobel Committee in the presence of the King of Norway.

What's the matter?

By Bob Beale

WILLIAM HENRY BRAGG
1862-1942
WILLIAM LAWRENCE BRAGG
1890-1971
Nobel Prize in Physics 1915

University of Adelaide

The Nobel Foundation

Getty Images

By working together, the father-and-son team of William and Lawrence Bragg founded a technique that revolutionised the way matter is analysed, earning them a Nobel Prize in the process.

William and Lawrence Bragg hold a unique place among Nobel Prize recipients, as a father-and-son team who received a joint award. Lawrence Bragg was also remarkable as the youngest person, at 25, to ever be awarded the prize.

Although the Braggs carried out their pioneering research in Britain, Australia lays legitimate claim to share in their glory by virtue of the formative years they spent in Adelaide, where Lawrence Bragg was born and educated. His father, William (to avoid confusion, the son was known as Lawrence even though his first name was also William) was born in 1862 in the English village of Westward, near Wigton in Cumberland. He was one of three sons of Robert and Mary Bragg. Robert was a farmer and Mary died when William was aged only seven.

He was sent to live with his uncle (another William Bragg), a druggist in Leicestershire. He was a bright child and became the youngest boy at the time to pass the local Oxford junior examinations in 1873.

At 13, he was sent to school on the Isle of Man, where another uncle – his mother's brother this time – was a teacher. William's autobiography recorded that he was bullied at first but went on to be reasonably happy. The religious background of his uncle's home and the school appear to have had a strong influence on his attitudes during his life, although as an adult he was a sceptical scientist – with great faith in fact, evidence and scientific method.

During his last school year, some type of religious storm swept the school, with a focus on terrifying stories of hellfire and damnation. In a lecture many years later he said: "I am sure that I am not the only one to whom when young, the literal interpretation of Biblical texts caused years of acute misery and fear." Yet he also said on another occasion: "From religion comes a man's purpose; from science his power to achieve it."

According to a biography of William Bragg, "William did well in maths and was interested in music, sport, chess and acting, and was appointed head prefect. He won a scholarship to go to Trinity College, Cambridge, in 1881, where he studied maths and physics, achieving first class honours."

He went on to work in the famous Cavendish Laboratory, which was headed at the time by J.J. Thompson, who won a Nobel Prize in Physics in 1906 for identifying the electron.

William played tennis with Thompson, and it was he who suggested that Bragg apply for the chair of maths and physics at the University of Adelaide in South Australia. From a field of 24 applicants, Bragg was shortlisted and interviewed in London. Thompson was among those on the selection committee.

Bragg got the job and sailed for Adelaide, arriving there early in 1886 at the age of 23, having consumed physics textbooks on the journey to expand his knowledge for his new role. The university, although a pioneering one in Australia, was short of apparatus, so he apprenticed himself to a firm of scientific instrument makers to enable him to build his laboratory equipment.

Although William was short on lecturing experience, he trained himself so well that he is remembered at Adelaide for being a gifted and enthusiastic teacher who gave popular practical demonstrations of physics phenomena to students and the public alike.

Musical effects

William also had a musical ear and a talent for playing the flute, which blossomed into an interest in acoustics and led him to becoming involved in helping to design the organ case at the university's conservatorium of music. Later, during two wars, both he and his son worked on military acoustic sensing and ranging devices.

William married Gwendoline Todd in 1889. He was 26; she was 18. Her father was Sir Charles Todd, who had risen from head of the Adelaide Observatory to become Postmaster-General of South Australia. Todd's wife, Alice, was the source of the name given to the famous town of Alice Springs. William and Gwendoline had three children: two sons – William Lawrence and Robert – and a daughter, Gwendolen. From all accounts, their marriage was a happy one.

As a scientist, William Bragg was a relatively late bloomer and he began to consider doing

The Bragg family (1902). L-R: Lawrence, Gwendoline, Robert and William.

Most pertinent to the later story of the Braggs, however, was the accidental discovery in 1895 by a German physicist, Wilhelm Roentgen, of a mysterious form of radiation that he named x-rays

Getty Images

The complexities of chess appealed to William Bragg who was a keen player.

original research only in his early 40s, around 1903 and 1904, when he became president of the physics section of the Australian Association for the Advancement of Science.

That burst of creativity was sparked in part by the many extraordinary inventions and discoveries that were happening in physics at the time – which included wireless radio, telegraphy and radioactivity.

Most pertinent to the later story of the Braggs, however, was the accidental discovery in 1895 by a German physicist, Wilhelm Roentgen, of a mysterious form of radiation that he named x-rays.

William kept abreast of these developments and pursued his own research, especially into radioactivity. He exchanged ideas with New Zealander Sir Ernest Rutherford (another Nobel Prize winner) and made a striking discovery that when alpha particles were emitted by a radioactive source, the distance they travelled could be used to work out what kind of atomic nucleus they came from.

International acclaim followed and he was honoured in 1907 with election as a Fellow of the prestigious Royal Society in Britain, having gained the support of Rutherford, among others.

After 21 years, Adelaide could no longer hold him and, thanks to his original radiation research, William was awarded the post of Cavendish Professor at the University of Leeds in 1908.

Here the focus of the Bragg story begins to turn to his eldest son. "Billy" was a quiet, dreamy child. One of his favourite pastimes was shell collecting, which he did on family holidays by the sea. He even chanced upon a new cuttlefish, which was named sepia braggii after him. Lawrence was educated at St Peter's College, Adelaide – the very same school that later nurtured another great Australian Nobel laureate, Howard Florey. Lawrence was so adept academically that he went to Adelaide University at the tender age of 15. At the age of 18, he was awarded a first class honours degree in maths while other boys his age were still in high school.

But while ahead of his peers academically, Lawrence found social activities quite arduous. He once said to his sister Gwendolen, "You and I find 'things' easier than people, Gwendy."

With the family's move to Britain, Lawrence transferred on a scholarship to Trinity College at Cambridge University, where in 1909 he changed his focus to physics. He graduated with first class honours in 1912.

It was an exciting and fast-moving time in physics. Since Roentgen's discovery of x-rays, a string of researchers had been probing their mysteries and had made important progress in understanding not only what they were but what they revealed about the nature of materials and their structure.

The year before, Lawrence graduated from Cambridge, the British physicist Charles Barkla had discovered that different metals scattered x-rays in different ways – in short, that each element makes its own peculiar kind of x-rays (an insight that won him the Nobel Prize in 1917).

However, it wasn't known at the time whether x-rays were streams of particles or a wave-like radiation, more like light. The German physicist Max von Laue reasoned that passing x-rays through crystals to see whether they could be diffracted – forced to change direction – might offer insight into that dilemma.

In 1912, von Laue's experiments revealed not only that the mystery rays could be diffracted but that they formed a pattern on a photographic plate, revealing them to have the properties of waves, not particles – for which he, too, received a Nobel Prize in 1914.

The newly-graduated Lawrence Bragg had started actively studying von Laue's findings as soon as they became known; his father having drawn his attention to them during a summer vacation.

Lawrence returned to Cambridge and by November that year had published his first paper on the topic in the *Proceedings* of the Cambridge Philosophical Society. Importantly, he had conducted some ingenious experiments and worked out what is now known as the Bragg equation – how to accurately calculate the wavelength of each type of x-ray from its diffraction pattern. It related the angle that x-rays are deflected by a crystal to the distance between its orderly arranged atoms.

William's instrument-making skills then came to the fore and soon the pair had a purpose-built x-ray spectrometer available to put the maths to work and measure the deflection angles, again during vacation time together.

These instruments went on to be developed from using photographic film to record the diffraction patterns to electronic counters, but until computers came along it was a tedious and time-consuming process to use them for

Gwendoline Todd at 16.

University of Adelaide staff (1906). William is seated front row, third from left.

analysing three-dimensional structures. William and Lawrence studied the structures of many crystals, including proteins.

The son, however, had a rare intuitive insight and a great capacity to conceptualise problems and express them mathematically.

He conducted some ingenious experiments and worked out what is now known as the Bragg equation – how to accurately calculate the wavelength of each type of x-ray from its diffraction pattern

Physicists soon jumped onto this marvellous new bandwagon for probing the secrets of matter. So did chemists, who were previously confined to studying chemical reactions between compounds to try to work out their structures.

This was especially important for methodically devising new pharmaceutical products, but it was laborious and slow. Now, with an x-ray spectrometer, they could study the compound of interest directly.

Complementary skills

The Braggs knew they had discovered something profound but they could have hardly foreseen the myriad uses their work would be put to. William and Lawrence Bragg brought complementary interests and skills to their collaboration.

The father's experience, investigative expertise and command of instruments were important, but his interest in diffraction was what it revealed about x-rays.

"In saying something about the work for which we were awarded the Nobel Prize, I feel that I cannot but speak for both my father and myself," Lawrence said in a speech in Stockholm in 1922. "It was with his inspiration and under his guidance that any contributions of my own were made, and it was one of the proudest moments of my life when I heard that you had associated my name with his and awarded the prize to us jointly."

That generous and inclusive public praise and credit for his father apparently masked tensions he felt privately about the issue. Even so, the Nobel Prize clearly meant a great deal to him and he accepted it at such a young age with due humility. He said: "That you should have given me, at the very outset of my scientific career, even the most humble place amongst their ranks, is an honour of which I cannot but be very proud."

He also credited von Laue, "whom we owe the great discovery which has made possible all

Picnic in Adelaide Hills 1887. William is seated fifth from left.

L-R: Brothers Robert, William and James in the late 1870s.

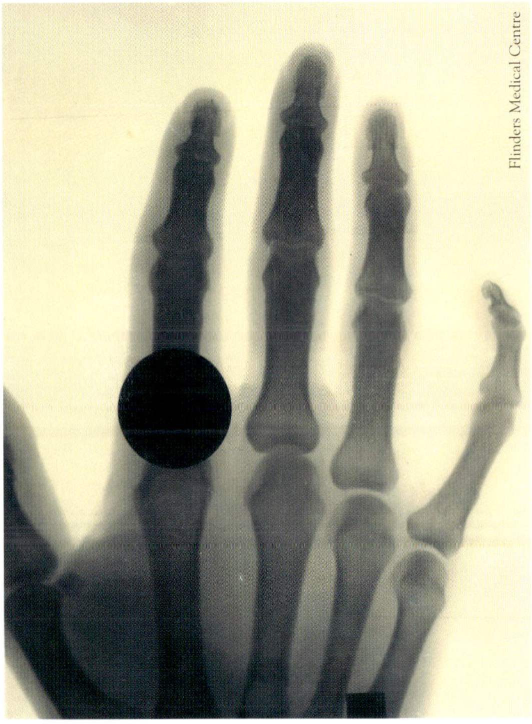

Flinders Medical Centre

The first x-ray taken in Australia: William Bragg's hand.

progress in a new realm of science, the study of the structure of matter by the diffraction of x-rays."

The pair won the Prize in Physics on November 12, 1915 for "a discovery of epoch-making significance" – as the Nobel citation speech put it – into x-ray crystallography. Put simply, it was a method for analysing the three-dimensional structure of crystals, atom by atom. Lawrence Bragg worked out the mathematics of how to do it and his father invented the instrument – the x-ray spectrometer – that made such precise measurements possible. Together, they founded a new branch of science that is still in everyday use in laboratories the world over.

It was a fundamental advance, an enabling technique with far-reaching consequences. It has since been used to unravel the structure of everything from diamonds to drugs, vitamins, insulin, bacteria, viruses and countless other organisms, compounds and materials. As such, it ushered in a revolution in physics, chemistry and biology.

In giving the citation speech, Professor G. Granqvist, the chairman of the Nobel Committee for Physics of the Royal Swedish Academy of Sciences, said: "Thanks to the methods that the Braggs have devised for investigating crystal structures, an entirely new world has been opened and has already in part been explored."

In turn, that brought more Nobel Prizes for many other researchers who followed in their footsteps.

Appropriately, in 1947 Lawrence Bragg helped to set up what became the Medical Research Council Laboratory of molecular biology at the Cavendish laboratory, Cambridge, and under his direction, Francis Crick and James Watson famously used x-ray crystallography to work out the double-helix structure of DNA.

William and Lawrence worked together from 1912 to 1914, investigating a range of crystal structures. They jointly published the results in abridged form in 1915 in a paper titled "X-rays and Crystal Structure", for which they were awarded the Nobel Prize.

From 1915 to 1919, Lawrence was involved in the war effort as a technical adviser on sound ranging or determining the distance of enemy artillery from the sound of their guns in the map section of British army headquarters in France. He was serving there when he heard the news about his prize.

In 1921, he married Alice Hopkinson of Cambridge, and they had two sons and two daughters (their eldest son became chief scientist with Rolls Royce). He moved back to Cambridge as Cavendish Professor of Experimental Physics from 1938 to 1953, and again served as a scientific adviser in Britain and Canada during World War II. There was also a string of other awards, honorary degrees and fellowships, and in 1941 he was knighted. When William heard the news, he was delighted. In a letter to his sister-in-law, Lorna Todd, on 5 January 1941, he wrote:

"You will learn by newspaper cable that Willie is knighted. Isn't that fine? … He will have to be Sir Lawrence: we can't have confusion worse than ever. I am so very glad for his sake. In spite of all care, people mix us up and are apt to give me first credit on occasions when he should have it. I think he does not worry about that all now, and will never anyhow have cause to do so now. I think I am more relieved about that than he is."

Because he won his Nobel Prize so young, Lawrence was one of the few laureates able to celebrate a golden jubilee

Because he won his Nobel Prize so young, Lawrence was one of the few laureates able to

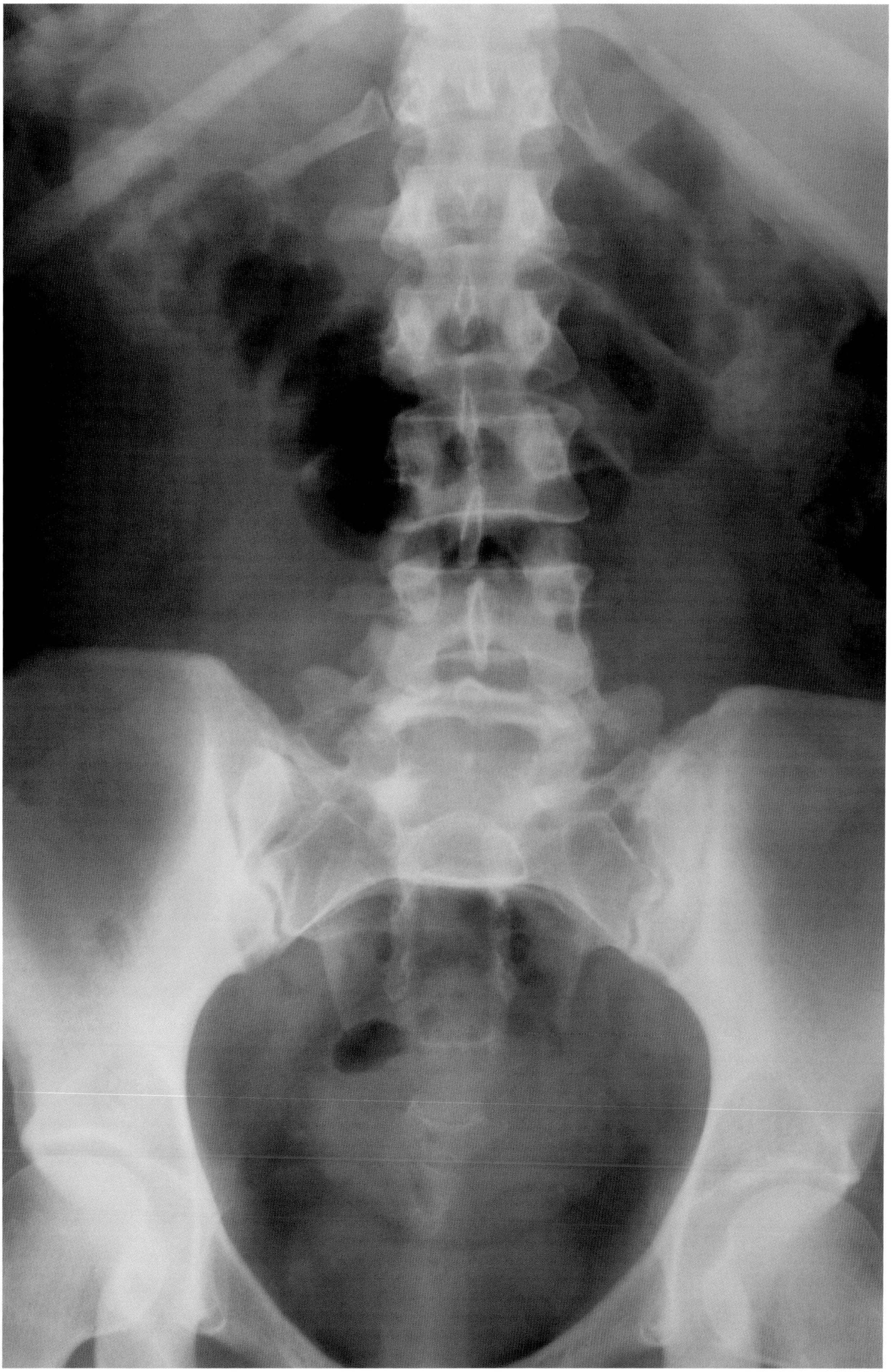

The Braggs founded a new branch of science with their studies of x-rays.

Swedish stamp from 1975 commemorating the Braggs' Nobel Prize in Physics 1915.

celebrate a golden jubilee, as a special guest at the awards ceremonies in Stockholm in 1965. While there he delivered the first Nobel Guest Lecture – reviewing developments in his field since 1915.

William's subsequent career and life were equally distinguished. After advising on anti-submarine matters during the war, he established a school of crystallographic research at University College, London, from 1915 to 1925.

He then served as director of the Royal Institution and of the Davy Faraday Research Laboratories in London, attracting and inspiring many young scientists who went on to great things. He maintained a broad and fruitful interest in research as well.

As one biographer put it: "His research interests embraced a great many topics and he was adept at picking up a subject, almost casually, making an important contribution, then dropping it again."

In 1923, William became head of the Royal Institution, serving in that post for two decades, only to be succeeded by Lawrence.

William was a popular lecturer and writer about science as well. He become well-known for giving "Christmas Lectures" for children, which also became best-selling books.

Sir William Bragg died in 1942. Those who knew him said he remained a simple and humble person, as well as a proud father.

When Lawrence succeeded his father at the Royal Institution, he developed that tradition by introducing year-round lectures that attracted 20,000 schoolchildren a year.

Lawrence also became a popular and successful lecturer who made many radio and TV appearances. He retired in 1965 and quit the Royal Institution the following year, but continued to lecture there until his death in 1971.

A fitting postscript to the lives of the two Braggs came from the winner of the 1964 Nobel Prize in Chemistry, Dorothy Hodgkin.

She recalled in a 1988 BBC Radio interview how the Braggs and x-ray crystallography were fundamental not just to her winning the prize but to her whole passion for and career in science. ■

Bob Beale is a freelance science writer.

Vital statistics

Name: William Henry Bragg
Born: Cumberland, England
Birthdate: July 2, 1862
Died: March 10, 1942
School: King William's College, Isle of Man
University: Trinity College, Cambridge
Married: Gwendoline Todd in 1889
Children: Lawrence, Robert, Gwendolen
Lived: Mainly Cambridge

Awards and accolades

1885: Elected to the Professorship of Mathematics and Physics in the University of Adelaide
1909: Cavendish Professor of Physics at Leeds (1909-1915)
1915: Nobel Prize in Physics
1916: Rumford Medal
1917: Awarded CBE
1920: Awarded a British knighthood
1925: Quain Professor of Physics at University College London (1915-1925)
1930: Copley Medal
1931: Order of Merit
1935: Elected President of the Royal Society

William Bragg was an honorary doctor of 16 universities and a member of the leading foreign societies. He was the author of many books, including *Studies in Radioactivity; X-Rays and Crystal Structure; The World of Sound; Concerning the Nature of Things; Old Trades and New Knowledge; An Introduction to Crystal Analysis,* and *The Universe of Light.*

Vital statistics

Name: William Lawrence Bragg
Born: Adelaide, South Australia
Birthdate: March 31, 1890
Died: July 1, 1971
School: St Peter's College, Adelaide
University: Adelaide University; Trinity College, Cambridge
Married: Alice Grace Jenny Hopkinson in 1921
Children: Stephen, David, Margaret and Patience
Lived: Mainly Cambridge

Awards and accolades

1914: Appointed as Fellow and Lecturer in Natural Sciences at Trinity College
1914: Barnard Medal
1915: Nobel Prize in Physics
1918: Awarded OBE and MC
1919: Langworthy Professor of Physics at Manchester University (1919-37)
1921: Elected Fellow of the Royal Society
1931: Hughes Medal of the Royal Society
1937: Director of the National Physical Laboratory
1938: Cavendish Professor of Experimental Physics, Cambridge
1941: Awarded a British knighthood
1946: Royal Medal of the Royal Society
1948: Roebling Medal of the Mineral Society of America
1958: Chairman of the Frequency Advisory Committee
1967: Companion of Honour

Lawrence was awarded many honorary fellowships and was an honorary or foreign member of American, French, Swedish, Chinese, Dutch, and Belgian Scientific Academies, besides being *Membre d'Honneur de la Société Française de Minéralogies et Cristallographie.* Together with his father, he published various scientific papers on crystal structure.

Why they were awarded the Nobel Prize

The work of William Bragg and his son Lawrence in 1913-1914 founded a new branch of science of the greatest importance and significance: the analysis of crystal structure by means of x-rays. The use of x-rays as an instrument for the systematic revelation of the way in which crystals are built was entirely due to the Braggs. They devised a method for analysing the three-dimensional structure of crystals, atom by atom. Lawrence worked out the mathematics behind it and William invented the instrument – the x-ray spectrometer. The branch of science they founded is still used everyday in laboratories all over the world.

Breaking the mould

By Bob Beale

HOWARD WALTER FLOREY
1898-1968
Nobel Prize in Physiology or Medicine 1945
Shared with Ernst Boris Chain and
Alexander Fleming

Photo Library

For a man who wasn't thinking much about relieving human suffering while he was experimenting with penicillin, Sir Howard Florey made one of the biggest contributions to the quest.

Of all of Australia's Nobel laureates, Howard Florey is unarguably the one whose work has provided mankind with the most profound benefits. Former prime minister Sir Robert Menzies once said Florey had more effect on the welfare of the world than any other Australian.

Through his central role in bringing the lifesaving miracle of penicillin to the world, and so ushering in the era of antibiotics, Florey's enduring legacy ranks him as a genuine giant in the world of public health. Without those drugs, millions of people – and countless domestic animals and livestock – undoubtedly would have died or suffered serious illnesses from bacterial infections.

It is easy to forget that before antibiotics became available, common infections took an appalling toll on human life. Many infants and mothers died in or soon after childbirth; tuberculosis and bacterial pneumonia were often deadly; and one bacterium alone – the infamous golden staph (staphylococcus aureus) – killed eight out of every 10 people infected after sustaining even the slightest wound.

Yet Florey claimed such concerns did not motivate his research. "People sometimes think that I and the others worked on penicillin because we were interested in suffering humanity," he told an interviewer in 1967. "I don't think it ever crossed our minds about suffering humanity. This was an interesting scientific exercise, and because it was of some use in medicine is very gratifying, but this was not the reason that we started working on it."

Sensitive scientist

This comment highlights the complexities of the individual behind the legend. His letters and the recollections of friends and colleagues point to a man who was a chain-smoker, inwardly shy, sensitive enough to be moved to tears by a musical performance, unpretentious, self-deprecating and possessed of great integrity. He hated nationalism and was deeply concerned for the plight of the dispossessed and hungry. Yet he was notorious for being gruff, driven, uncompromising, blunt, overbearing and, at times, callous to others.

He once unceremoniously sacked a technician for theft, yet quietly found the man another job. For all his achievement, he humbly attributed much of it to luck and to the hard work of those around him. A long-time colleague once commented that he never heard Florey utter a word in praise of himself.

Florey was born in suburban Adelaide on 24 September, 1898, the only son (he had four older sisters) of Joseph and Bertha Florey. Joseph was widowed when his first wife Charlotte died from tuberculosis but his early business success as a footwear manufacturer meant that Howard had a comfortable start in life. In 1906, his father's growing wealth enabled the family to move into "Coreega", a two-storey, 16-room sandstone house in the well-to-do suburb of Mitcham.

It was obvious from an early age that the young Howard was an exceptional individual. Between 1908-1910, he attended Kyre College preparatory school – now Scotch College – and later St Peter's College. He soon emerged as a genuine all-rounder, with diverse intellectual and physical talents, a strong character with leadership qualities. He won medals in gymnastics, sprinting and hurdles, and he captained tennis, cricket and football teams. At high school he was secretary of the debating society and was made head prefect in 1916. He regularly topped his classes academically and excelled in both science and the humanities, winning prizes in chemistry and history among other awards.

He was inspired by his chemistry teacher – known as "Sneaker" Thompson – the same man who had earlier sparked the intellect of another outstanding student from the same school who also went on to become a Nobel laureate: Lawrence Bragg.

Florey decided to study medicine at the University of Adelaide in 1917 but in the following year tragedy struck the family: not only did Joseph Florey lose all his wealth when his business failed but he died soon after of a heart attack. Despite more straitened circumstances, Florey was able to continue his studies with the help of scholarships and he again excelled in sport and academically at university. It was there that he met Ethel Reed, the bright, attractive and outgoing young woman who was to become his wife and a key member of his research team. She had begun studying medicine in 1919 and was the only female in her year.

Florey took his final exams in 1921 and was awarded a Rhodes scholarship. He left Australia in December of that year for Oxford University – working his passage as a ship's doctor – where he studied physiology and was awarded a Bachelor

Florey (back row, second from left) with fellow science students at St Peter's College, Adelaide, 1911.

Florey, the prefect, 1915.

Florey as a baby.

of Science degree in 1924. He did not make friends easily, writing home in one letter: "Sometimes I think I am liked here. At others, I get most depressed about it, and revile the English."

He knew himself well enough, though, to concede that: "I may not be able to throw off my selfishness and domineering manner." Those doubts were later scotched by his success in forming and leading an exceptional research team.

His adventurous spirit saw him join a university expedition to Antarctica in 1924 – by repute, the first to fly and crash-land an aircraft on the continent. Switching to Cambridge University, he became a research student and instructor while he completed his doctoral studies. A Rockefeller grant in 1925 took him for a time to America, where he rubbed shoulders with prominent scientists and honed his intellect and research skills before returning to Cambridge.

He did not make friends easily, writing home in one letter: Sometimes I think I am liked here. At others, I get most depressed about it, and revile the English

All the while he corresponded with Ethel, writing more than 150 letters to her up to 1926, when she began working at Royal Adelaide Hospital. Having suffered pleurisy and tuberculosis, she was steadily losing her hearing but their letters at the time reveal that they were discussing marriage.

She joined him in England in September 1927, and they were married the following month, a haste that later proved ill-advised since the marriage became a rocky one. Nevertheless, they remained together and had a daughter, Paquita, two years later, and a son, Charles, in 1939.

From 1930-34, Florey was Professor of Pathology at Sheffield University and in 1935 he was appointed director of the Sir William Dunn School of Pathology in Oxford, which he came to lead and dominate for the next three decades.

There he began to assemble the team – in itself an unusual approach in those more individualistic days – which would crack the bacterial infection problem. As one of his biographers, the late Gwyn McFarlane, said: "Florey was a rough, tough Australian, completely uncompromising, rather prickly, very energetic and tense as a coiled spring. And he brought to his work this extraordinary dedication which was very infectious, in such a way that he really could collect a team of people who became almost as dedicated and enthusiastic as himself."

Often forgotten behind the drama of the

Penicillin production at the Commonwealth Serum Laboratories in the 1940s.

A crucial discovery: mould, seen here magnified, and its effect on bacteria.

penicillin story is that Florey's other scientific contributions were significant, including important research into immune responses that helped pave the way for vaccination.

Indeed, at 43 he had published so many important research papers that he was elected to the prestigious Royal Society. Two decades later, Florey was elected its president, the first pathologist and the first Australian to obtain that post.

His interest in penicillin was born amid a broader surge of scientific inquiry into the drug-making potential of biological compounds produced by the human body, animals and plants.

At Oxford, Florey hired Ernst Chain, a German biochemist, to supplement his own technical skills. Chain was working on snake venom but Florey and others attracted his interest to a substance known as lysozyme, an antibacterial enzyme found in tears and nasal secretions. Together, Chain and Florey decided to systematically survey all known natural antibacterial substances.

Among those substances was a simple mould discovered years earlier in London by the Scottish researcher Alexander Fleming, in one of the greatest and most famous pieces of luck in medical research.

Fleming returned to his laboratory from a two-week holiday in 1928 to find that a petrie dish with a culture of bacteria left uncovered on his desk had been settled on by a mould spore that had begun to grow to about the size of a 20-cent coin. Around the rim of the mould the bacteria were dead.

Mouldy musings

At first, Fleming merely noticed it then threw it into a bucket. He later retrieved it and tinkered with it experimentally. He had earlier made a study of different antibacterial substances and discovered lysozyme, so the mould interested him enough to cultivate it. The green mass that grew on the surface of a broth had such a strong effect on bacteria that even when diluted 500 to 800 times, it completely prevented the growth of staphylococci bacteria. It turned out to be a species of the penicillium group of moulds – penicillium notatum – described for the first time only in 1911.

Fleming discovered that penicillin was highly effective against many different kinds of bacteria. He performed basic tests on white blood cells and mice to show that it was not toxic and had a little success testing its effect on infected wounds. He published a brief scientific report about it in 1929 but, believing it to be too difficult to work with and not showing enough promise beyond being a possible disinfectant of skin wounds, he moved on.

However, luck again intervened, as Lennard

Often forgotten behind the drama of the penicillin story is that Florey's other scientific contributions were significant, including important research into immune responses that helped pave the way for vaccination

Bottles for the surface culture of penicillin are filled with growth serum.

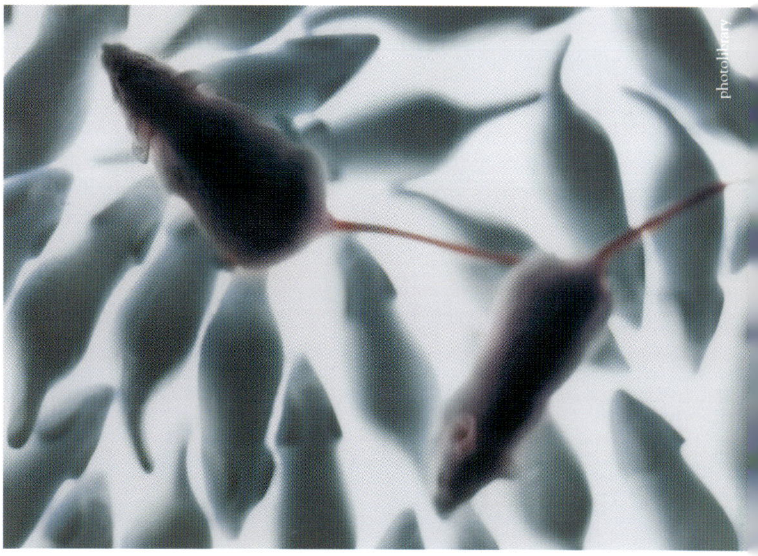

Florey tested penicillin on mice to assess its antibacterial properties.

Bickel, a Florey biographer, recalled in a 1998 ABC radio interview. "Chain was reading Fleming's old paper of 15 lines when he had a vision of a woman, one of the ladies who worked at the Dunn School, walking along a corridor with a dish in her hand on which a mould was growing. He went to see this lady and said to her, 'This mould that I saw you with …', she said, 'Yes', he said, 'What is it?' She told him it was penicillium notatum, this species of mould, and he said, 'That is the very mould that Fleming found in 1928'. She said, 'Yes, he cultured it and he gave us a piece of it, and we've kept it alive ever since'. Science is so full of pieces of serendipity."

In 1938, Florey and Chain began to study the mould more closely and found that it grew slowly and had special needs. They experimented with different growing mixtures, including Marmite, malt extract, meat and yeast extracts and studied its growth rate. They began to work out how to extract the active penicillin compound from the brown juice that accumulated beneath the surface layer of green mould.

Funds were short despite Florey's many urgings for more financial support, so in the Australian tradition of making do with whatever materials were to hand, he encouraged his team to experiment with growing the mould in biscuit tins and various dishes and pans. Hospital bedpans turned out to be the best-shaped growing containers. Old dairy equipment, a letterbox and an aquarium pump were among the items pressed into service to make the first penicillin.

It turned out to be highly unstable stuff. The scientists quickly moved on from filtering the juice through parachute silk to a more sophisticated extraction process using solvents. Chain worked out a series of steps to isolate, purify and concentrate the penicillin in the liquid but another member of the team improved the process so that soon they were able to extract and produce penicillin as a brown powder in small but useful quantities.

Now the team could experiment with the powder to test its effects. Chain dissolved some in water and injected it into two mice that survived the experience despite its highly impure state.

Working with increasing strengths of the powder, Florey's team tested it on blood, hormones and living cells. The turning point came, however, one morning in May 1940, when Florey injected a lethal dose of streptococcus pyogenes bacteria into eight mice but injected four of them with penicillin as well. The team anxiously watched the mice and by the middle of that night – just 16 hours later – all the unprotected mice were dead while those that had received penicillin remained alive and well.

Chain is said to have almost danced with glee and even the understated Florey gave way to

Assaying penicillin in the 1940s.

Penicillin autoclave unloading, 1942.

Installation of the penicillin seed tanks, 1946, at Commonwealth Serum Laboratories.

excitement when he later called his assistant Margaret Jennings to report the outcome of the experiment and said: "It looks like a miracle."

The team duly reported to other scientists that penicillin was a therapeutic agent able to kill sensitive germs in a living body and, with so many people suffering war wounds, the race began to work out ways to make enough of the stuff to test it on people.

Not until February 1941 was there sufficient funding to do that. A policeman named Albert Alexander, near death from an infection sustained when he pricked himself on a rose thorn, was selected as the first recipient of penicillin.

Within a day, Alexander was showing dramatic signs of recovery but there was no more penicillin available to give him further treatment. Despite heroic efforts to filter his urine to recapture any excreted penicillin, Alexander's infection took hold again and he died. It became apparent that a course of several treatments was needed to thoroughly kill off germs and Florey firmly decided that not until supplies of penicillin were great enough to effect a cure would it be tested on anyone else.

With British funds limited due to the war effort, Florey and a colleague took the risk of flying a blacked-out aircraft to the United States a few months later to seek help in boosting production of the drug.

Better ways of growing the mould were soon found and another species of it – penicillium chrysogenum – was discovered by Mary Hunt (known as "Mouldy Mary"). It turned out to be a much better producer of penicillin.

Florey returned to Britain and the first penicillin to be used in a war zone – in Cairo – was administered in 1942. The next year, in belated recognition of the significance of his work, the British Government finally gave Florey's team full funding.

The team anxiously watched the mice and by the middle of that night – just 16 hours later – all the unprotected mice were dead while those that had received penicillin remained alive and well

His wife Ethel was a vital member of the team during this time, working long hours with Florey on experiments and testing the drug and publishing the results jointly with him. His laboratory now became a makeshift penicillin factory with teams of workers tending hundreds of jars of mould. By October that year, the first supplies went to the British Army. With better extraction and greater production came increasing success in human use, with most war wounds now being cured of infection.

By 1945, enough penicillin was being produced, including by drug companies in the United States and Australia, notably the Commonwealth Serum Laboratories in Melbourne – to take penicillin to the wider community. In 1944, Australia was the first country to offer it for civilian use.

Resistance forms

Within a few years, in a foretaste of today's global problem with antibiotic resistance, several strains of bacteria started to become resistant to natural penicillin and the pharmacological spotlight soon turned to making synthetic forms which were developed in the 1950s and '60s.

n 1944, Australia was the first country to offer penicillin for civilian use

Florey's remarkable achievement was quickly recognised. He was knighted by Britain in 1944 and honoured as well by France, the United States and Australia for having such an unforeseen but undeniable influence on the course of the war and on medicine.

In December 1945, Florey, Chain and Fleming were jointly awarded the Nobel Prize in Physiology or Medicine for "the discovery of penicillin and its curative effects in various infectious diseases".

In 1966, Lady Ethel Florey died at home. In London on June 6, 1967, Lord Florey married again, to Dr Margaret Jennings, a member of his team and his special assistant for more than three decades. Just eight months later, on February 21, 1968, he died in Oxford of a heart attack, aged 69 (Lady Margaret Florey died in 1994).

In the eulogy at Westminster Abbey, Lord Adrian said of him: "Florey inspired the research and made it succeed. We have still to adjust our ideas to the extent of the revolution in medical treatment that Florey brought to success."

Norman Heatley, a key worker in the Oxford team, wrote of Florey's "genius in leading his colleagues and providing at the right times the encouragement, advice, inspiration and realism which kept the team working constructively throughout".

Perhaps the last word should go to one of

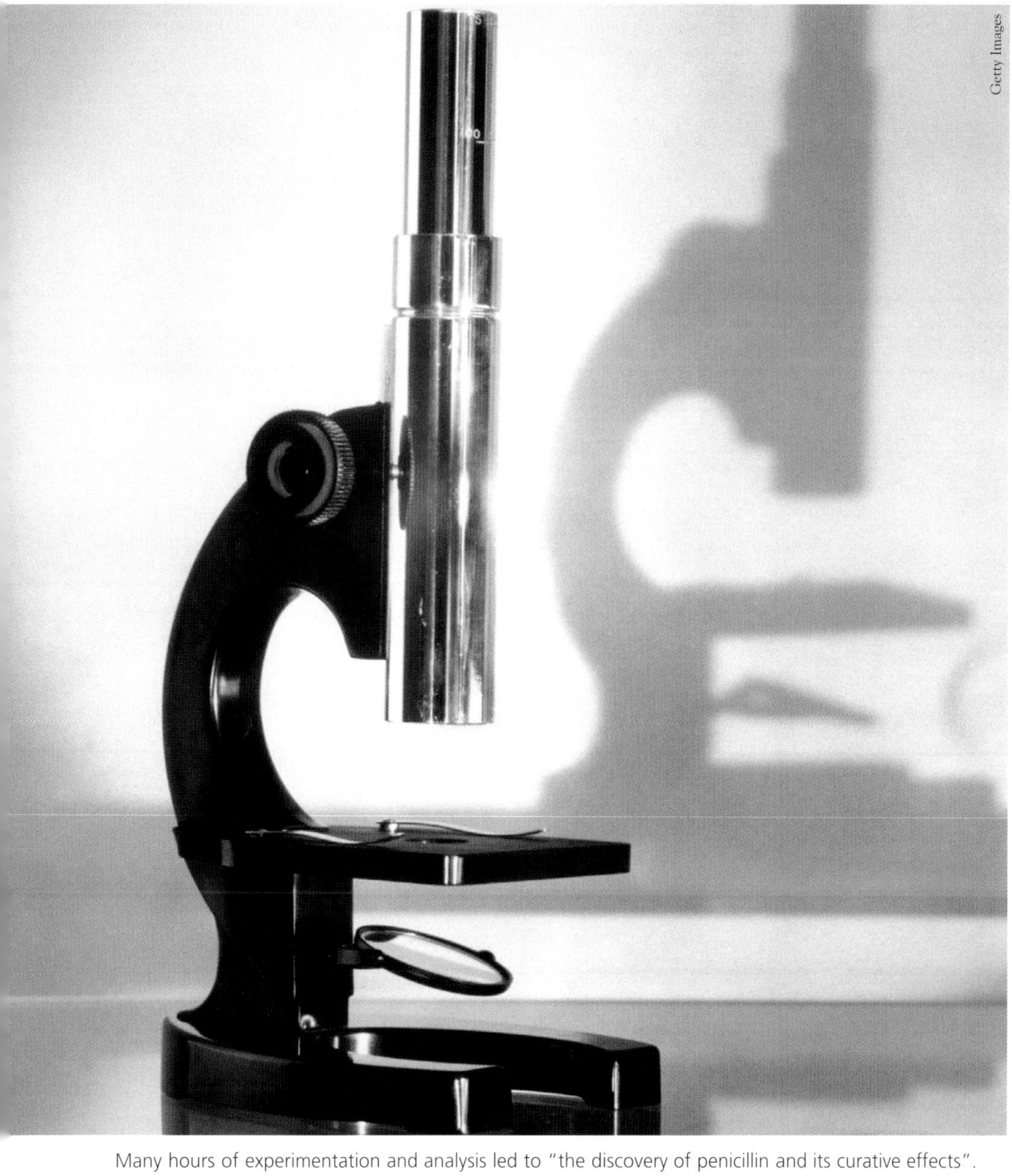

Many hours of experimentation and analysis led to "the discovery of penicillin and its curative effects".

ICSMR, ANU

Portrait of the young man as a scientist: Howard Florey went on to win the Nobel Prize for his work on penicillin.

his friends, Sir Alan Drury, who recalled Florey coming to Britain: "His drive and ambition were manifest almost from the day he arrived. A great fire seemed to burn within him, and his many-sided character was never concealed. We could all see the power in him and wondered whether he would ever find the right outlets for his greatness."

Florey recognised that his success was due to a team effort and for this reason he always acknowledged the team in his publications, usually listing them in alphabetical order. He disliked the public spotlight and, unlike Fleming, shunned media attention to the extent that Fleming's name is still far better known for the discovery of penicillin than is Florey's.

In his oration speech, Fleming acknowledged the role of luck in his discovery. He said: "My only merit is that I did not neglect that chance observation and pursued it as a bacteriologist. The first practical use was to sort out bacteria that were sensitive from those that were not. We tried to concentrate it but found, as others did later, that it was easily destroyed – and, to all intents and purposes, we failed".

In his own oration, Florey lauded his team and its scientific approach, and looked forward to the revolution that penicillin and other probiotics promised to bring. Other scientists followed in his path, discovering new moulds and other types of antibiotics, some of them with the help of Florey's team.

In his Nobel Prize presentation speech, Professor Goran Liljestrand, of the Royal Caroline Institute, saw Florey's work in stark contrast to the bloody war that had just ended. "To overcome the numerous obstacles, all this work demanded not only assistance from many different quarters but also an unusual amount of scientific enthusiasm and a firm belief in an idea. In a time when annihilation and destruction through the inventions of man have been greater than ever before in history, the introduction of penicillin is a brilliant demonstration that human genius is just as well able to save life and combat disease." ■

Vital statistics

Name:	Howard Walter Florey
Born:	Adelaide, South Australia
Birthdate:	September 24, 1898
Died:	February 21, 1968, Oxford, England
School:	St Peter's College (Adelaide)
University:	University of Adelaide; Oxford University; Cambridge University.
Married:	Mary Ethel Hayter Reed in October 1927. Ethel died in 1966. Married Dr Margaret Jennings in June 1967
Children:	Paquita, Charles
Lived:	Mainly in Oxford, England

Awards and accolades

1935:	Professor of Pathology and a Fellow of Lincoln College, Oxford
1944:	Nuffield Visiting Professor to Australia and New Zealand
1944:	Created a Knight Bachelor
1946:	Honorary Fellow of Gonville and Caius College, Cambridge
1949:	Contributor to and editor of *Antibiotics*. He also had many papers published on physiology and pathology
1952:	Honorary Fellow of Magdalen College, Oxford
1960:	President of the Royal Society (first Australian to be elected)
1962:	Provost of the Queen's College, Oxford
1965:	Made Baron Florey of Adelaide and accepted the Chancellorship of the Australian National University that same year.

Other honours include: the Lister Medal of the Royal College of Surgeons, the Berzelius Medal of the Swedish Medical Society, the Royal and Copley Medals of the Royal Society, the Medal of Merit of the US Army, and many others. He was a Fellow of the Royal College of Physicians and held an honorary fellowship of the Royal Australian College of Physicians. He has been awarded honorary degrees by 17 universities and was a member or honorary member of many learned societies and academies in the field of medicine and biology.

Why he was awarded the Nobel Prize

Howard Florey, Ernst Chain and Alexander Fleming were jointly awarded the Nobel Prize in 1945 in Physiology or Medicine for the discovery of penicillin and its curative effects in various infectious diseases. Penicillin had been discovered by Fleming in 1928 but the active substance was not isolated. In 1939, Florey and Chain headed a team of British scientists, financed by a grant from the Rockefeller Foundation, whose efforts led to the successful small-scale manufacture of the drug from the liquid broth in which it grows. Without penicillin, millions of people would have died or suffered serious illnesses from bacterial infections.

In the name of science

By Brad Collis

The Walter and Eliza Hall Institute

FRANK MACFARLANE BURNET
1899-1985
Nobel Prize in Physiology or Medicine 1960
Shared with Peter Brian Medawar

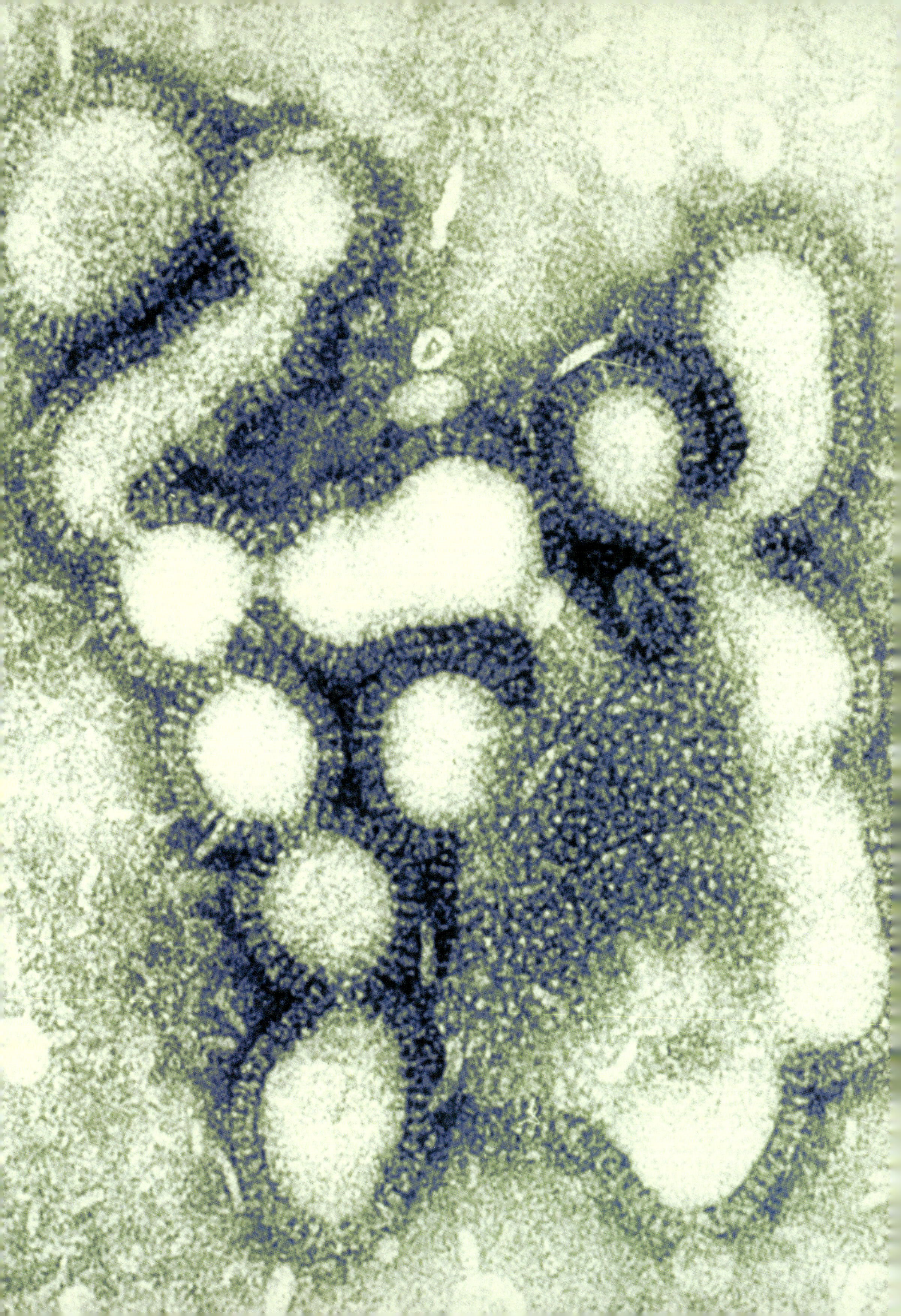

Sir Frank Macfarlane's study of viruses formed the basis of immunity that has become vital in today's field of organ transplants.

It was Christmas 1950, when an unusually warm, wet weather front moved across the Murray Valley in northern Victoria and created ideal conditions for an explosion in mosquito numbers.

Within days, the surrounding farmland was strewn with dying rabbits. Scientists, by luck, had discovered the carrier needed to spread the myxoma virus they had developed which they had been trying for years to use on the country's devastating rabbit plague. Until now it had been assumed the virus could only spread very slowly by direct contact among rabbits. However, the warm, wet weather and the mosquitoes revealed there was a highly effective natural carrier after all.

Farmers, whose landscapes and livelihoods were being destroyed, were ecstatic. But their jubilation quickly turned to horror. People also began to fall ill and die from a deadly new brain disease. The media quickly blamed the new rabbit virus and the scientists who had unleashed it.

Virologists knew that myxoma, which causes myxomatosis in rabbits, had no effect on humans, but were struggling to be heard or seen in the outcry and hubbub of "Death Valley" headlines.

Something or someone was needed to calm a panicking population – someone who was not only eminent in medicine but who was also respected and trusted by the community. The man the government turned to was noted virologist Dr Frank Macfarlane Burnet, director of the Walter and Eliza Hall Institute in Melbourne.

Burnet, known to friends and colleagues by his childhood nickname "Mac", decided the best way to calm everyone was to simply inject himself with a large dose of the myxoma virus and show people, in the most dramatic way possible, that the deadly new human disease – Murray Valley encephalitis – was not myxomatosis.

After some discussion it was decided that both Burnet and an associate, Frank Fenner (who would later achieve world renown for his part in eradicating smallpox from the planet) would inject each other.

Fenner had just been appointed as a professor of microbiology at the John Curtin School of Medicine at the Australian National University in Canberra and was working at the Walter and Eliza Hall Institute until his new laboratories were built. By chance, he and Burnet were working side by side. Fenner was filling in time by studying the myxoma virus and Burnet on the mystery virus that was killing people.

Human guinea pigs

Burnet prepared a compound containing the myxoma virus – a dose 100 times greater than a rabbit or human would receive from a mosquito: "He injected me and I injected him," recalled Fenner, who, with CSIRO scientist Francis Ratcliffe, later wrote the definitive textbook *Myxomatosis*.

Just after the pair had finished injecting themselves, the head of the CSIRO, Ian Clunies Ross, walked in demanding that since myxomatosis was a CSIRO project, he should also be injected. This was done and after three weeks when all tests showed the scientists were free of any myxoma antibodies (as they knew would be the case), the myxomatosis scare faded into history.

What didn't fade, however, was the lasting impression in the public's mind of the scientists' perceived bravery. In becoming human guinea pigs, Burnet, Fenner and Clunies Ross had perpetuated an important, although fading tradition in which scientists take the unknown risks as the first volunteers in their experiments.

> Burnet decided to simply inject himself with a large dose of the myxoma virus and show everyone that the new deadly disease was not myxomatosis

Meanwhile, the mystery human virus was identified as a new strain of encephalitis, triggered by the same warm, wet weather that got myxomatosis off to a flying start. Murray Valley encephalitis, in which Burnet became the world authority, remains to this day.

The approach to the myxomatosis scare typified Burnet's practical, sleeves-up attitude. In much the same manner as Pasteur, Burnet was committed to tackling practical problems. That same practical attitude played a role in his momentous decision a few years after the myxoma episode to move away from virology and concentrate on

Scientists catching mosquitoes, carriers of the myxoma virus, just after the disease outbreak.

Burnet, under rug, with his mother and sister Doris, 1903.

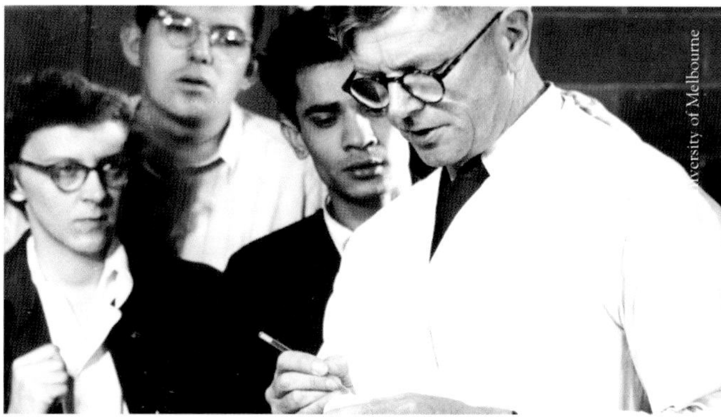

Burnet demonstrates his technique with an egg shell and needle, 1944.

Melbourne Hospital (Burnet, front row, second from left), 1924.

immunology – the decision that culminated in him winning the Nobel Prize in 1960.

By the mid-1950s, Burnet had realised that the nature of virology research was changing: modern laboratories were turning to the new fields of molecular biology and DNA coding. Against this, Burnet's main research tools – a chicken egg and a microscope – looked old-fashioned. Throughout his career, Burnet favoured an experimental technique developed while working at the National Institute of Medical Research in London in the 1930s. An opening was made in the shell of an egg and a virus sample was injected into the membrane that surrounded the chick embryo. In these conditions most viruses multiplied, allowing closer study.

Yet rather than learn new methods – such as the use of tissue culture – in virology, in which he had already achieved acclaim, Burnet decided to move on. He had always been interested in immunology and directed his own and the Institute's work into this area.

While the radical change in direction was later justified with the awarding of the Nobel Prize, others felt Burnet's move was dictatorial. Gustav Nossal, who would eventually succeed Burnet as director of The Walter and Eliza Hall Institute, arrived at the Institute in 1957 expecting to work on viruses. He was reportedly taken aback when told of the change in direction.

For a man regarded as one of the great scientific thinkers of the 20th century, Burnet's propensity for gloomy – and often wrong – predictions about science, especially after his formal retirement in 1965, often puzzled associates.

Professor Jim Goding of Monash University's Department of Pathology and Immunology recalls: "On the occasions that I got to know Burnet socially, I challenged him concerning his predictions. When confronted with the spectacular progress since his predictions, he would say, 'I stand corrected', but then immediately reiterate his belief that most of the really important work had been done and our generation was just tidying up loose ends."

Gustav Nossal believed Burnet's stance simply reflected: "the real pain that he felt in having to leave the scene of discovery". It showed up one of Australia's greatest scientific heroes as being very human.

Innumerable records and memoirs tell the story of Frank Macfarlane Burnet. He was born in the Gippsland town of Traralgon on September 3, 1899, the second of seven children. His father, also Frank, was the branch manager of the Colonial Bank and had emigrated from Langholm, Scotland, in 1880. In 1893, he married Hadassah Mackay, the daughter of a local school teacher, George Mackay, who had also emigrated from Scotland.

From the time he was old enough to run around the district, which was still lush with bushland, the young Burnet revealed a fascination for biology, assiduously compiling the usual boyhood collections of butterflies, birds' eggs, rocks and beetles. His interest in the natural world became even more marked at 10, when his family moved to Terang in western Victoria, where the wildlife around Lake Terang was a source of unlimited discovery.

Those who have written monologues and memoirs on Burnet's life, including his friend and colleague Frank Fenner, discovered the adolescent Burnet fitted a familiar profile of boys who were later drawn to an academic career. They were shy, socially late-maturing, and had a strong devotion to their hobbies.

Beetle mania

In Burnet's case, his passion from boyhood was beetles, which he began to record and draw. He read all the biological sections of an old *Chambers Encyclopaedia* published in the 1860s, which introduced him to Charles Darwin. He wrote to Melbourne for a book about beetles, and was sent a translation of Fabre's *Souvenirs Entomologique*. Later he acquired Froggatt's *Australian Insects*, and covered its pages with his own entries on beetle collecting. His keen interest prompted the local Presbyterian minister to encourage his parents to send Frank to university.

He was sent to Geelong College for four years – an experience he later revealed was not a particularly happy one – but in his final year he gained a scholarship to Ormond College, at the University of Melbourne. There he pursued medicine, mainly because it appealed more than what seemed to be his only other professional options – a career in law or in the church.

Frank Fenner wrote that Burnet's early years at university were accompanied by wide reading and horizon-broadening, and a sorting-out of his ideas on religion. Despite growing up in a Presbyterian household in which church every Sunday was an unbending ritual, he soon became agnostic. Darwin's writings on evolution strongly influenced his early scientific work, while the incisive work of writer H.G. Wells helped form his views on society.

At the end of a medical course that was shortened to five years because of World War I and the perceived need to produce graduates quickly, Burnet graduated MB, BS in April 1922, coming second in a class that contained four others who later achieved fame in science and medicine: Sir Roy Cameron (pathology), Professor R.A. Willis (medical research), Dame Jean Macnamara (poliomyelitis research and campaigner for the introduction of myxomatosis) and Dame Kate Campbell (paediatrics).

After graduating, Burnet spent two years gaining his MD, then spent a year as resident medical officer in the Royal Melbourne Hospital. In the surgical wards, he came to know two eminent surgeons, Sir Alan Newton and Sir Victor Hurley, each of whom later served as a chairman on the board of The Walter and Eliza Hall Institute while he was director.

However, it was while serving as house physician to Melbourne's leading physician at the time, the neurologist Dr R.R. (later Sir Richard) Stawell, that Burnet became convinced his future lay in clinical neurology.

He applied for the post of medical registrar at Melbourne Hospital, intending it as a stepping stone, yet the medical superintendent had other ideas. Convinced that Burnet's shy and awkward personality was more compatible with a career in a laboratory, the superintendent steered him into the job of pathology registrar and a few months later, senior resident pathologist. This became the start of his brilliant scientific career and his long association with The Walter and Eliza Hall Institute (the pathology laboratories of the hospital operated as part of the Institute).

In 1924, the Institute was given a fresh intellectual stimulus with the arrival from University College, London of Dr Charles Kellaway (later Sir Charles) who, as director, transformed the Institute into a research facility specialising in physiology, biochemistry and bacteriology.

Kellaway saw Burnet as the potential leader of the small bacteriology section but decided he should first have overseas training. In 1925, Burnet left for England as a ship's surgeon. He took a position at the Lister Institute in London because there was a paid position that still gave him time for research. Under the supervision of Professor J.G. Ledingham, he gained a PhD from the University of London (1928).

Burnet's time in London wasn't all work and staring through a microscope. He returned to Australia in 1928 with his fiancée Linda Druce, an Australian who had been living in London. They married later that year on July 10, 1928, and spent their honeymoon skiing at Mt Buffalo in northern Victoria.

Yet it was another posting to London, this time to the National Institute of Medical Research at Hampstead in London in 1932-33, that changed the course of Burnet's scientific life. The post placed him in the thick of research that was opening up science's understanding of animal virology.

Getty Images

The young Burnet revealed a fascination for biology, assiduously compiling the usual boyhood collections of butterflies, birds' eggs, rocks and beetles.

During this visit, Burnet developed his work on the use of the chick embryo to study animal viruses. He also acquired a powerful friend in Sir Henry Dale, who offered him a permanent position at the Institute.

However, Burnet decided to return to Melbourne, where he became assistant director of The Walter and Eliza Hall Institute, in charge of the virus section.

Back in Melbourne, Burnet continued to study the behaviour of a variety of viruses in the developing chick embryo. Seizing opportunities as they arose, he worked on psittacosis (parrot fever), carried out studies on poliovirus, and recognised a rickettsia (a micro-organism that resembles bacteria but can be as small as a virus and reproduces only inside a living cell) to be the cause of Q fever. Q fever, characterised by fever, chills, headache and weakness, is often mistaken for flu and is a zoonotic disease that spreads from animals, such as sheep, cattle and ticks, to people. The causative organism of Q fever was later named Coxiella burneti in Burnet's honour.

Influenza interest

However, his major interest from 1939 was the influenza virus, prompted by the discovery of methods of growing the virus in chick embryos. With the onset of World War II, his attention was focused on methods of immunising against influenza in case there was ever another pandemic like that of 1918-19, which killed more people than the entire First World War.

In 1942, he was offered a professorship at Harvard University in the United States and was strongly tempted, but he later explained that he felt a strong sense of loyalty towards The Walter and Eliza Hall Institute and Australian science. The patriotism he felt towards his country was reflected in the fact that many of his papers were first published in Australian scientific and medical journals. He even felt that it was an advantage for scientists to be relatively isolated in Australia, since it protected them from being too influenced by fashions in scientific thinking.

In 1944, Burnet succeeded Kellaway as director of the Institute – even though Kellaway and Burnet himself had reservations about the appointment. Kellaway was worried that Burnet lacked the necessary leadership skills, and that the post might hinder Burnet's research work. The Institute's board overruled Kellaway, and Burnet immediately showed he was capable of making big decisions.

He decided that the Institute's activities should be concentrated on animal virology, especially the influenza virus. His team was to eventually unravel the nature of the influenza viral enzyme (neuraminidase), which became the starting point for four decades of Australian research that culminated in 1999 with the release of the world's first anti-influenza drug, Relenza.

During that time, because Burnet kept the Institute at the forefront of influenza research, it attracted numerous overseas scientists. Those who worked at the Institute had no doubt that they were privileged to be working with a man of genius. Burnet also had a reputation as a hands-on director.

Until 1955, when he started travelling overseas more often and had to juggle increasing demands on his time, Burnet spent more than half of each day at the laboratory bench. He usually worked alone, apart from one or two graduate assistants and one or two technicians.

Few of Burnet's research papers list a co-author other than his graduate assistant of the day. Fenner says that Burnet was careful in selecting his graduate assistants – and the rollcall of highly competent women assistants (including his daughter Deborah) repaid him with many years of service. One, Patricia Lind, worked in the lab with Burnet from 1944 until his retirement in 1965.

Burnet and his second wife, Hazel, on their wedding day in 1976.

Ormond College, 1917. Burnet, front row, second from left, lived there as a medical student and, later, after the death of his first wife Linda.

During the 1940s, Burnet revised his interest in the body's immune response to attack by micro-organisms such as potentially harmful bacteria and viruses. This led to the work on acquired immunological tolerance, for which he and Dr Peter Medawar would be jointly awarded the Nobel Prize in Physiology or Medicine.

When Burnet and Medawar travelled to Sweden in 1960 to receive their laureates, Professor Bertil Lindblad, the president of the Royal Academy of Sciences, articulated their achievement during the award ceremony: "In your discovery of immunity produced in the embryonic stage and of actively acquired tolerance you have found a new biological law, opening up new vistas in experimental biology. The phenomenon of immunological tolerance which you have discovered will most certainly be of direct practical importance for the treatment of various kinds of injuries and diseases."

When he travelled to Sweden to receive his laureate, Burnet was clearly conscious of the historical significance of his achievement, and was keen to place it into a wider educational context. "I think that this occasion has a rather special significance for my own country, a middling small country a little bigger than Sweden but only now beginning to create an image of its own in the eyes of the world," he said. "Some day I hope that we will take our place along with Sweden as one of the centres where knowledge can go along with social progress to the good life we all seek."

Even though Burnet was still a shy and withdrawn character, those who worked at the Institute had no doubt that they were privileged to be working with a man of genius

He then had a message for all students, not just those who might embark on scientific careers: "To advance science is highly honourable but other things are equally honourable. Perhaps when you are 20 to 30 years older, research as we know it may be less important than it is today but there will always be an obligation to pass on to the new generation the tradition of liberal scholarship. I hope that when you are as old as I am, skill and success in education will be as highly rewarded as success in scientific discovery is today. But whether your career is in research, in education, or in seeing that some of the wheels of our complex civilization turn as they should, I wish you luck."

Although Burnet won the Nobel Prize for his work on acquired immunological tolerance, he was dismissive of its wider implications (one headline quoted his comments under the headline, "No great value in transplants, says top scientist"). Burnet considered his greatest contribution to biology to be his clonal selection theory. This theory, which sought to explain how animals could produce such a wide variety of antibodies, opened up vast new areas of research. However, it would take nearly another 20 years for the actual mechanism of clonal diversification to be discovered by Hozumi and Tonegawa in 1976.

Burnet had an unforgettable impact on the stream of scientists who came to The Walter and Eliza Hall Institute, especially between 1944 and 1965. Although his management style could seem like a benign dictatorship, the Institute's scientists were granted many freedoms. Unlike others, Burnet did not put his name on papers to which he had not contributed. Highly animated discussions over morning and afternoon tea were a feature of Institute life. He read every paper from staff, scribbled clarifications over any points he felt were unclear, and returned the manuscript the following day.

Reluctant spokesman

Burnet never courted interviews but he recognised the importance of talking to the media if the public were to understand the role of science and scientists. He would interrupt his precious time at the laboratory bench to take inquiries from the press and occasionally appeared on television.

After Burnet notionally retired in 1965 he broadened his interests even further, becoming a public commentator on issues as disparate as cigarette smoking and uranium exports.

In 1966, he returned to the University of Melbourne where, over the next 12 years, he wrote 13 books on subjects ranging from immunology and human biology to ageing and cancer. He also produced a fourth edition of his first book, *Biological Aspects of Infectious Disease.* Most of Burnet's writings were not technical monographs, but written for the doctor or biologist who was not a specialist in virology, immunology or gerontology. Many of his popular books are highly readable.

In 1973, Burnet's life was turned upside down when his wife Linda died of lymphoid leukaemia. She had been his companion on many of his overseas trips and when separated by distance, they had corresponded devotedly. Many of their letters are held at the University of Melbourne Archives, along with Burnet's photographs, diaries, notes, press clippings and early watercolours and sketches.

After his wife's death, Burnet lived for a while at Ormond College at the University of

Burnet, one of Australia's most acclaimed scientists.

Melbourne, which had been his home as a medical student. It was here that he renewed his friendship with the Master, Dr Davis McCaughey (later appointed Governor of Victoria).

While devastated by his wife's death, Burnet was to find love again. In 1976 he married Hazel Jenkin, a widow who had endowed the library in the School of Microbiology to commemorate her only daughter Heather, who had died while a graduate student. In November 1984, Burnet was operated on for cancer and initially seemed to have made a good recovery, but secondary lesions were discovered in August 1985. As a biologist, Burnet accepted the inevitability of death and was impatient with proposals designed to prolong the life span.

He had earlier recognised the scope for research "on the best means of minimising the depression and misery of pre-death", so it is perhaps fitting that he was spared those things. He remained mentally acute until he lost consciousness shortly after his last illness took hold. He died on August 31, 1985, aged 86, at his son Ian's home at Port Fairy, near where he had spent his boyhood.

He was given a State funeral by the Federal Government, and was buried at Tower Hill Cemetery, near Port Fairy. His wife Lady Burnet survived him, along with his children from his first marriage, Ian, Elizabeth and Deborah, and eight grandchildren.

Sir Frank Macfarlane Burnet remains one of Australia's most acclaimed scientists; a man who made an extraordinary contribution to human medicine and well-being, and who also stood up as a proud Australian. Greatness became a mantel that he appeared to wear quite comfortably, yet at heart he didn't move that far from the shy, curious child who used to wander through the bush, hunting down butterflies and drawing beetles in a sketch pad. ■

Brad Collis is a Melbourne-based author and journalist.

References: Australian Academy of Science, Biographical Memoirs of Deceased fellows,
Frank Macfarlane Burnet 1899-1985, by Frank Fenner: http://www.asap.unimelb.edu.au/bsparcs/aasmemoirs/burnet.htm
Tim Sherratt, From Beetles to a Nobel Prize, http://www.asap.unimelb.edu.au/bsparcs/exhib/journal/as_burnet.htm
Brad Collis, *Fields of Discovery – Australia's CSIRO*, Allen & Unwin, 2002
Alex Webling, *Nobel Prize Laureates and Australia to 2000*, private monograph
Prof Jim Goding, profile of Frank Macfarlane Burnet, http://www.wehi.edu.au/collegiate/ASI/burnet.html

Vital statistics

Name: Frank Macfarlane Burnet
Born: Traralgon, a small town in the industrial Latrobe Valley, Gippsland, south-eastern Victoria
Birthdate: September 3, 1899
Died: August 31, 1985, Port Fairy, Victoria
School: Geelong College
University: University of Melbourne; University of London
Married: Edith Linda Marston Druce in 1928, widowed in 1973. Married Hazel Gertrude Jenkin in 1976
Children: Three from his first marriage – Ian, Elizabeth and Deborah
Lived: Mainly in Melbourne, but also in London.

Awards and accolades

1942: Fellow of the Royal Society of London
1946: Honorary doctorate from Cambridge
1947: Royal Medal, Royal Society of London
1948: Fellow of Royal Australasian College of Physicians
1951: Awarded a British knighthood
1953: Fellow of Royal College of Surgeons
1958: Order of Merit
1959: Copley Medal, Royal Society of London
1960: Honorary doctorate from Harvard University
1961: Australian of the Year
1968: Honorary doctorates from Oxford and Monash universities
1978: Order of Australia
1981: Australian of the Year

Burnet was the most highly honoured scientist to work in Australia. Following his death in 1985, two Melbourne research units and an annual oration at the Australian Society of Immunology were named in his honour.

Why he was awarded the Nobel Prize

The Nobel Prize was awarded with Sir Peter Brian Medawar of Great Britain for the discovery of acquired immunological tolerance. How does the human body recognise the difference between existing micro-organisms and dangerous invaders? This fundamental question worried away at Burnet during the 1940s. He studied every piece of literature he could find on the subject – and it was his ability to think laterally across all this information that produced the breakthrough hypothesis.

During Burnet's study of influenza viruses in the 1930s and 1940s, he had developed methods for growing viruses in chicken embryos inside eggs. He noticed in his research that although normal hens could be infected with influenza and develop antibodies, chickens born from eggs infected with the virus did not develop antibodies when exposed to influenza. The chickens had an "acquired immunological tolerance" and did not raise an immune defence because their immune systems didn't recognise the virus as foreign. This showed that when the immune system developed during the foetal stage, it "recorded" all proteins and micro-organisms in this enclosed environment as "self" – including the virus injected by the scientist.

Burnet's theory was confirmed in experiments by English researcher Peter Medawar, and it was for their collaboration in developing a new understanding of the immune system that they were awarded the Nobel Prize in 1960. This understanding of the basis of immunity has become vital in the field of organ transplants. Prior to Burnet and Medawar's work, it was thought that immunity was in some way related to the nature of an individual's blood.

Unlocking mysteries of the mind

By Brad Collis
Additional research by Katrina Lobley

JOHN CAREW ECCLES
1903-1997
Nobel Prize in Physiology or Medicine 1963
Shared with Alan Lloyd Hodgkin and
Andrew Fielding Huxley

Australian Academy of Science

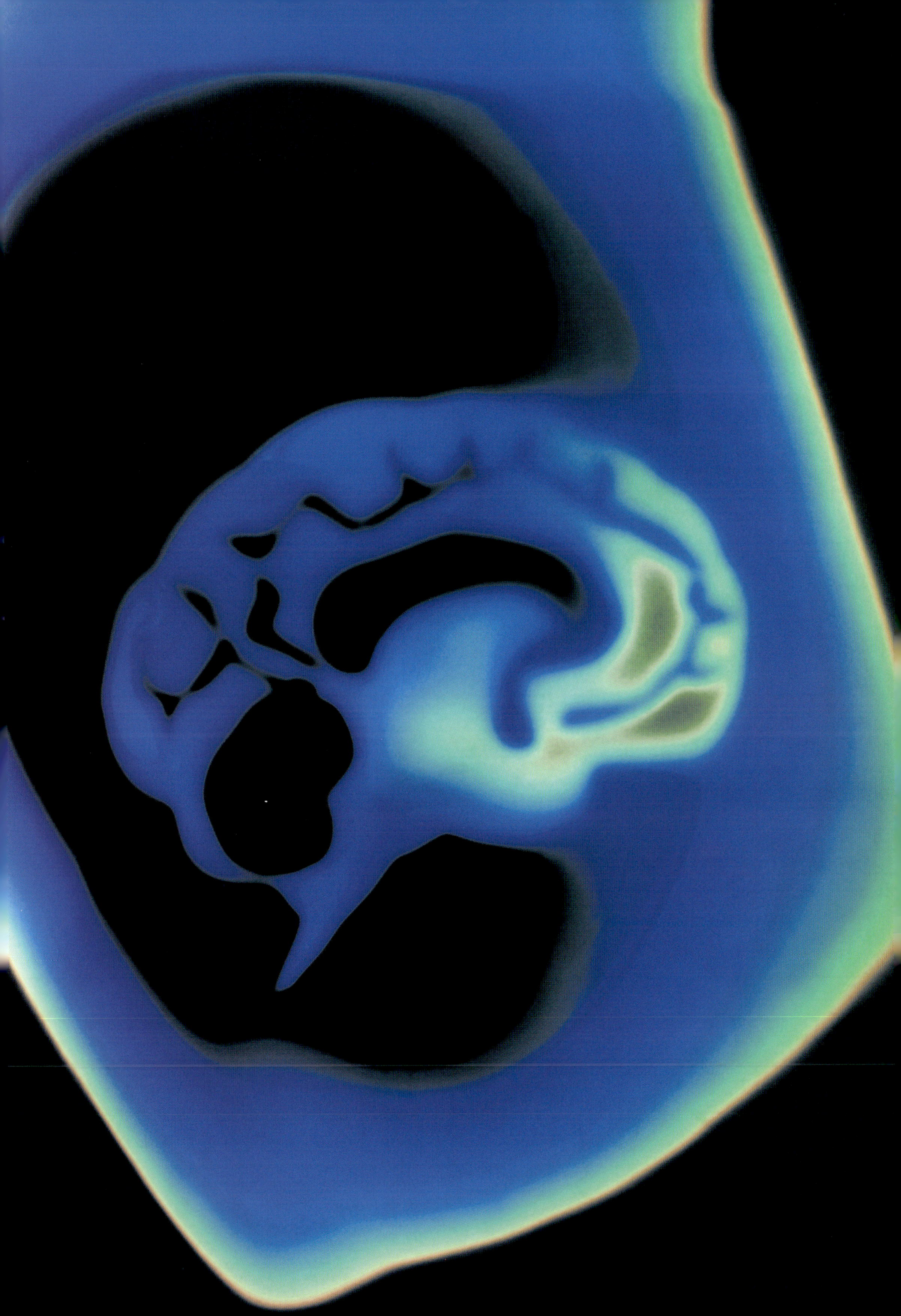

The meaning of life is a question we have all pondered at some time. John Eccles' research provided some vast insights into the workings of the mind.

Are you your brain or your mind? Is the essence of who you are controlled by the brain – an organ – or the mind, which has no known physical dimension? Is your mind a mysterious, ethereal controller of your brain, or a complex but wholly explicable product of the brain's chemical, electrical and neurological processes? Indeed, is the mind real or imagined?

Start exploring questions like these and you plunge into a maelstrom of conflicting philosophies and sciences; the same ferment of human thought that has pitted atheists and spiritualists against each other throughout the ages.

Ponder the existence and function of the mind and you enter into the world of Sir John Carew Eccles, whose obsession with the brain and central nervous system reshaped science's understanding of neurological processes.

His research and insights into these complex and controversial mechanisms that define humanity culminated in him being awarded science's highest honour, the Nobel Prize, in 1963. The honour was shared with two British neurology specialists, Sir Alan Lloyd Hodgkin and Sir Andrew Fielding Huxley.

Their citation read: "For their discoveries concerning the ionic mechanisms involved in excitation and inhibition in the peripheral and central portions of the nerve cell membrane".

In plain English, the three scientists, individually and collectively, significantly expanded medical researchers' understanding of the central nervous system, including how it works and controls the body.

For Eccles, winning the Nobel Prize was the culmination of his search for the mechanisms by which the mind controls the body.

He always said his quest began as an 18-year-old medical student, when he was struck by an awesome feeling of his own uniqueness; he marvelled at his brain's capacity for thoughts and emotions.

Yet perhaps his obsession with the brain started at a much younger age. A memoir entitled *Hello Eccles*, written by his family, states: "He was born with an uncommonly large head and the doctor told his parents bluntly that this indicated the child would either be an idiot or be brilliant. Anxiously they awaited his development."

One of Eccles' daughters, Mary Mennis, said her father once confessed that he'd done an IQ test and was off the top of the scale, that his IQ couldn't actually be tested.

Prior to Eccles' groundbreaking neurological studies that greatly increased knowledge of the mammalian nervous system, the brain was regarded by many neuroscientists as a powerful, albeit poorly understood, biological machine – a haphazard collection of circuits that "wire" bodily functions via the central nervous system.

When processed by the brain, the combined action of all this circuitry, feeding information into and out of the brain, was thought to somehow create the sense of personal control and individual identity.

Eccles, also a neuroscientist, considered such conventional views as ridiculously simplistic. In fact, he described many of the more popular theories of the day as "impoverished and empty".

While every bit a disciplined scientist with an obsession for rigidly objective experimentation, Eccles nonetheless believed that if the science was going to be credible it couldn't simply dismiss what he called "the mystery of human creativity and uniqueness".

He wrote: "Extensive experimental studies have shown that mental acts of attention and intention activate appropriate regions of the cerebral cortex. An intention to move, for example, initiates the firing of a set of neurons of the supplementary motor area about 200 milliseconds before the intended movement takes place. If the mind is the brain, this would mean either that one part of the brain activates another part, which then activates another part, or that a particular region of the brain is activated spontaneously, without any cause, and it is hard to see how either alternative would provide a basis for free will."

Free will or mind machine

Eccles believed that the notion of the brain as nothing more than a "virtual machine" could not stand up to even basic scrutiny as soon as a scientist acknowledged the mysteries that the "machine" theory couldn't explain.

For example, to sidestep the mystery of the uniquely human capacity to "wonder" as something that would be learned in the course of time as the brain became better understood was, to Eccles, shoddy science.

"I maintain that the human mystery is incredibly demeaned by scientific reductionism, with its claim in promissory materialism to account eventually for all of the spiritual world in terms of patterns of neuronal activity," he wrote.

"This belief must be classed as a superstition

He was born with an uncommonly large head and the doctor told his parents bluntly that this indicated the child would either be an idiot or be brilliant

Eccles, the Rhodes Scholar.

Eccles, seated in the middle, with colleagues at Oxford.

Young Eccles, fourth from left, with his family.

and we have to recognise that we are spiritual beings with souls existing in a spiritual world as well as material beings with bodies and brains existing in a material world."

Eccles was never shy about including both the philosophical and the anatomical in the course of his research, which went further than anyone had previously gone into concerning whether, or how, both physical and mental states might exist or interact in the nervous system.

This was largely due to him being – at that time in his life – a practising Catholic with a strong belief in God, which he took to his research without any sense of contradiction as he peeled away at the mechanics of the human brain and its neurological processes – processes that other scientists believed to be the inner and wholly human origin of spiritual awareness.

In an undated interview with ABC Radio he once explained his scientific obsession this way: "How we come to be and how we are what we are is beyond any understanding. I have been obsessed by this; trying to understand the very nature of my existence. Somehow or other, what is going on in your brain turns into a perception. This is a tremendously important problem beyond any solution, but it is a real problem."

John Eccles, the scientist who defined some of science's most imponderable questions during the 20th century was born on January 27, 1903 at Northcote, a suburb of Melbourne. His father, William James Eccles, and his mother, Mary (née Carew), were both school teachers, born in Victoria.

At the age of 12, Eccles began his secondary schooling at Warrnambool High School. Before entering the University of Melbourne, he spent a year at Melbourne High School studying science and mathematics. He headed the school at the final state-wide exam, shared the state geometry prize and gained a senior scholarship to the University.

Although interested in mathematics, Eccles chose to study medicine and started his five-year course in 1920 at the age of 17. He was active in various university societies and sport, gaining a full blue in athletics after setting an Australian universities' record in pole vaulting.

Like many students whose courses included the biological sciences, he was strongly influenced by Darwin's *Origin of Species.* It encouraged Eccles to read widely on both physiology and philosophy and it was when he couldn't find a satisfactory explanation of the interaction between the mind and the brain that he decided, while still a student, to become a neuroscientist.

After reading Sir Charles Sherrington's 1906 book, *The Integrative Action of the Nervous System*, Eccles applied himself to achieving a Rhodes Scholarship just to be able to work with the pioneering neurophysiologist at Oxford University.

"I, at the age of 17 or 18 as a medical student, suddenly came up against a problem. What am I? What is the meaning of my existence as I experience it? And I proceeded to read quite a lot about what the philosophers had said and the psychologists and was profoundly dissatisfied with it. So I decided that they didn't know enough about the brain and the brain was the essence of all my consciousness and everything that I knew myself to be. I decided that I would learn something myself about it."

Eccles completed his medical course in February 1925, gaining first class honours, first place and several clinical prizes. He graduated with the degrees of Bachelor of Medicine and Bachelor of Surgery, and had already been awarded the Rhodes Scholarship for Victoria.

After six months as a resident medical officer at St Vincent's Hospital in Melbourne, he left Australia in August 1925 and arrived in Oxford in October. Eccles then worked closely with Sherrington until 1931 in the Oxford Neurophysiology School, which produced major advances in neural understanding.

Sherrington, also a published poet, was 74 and his experiments with Eccles were the last in which he actively participated. However, the pair formed a deep friendship that would last until Sherrington's death.

Later in 1952, when Eccles was in England to deliver the eight Waynflete Lectures, he visited Sherrington at his Eastbourne nursing home just days before his great mentor died.

In 1932, Eccles became a Staines Medical Research Fellow and two years later, lecturer in Physiology at Magdalen College, Oxford University.

It was when he couldn't find a satisfactory explanation of the interaction between the mind and the brain that he decided, while still a student, to become a neuroscientist

In the 1920s, the Oxford Neurophysiology School, under Sherrington, led the world in the field of mammalian central nervous system physiology. It was also a time of controversy between the exponents of rival chemical and electrical theories on the nature of the process of transmission at the terminal junctions (synapses) of nerve cells (neurones). Although there was strong evidence of chemical transmission at excitatory and inhibitory synapses in the peripheral nervous system, synaptic transmission in the central nervous system was widely considered to be an electrical process.

Chemical resistance

Eccles resisted many aspects of the chemical transmitter theory. The techniques then available for studying central synapses were simply inadequate for determining the nature of the transmission process, but the debate was important in defining problems and stimulating considerable experimental work. The eventual victory of the chemical theory was another 20 years away.

For a while Eccles looked like he would stay at Oxford for the duration of his academic career, but in

Eccles' passion for his research led beyond science to philosophy.

Eccles and friend in Lindau, Germany.

1937 he was drawn back to Australia by a position at the Kanematsu Memorial Institute of Pathology at Sydney Hospital. His mentor, Sherrington, had retired from his post in 1936 and Eccles was disappointed at the direction that research was taking at Oxford. He was also increasingly worried about the political uncertainty in Europe.

"There was the ominous rise of Hitler against the unprepared Western Alliance, so I decided, perhaps unwisely, to return to what seemed the security of Australia. There was the opportunity to create in Sydney a research institute matching the Hall Institute at Melbourne Hospital with Kellaway and Burnet. I accepted the directorship of the Kanematsu Institute at Sydney Hospital."

Eccles said that in retrospect, he felt he should have stayed in England and weathered the storm but instead, "I embarked on my odyssean journeyings, never to return to my beloved England. It was a fateful choice".

He found the academic isolation severe. "The Sydney University Medical School was a very dim place, being little more than a teaching institution. Unbelievably, it was completely locked up by guards at 5pm. Even the professors had to scurry out to avoid imprisonment for the night!"

At the Kanematsu Institute he led a team studying the actions of chemical substances on the transmission of nerve impulses to muscles. One of the team members was Bernard Katz, a refugee from Nazi Germany, and a key member of the research team that eventually proved the chemical theory in nerve transmission.

It was Katz who was eventually able to show how chemical transmitters released from nerve endings produce electrical currents, and that the chemical alters the configuration of molecules in the cell membrane. This allows ions to flow into muscle fibres and generate electrical currents. Katz was awarded the Nobel Prize in Physiology or Medicine in 1970. However, Katz's collaboration with Eccles was interrupted by Japan entering the war – Katz joined the Royal Australian Air Force in 1942 and Eccles found himself working on the problems of communication in the noisy environment of aircraft and tanks.

The last straw, however, came when the hospital board decided to add two floors to the Kanematsu Institute to accommodate resident medical staff, effectively curtailing future expansion of the research laboratories. Eccles resigned in late 1943 and moved to Dunedin, New Zealand as professor of Physiology at the Otago Medical School.

His new job required him to take on a heavy teaching load, so Eccles, who would be called "Prof" and eventually "Jack" by younger colleagues, conducted many of his experiments late at night and in the early morning. Many people, when describing what Eccles was like, mentioned his enormous energy, along with a powerful personality.

It was in New Zealand that Eccles met science philosopher Karl Popper, who became a pivotal influence on Eccles and his research.

"The year 1944 was important in my scientific life above all my post-Sherrington years, because my intimate association with Karl Popper dates from that time," Eccles said. "Many people, including myself, had our scientific lives changed by the inspiring new vision of science that Popper gave us."

Popper was a vigorous proponent of the "falsifiability" of hypotheses as the only true test of validity for a scientific theory. In other words, he argued that proving a proposition to be valid was not good enough – just as much effort had to go into deliberately trying to disprove it as well.

Eccles found this a liberating way to approach experiments. It meant he could be

Eccles (far right) with 1945 Nobel Prize winner, Howard Florey, next to him.

Eccles with his family: he had five daughters and four sons with his first wife.

absolutely daring in developing any hypothesis, and rejoice even when he proved it wrong because this in itself was a scientific success. Eccles would later say this greatly lifted his conceptual power because in many ways the shackles were off. It removed any hesitation about proposing a theory that might later be proved flawed.

Eccles and Popper eventually collaborated on the influential 1977 book *The Self and Its Brain*, a fascinating probe into the body-mind, self and soul puzzle. It remains the most cited of all of Eccles' philosophical writings.

Concerned that his heavy teaching load in Dunedin was limiting his competitive edge, Eccles returned to Australia in 1952 to take up the foundation chair of Physiology in the John Curtin School of Medical Research at the recently established Australian National University in Canberra.

He regarded his "14 golden years" there as the most productive and enjoyable period of his career. Colleagues remember him as honest and occasionally terse. However, Eccles would unreservedly heap praise upon anyone who had made significant discoveries or progress.

In Canberra, Eccles made significant contributions to the understanding of the ionic basis of central nervous system excitation and inhibition, and the nature of chemical transmitters involved in this. It was for this work that he received the Nobel Prize. His daughter Rose was working with him as a researcher in Canberra when he received the Nobel Prize news. She recalled the momentous occasion, which Eccles downplayed in his own writings: "We had a wonderful celebration in the lab and we presented him with a candlestick," she said.

Forced retirement

In 1966, despite a brilliant career and international acclaim, Eccles crashed head-on into the cultural apathy that dogged academic achievement in Australia. Despite his knowledge, experience and zest for science and life, he found absolute disinterest within the nation's bureaucracy and university hierarchies in finding a way around retirement at the compulsory age.

Reluctantly, he left Australia again and was welcomed with open arms by the American Medical Association's Institute for Biomedical Research in Chicago. It wasn't a happy or successful period for Eccles, and he moved three years later, this time to be distinguished professor of Physiology and Biophysics at the State University of New York.

His departure from his home country had also been coloured, in the eyes of some, by his divorce from his wife Irene, with whom he had nine children, and his marriage shortly afterwards to Dr Helena Taborikova, a Czechoslovakian neurophysiologist whom he had met at a scientific congress in Prague in 1963.

Eccles' daughter Rose Mason felt her father's decision to leave Australia again had more to do with personal than professional circumstances. When interviewed for a Radio National program celebrating her father's centenary, she said her parents had "drifted miles apart".

"My mother had become very religious and she went to religious meetings. She got up and went to daily mass and this used to irritate him because he'd get to bed at two or three and the alarm would go at six and, you know, wake him up. Both were incompatible in many ways, their characters, and he was much more open – he loved meeting people."

Despite the incompatibilities, daughter Mary Mennis said her mother was shocked to hear – via a friend who heard it on the evening news – that Eccles, who was in Chicago, was filing for divorce.

Eccles ceased to be a practising Catholic. When his second wife Helena was interviewed on Radio National, she said he could no longer accept many of the dogmas of the church.

When Eccles did finally retire in 1975 at the age of 72, he moved to the "idyllic mountain surroundings" of Contra in Switzerland. Both he and Helena, who outlived him, were sad that Australia had not accommodated his willingness to keep experimenting and writing well beyond retirement age.

He travelled widely throughout Europe and the US and played a prominent role in the International Physicians for the Prevention of Nuclear War organisation.

Eccles also continued to wrestle intellectually with a new generation of researchers. In Frankfurt in 1993 during a celebration of his 90th birthday, he was still pushing advanced ideas on cerebral function and the nature of consciousness and the self-psyche-mind-body conundrum.

Mental interactions

Eccles came to believe that humans have a non-material mind or self that acts upon, and is influenced by, our material brains – a mental world in addition to the physical world, and that the two interact.

As for what the mind actually is, Eccles conceded that it couldn't be pure nothingness, otherwise it couldn't exist, in which case he reasoned it to be composed of finer grades of energy-substance. Indeed, he suggested our inner constitution might comprise several non-physical levels.

Eccles always had plenty of opponents, however. In his last book, 1994's *How the Self Controls Its Brain*, he was buoyed by advances in quantum

physics and the latest discoveries about the microstructure of the neocortex, which he felt were on the way to supporting his proposition.

Eccles called the fundamental neural units of the cerebral cortex dendrons, and he proposed that each of the 40 million dendrons is linked with a mental unit, or psychon, representing a unitary conscious experience. In willed actions and thought, he proposed that psychons act on dendrons and momentarily increase the probability of the firing of selected neurons.

Eccles remained in basic agreement with the neo-Darwinian theory that evolution is driven by random genetic mutations followed by the weeding out of unfavourable variations by natural selection. However, he added to this his belief in a divine providence operating over and above the biological evolution.

As for what happens after death, Eccles had little option but to fall back onto his personal perspective: "We did come to be by something we do not understand at all, and therefore in our apparent ceasing to be, with death, it is the same problem. So I should say we have hope because we know nothing, and we should not dogmatise it. This is part of the message I have for humanity, that there is such tremendous mystery, so little is known, and the wonder is so enormous that we must have hope."

Eccles took the scientific exploration of the mind and brain as far as anyone had been able to push, but even he was forced to concede that perhaps the meaning of life might always be a mystery. He said: "For me the one great question that has dominated my life is: what am I?"

After suffering ill health from 1994, John Eccles died on May 2, 1997 at Locarno, Switzerland.

At Eccles' request, he was buried in Contra near his last home. ■

Eccles and his first wife Irene with the Nobel medal.

Vital statistics

Name:	John Carew Eccles
Born:	January 27, 1903, in Northcote, Melbourne
Died:	May 2, 1997, in Locarno, Switzerland
School:	Warrnambool High School, Melbourne High School
University:	University of Melbourne, Oxford University
Married:	Irene Frances Miller, of Motueka, New Zealand in 1928, divorced 1968; married Helena Taborikova, of Prague, Czechoslovakia, 1968
Children:	Five daughters, including Rosamond who worked at the Australian National University with him as a researcher from 1955-66, and four sons with Irene Miller
Lived:	Australia, New Zealand, the United States, England and Switzerland

Awards and accolades

1927-29:	Christopher Welch Scholar, Oxford
1932-34:	Staines Medical Fellow
1952-66:	Professor of Physiology, Australian National University
1952:	Waynflete Lectures, Magdalen College, Oxford
1958:	Knight commander
1962:	Royal Medal, Royal Society
1963:	Australian of the Year
1966:	Foreign Associate, National Academy of Sciences
1968-75:	Distinguished Professor of Physiology and Biophysics, State University of New York at Buffalo
1990:	Companion in the Order of Australia
1993:	90th birthday marked by the establishment of the Sir John Eccles Lecture, University of New South Wales

Why he was awarded the Nobel Prize

Sir John Eccles' work with the brain and central nervous system reshaped science's understanding of neurological processes. His discoveries expanded the knowledge available surrounding the central nervous system including how it works and how it controls the body.

In particular, his work concerned the electrical changes, which the nerve impulses elicit when they reach another nerve cell. He made significant contributions to the understanding of the central nervous system including excitation and inhibition, and the nature of chemical transmitters involved in this. Eccles showed how excitation and inhibition are expressed by changes of membrane potential.

When the response is sufficiently strong to cause excitation, the membrane potential decreases until a value is reached at which the cell fires off an impulse. This impulse travels through the nerve fibre of the cell and causes contraction in a muscle. He showed how excitation and inhibition correspond to ionic currents, which push the membrane potential in opposite directions.

By understanding the nature of the events in the peripheral and central nervous system, a greater understanding of nervous action has been achieved. Eccles believes that the origin of each of us stems from codes of genetic inheritance and the most significant questions we can ask scientifically concern the working of our nervous systems – the reception, communication and storage devices that subserve all our perception, our thoughts, our memories, our actions, our creative imaginations, and our ideals.

References: David Pratt, John Eccles on Mind and Brain, http://ourworld.compuserve.com/homepages/dp5/eccles.htm Original article published in *Sunrise*, June/July 1995.
Australian Academy of Science, Biographical Memoirs, John Carew Eccles 1903-1997, http://www.science.org.au/academy/memoirs/eccles.htm#1
Royal Society of New Zealand 1999 Academy Yearbook, Sir John Scott, Sir John Carew Eccles, http://www.rsnz.govt.nz/directory/yearbooks/year99/eccles.php
A.D'A. and D.D'A Webling, *Nobel Prize Laureates and Australia to 2000*, private monograph.

A quantum leap

By Kathy Graham

The Nobel Foundation

ALEKSANDR MIKHAILOVICH PROKHOROV
1916-2002
Nobel Prize in Physics 1964
Shared with Charles Hard Townes and
Nicolay Gennadiyevich Basov

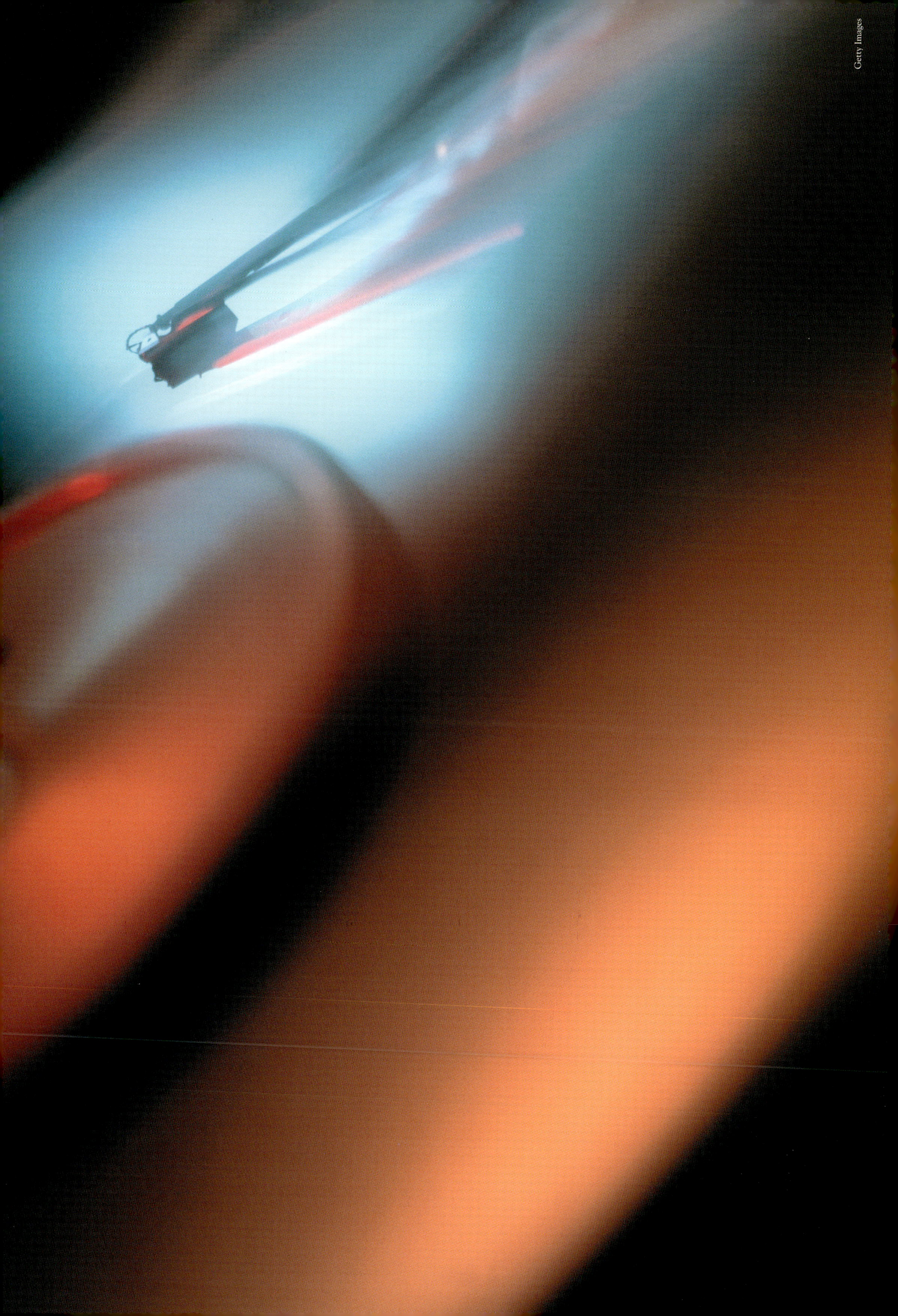
Getty Images

Aleksandr Prokhorov's invention of laser technology laid the groundwork for many scientific advances including compact discs and modern surgery.

Today's laser applications which range from supermarket checkouts to global communications and broadcasting networks can be traced back to the work carried out by Aleksandr Mikhailovich Prokhorov. His work provided the first practical demonstration of quantum electronics, and the origin of today's vast industry for lasers. Other applications include high fidelity recording on compact discs, surgery, printing and astronomy and space research.

Prokhorov was born in 1916 at the family farm in Russell Road, Peeramon in the Atherton Tablelands, Queensland. His Russian émigré parents, Mikhail Ivanovich Prokhorov and Mariya Ivanovna Mikhaelovna (Prokhorov) had fled from Siberia to Australia in 1911 because of Mikhail's involvement in revolutionary activities. As an opponent of the Czarist regime, he had been sent to Siberia following the abortive 1905 revolution.

After the hoarfrost, howling wolves, frozen mountains, salt mines and human chain gangs of Siberia, the young couple's new home in Australia's tropical far north was literally a world away. Early last century, the Cairns district was actively attracting migrants from all over Europe to develop tin and gold mines, cut timber and work on the railways. Among the new arrivals looking for a fresh start were a small group of Russian families, many highly educated and, like the Prokhorovs, escaping the serious social upheavals taking place in their homeland.

Looking for some sense of community, many took up new land selections in the same area of lush rainforest country characteristic of the Atherton Tablelands. This part of the world is still known as "Little Russia". As for Peeramon, little remains of its pioneering past. Today the town's main attraction is Peeramon Hotel, the oldest pub in Queensland.

Prokhorov was the third and youngest child. His parents also had two daughters Jane and Lila. When he was seven years old, Lila – the eldest sister – died unexpectedly. That same year, 1923, the family decided to return to Mother Russia, a move probably influenced by the propaganda filtering out of the country following the October Revolution of 1917.

When the Communists took power in the country, they promised a "worker's paradise", raising the hopes of many that the horrors visited on Europe by the First World War would be expunged by a new utopian world order.

Life under Stalin

Life in Russia as a teenager was not easy for the young Prokhorov. The civil war had been a disaster for the country's economy, culminating in the enormous famine of 1920-21. The following year, the Union of Soviet Socialist Republics (USSR) was established and when Vladimir Lenin died soon after, his successor Joseph Stalin began his reign of terror.

It was during this period that Stalin introduced farm collectivisation, effectively wiping out the peasantry both as a class and as a way of life. He also oversaw the execution or exile of millions to Siberian concentration camps.

Against this bloody and turbulent background, Prokhorov grew into adolescence. In 1934, he was accepted into the Leningrad State University and graduated four years later with an honours degree in physics. This led to postgraduate work at the P.N. Lebedev Physical Institute, Russian Academy of Sciences in Moscow. There he studied the propagation of radio waves over the earth's surface and with one of his research directors, the physicist V.V. Migulin, proposed a novel technique for using radio interference to explore the ionosphere, one of the layers of the upper atmosphere.

In June 1941, the German invasion of Russia interrupted Prokhorov's studies and he entered military service, despite eligibility for exemption because of his academic research. In the same year, he married Galina Alekseyevna Shelepina, a geographer with whom he later had one son, Krill.

However, instead of a life of simple domesticity with his new bride, Prokhorov went on to serve with the Red Army in a period of warfare that would eventually kill one sixth of the population.

The battles for Leningrad (Prokhorov's old stomping ground) and Stalingrad were particularly protracted and obscene. While defending his country, Prokhorov was wounded twice.

After his second injury in 1944, he was demobilised. He was awarded the Order of the Patriotic War, 1st class, and the Medal for Valour for his bravery.

After taking off his uniform, Prokhorov wasted no time in resuming his studies as a senior research associate in the Oscillations Laboratory of the P.N. Lebedev Physical Institute. His main passion by this time was nonlinear oscillations about which he devised a theory that became the subject of his thesis. This earned him his candidate's degree (equivalent to a master's degree) in 1946. For this work, he and two other physicists were awarded the prize named in honour of the academic Leonid I. Mandelshtam, a Soviet radio physicist of considerable repute.

From 1947, he studied the radiation from electrons produced in a synchrotron (a device that accelerates charged particles such as protons and electrons in widening circles to very high energies). He demonstrated that the electrons radiate at wavelengths of the order of centimetres in the microwave region. As a result of these investigations, he wrote his PhD thesis "Coherent Radiation of Electrons in the Synchrotron Accelerator" in 1951.

The year before, Prokhorov was appointed assistant director of the Oscillation Laboratory at the P.N. Lebedev Physical Institute. In 1954, he was appointed head of the Institute, a post he held until his retirement. The department of oscillations soon became a research nursery. In 1959, the laboratory of radio astronomy was organised from one of the groups in the department, and in 1962 another specialist laboratory was dedicated to quantum physics.

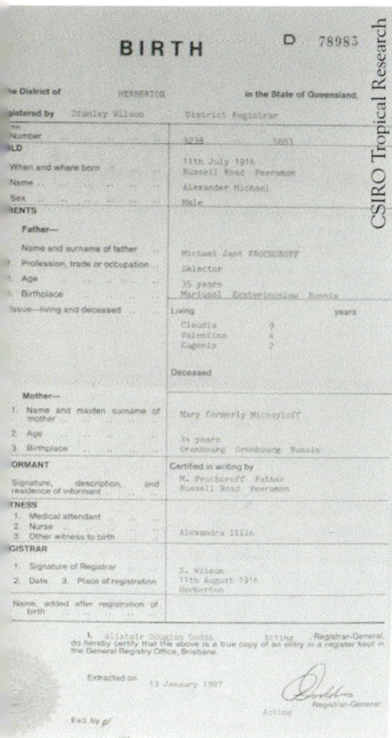

BIRTH D 78983

...e District of HERBERTON in the State of Queensland.
...gistered by Stanley Wilson District Registrar

Number	9238 5881
...LD	
When and where born	11th July 1916 Russell Road Peeramon
Name	Alexander Michael
Sex	Male
...RENTS	
Father—	
Name and surname of father	Michael Jant PROCHOROFF
Profession, trade or occupation	Selector
Age	35 years
Birthplace	Mariupol Ekaterinoslaw Russia
Issue—living and deceased	Living years: Claudia 9, Valentina 4, Eugenia 2; Deceased
Mother—	
1. Name and maiden surname of mother	Mary formerly Michayloff
2. Age	34 years
3. Birthplace	Orenbourg Orenbourg Russia
...ORMANT	Certified in writing by
Signature, description, and residence of informant	M. Prochoroff Father Russell Road Peeramon
...TNESS	
1. Medical attendant	
2. Nurse	
3. Other witness to birth	Alexandra Illin
...GISTRAR	
1. Signature of Registrar	S. Wilson
2. Date 3. Place of registration	11th August 1916 Herberton
Name, added after registration of birth	

I, Alistair Douglas Dodds Acting Registrar-General do hereby certify that the above is a true copy of an entry in a register kept in the General Registry Office, Brisbane.

Extracted on 13 January 1987

Registrar-General
Acting

Exd. by

CSIRO Tropical Research

Prokhorov's birth certificate.

Eacham History Society

Logging at Peeramon, Prokhorov's birthplace.

Prokhorov (left) and Basov (right) showing their lab to Charles Townes.

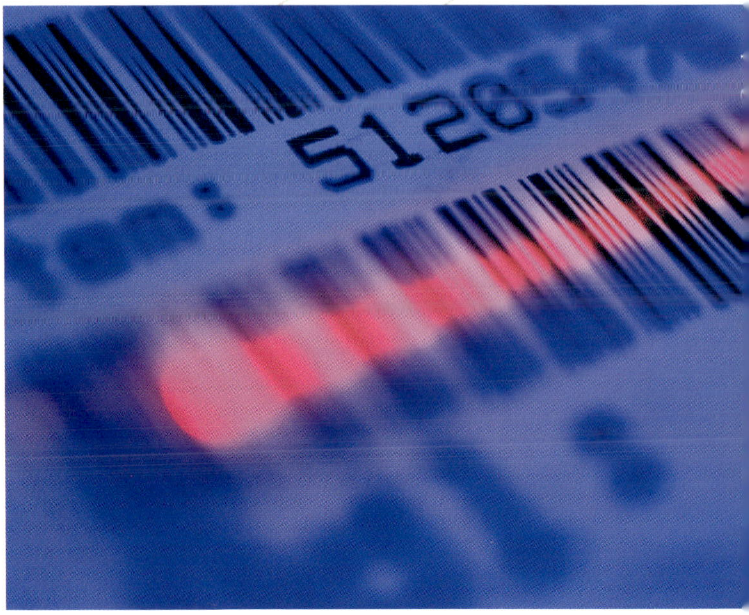

Laser on a barcode.

These days, the laser is taken for granted as a common component in domestic, industrial and scientific equipment. Yet for the first 10 years of its existence, it was regarded as a genius invention with little practical use. Its potential only began to emerge in 1964, when three scientists shared the Nobel Prize in Physics for its discovery.

For the first 10 years of its existence, the laser was regarded as a genius invention but with little practical use

After completing his doctorate, Prokhorov began to investigate the question of radio spectroscopy on a wide scale. Somewhat later, he explored quantum electronics. To assist in this field of inquiry, he organised around him a team of young research scientists. Together they applied radar and radio broadcasting techniques, developed primarily in the US and England during and after the Second World War, to study the vibrational and rotational spectra of molecules.

Prokhorov focused his research on a class of molecules called asymmetric tops, which have three different rotational axes of symmetry and are the most difficult to analyse in terms of their rotational spectra. In addition to purely spectroscopic research, he carried out a theoretical analysis of the application of microwave absorption spectra to improving frequency and time standards. This latter work led to his collaboration with Nikolay Basov.

While the pair was searching for a technique to amplify microwave signals in spectroscopic experiments, they hit upon the idea of using a gas-filled cavity with reflectors at either end, in which the microwave beam would be intensified. They then discovered that this method produced microwaves with an extremely narrow range of frequencies.

This laid the theoretical groundwork for the eventual construction of a molecular oscillator (or "maser") operating on ammonia. This is a device emitting microwave radiation of a single wavelength, and was a precursor to the laser. In 1955, Prokhorov studied the electronic para-magnetic resonance (EPR) spectra of ruby with A.A. Manenkov, which actually made it possible to suggest it as a material for lasers in 1957.

Prokhorov at his desk at home.

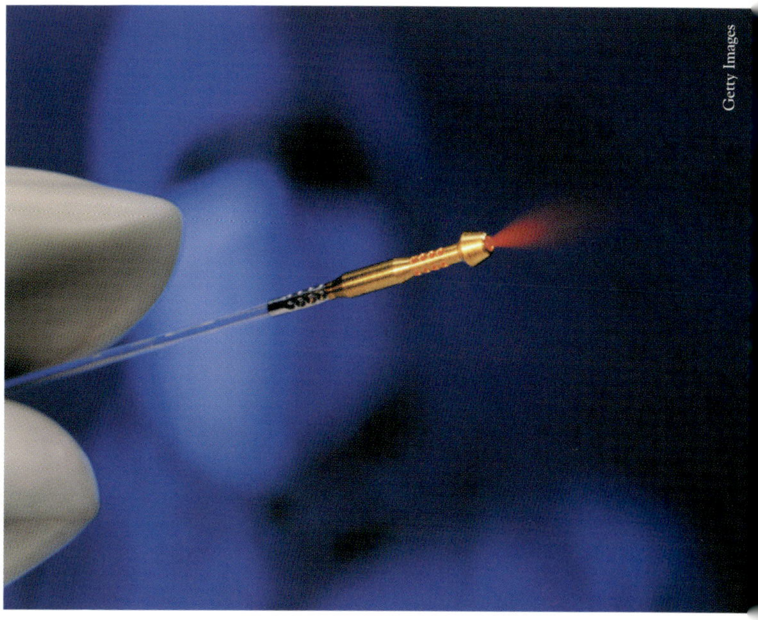

His work preceded modern applications of laser technology.

The two announced the discovery of their molecular generator in a paper read before the All-Union Conference on Radio Spectroscopy in May 1952. However they held off publishing their results for more than two years, by which time the American physicist Charles H Townes had built a working maser and published his conclusions in *Physical Review*.

Prokhorov's research came from a combined understanding of the science of optics and radio engineering. In 1964, he won the Nobel Prize in Physics for discoveries leading to the development of the laser. He shared this with compatriot Basov, also at the P.N. Lebedev Physical Institute in Moscow and US scientist Charles Townes whose discoveries were made independently at Columbia University and Bell Telephone laboratories.

"Many believed that we had gone crazy, that it was impossible," Prokhorov said in a television interview in 2001. "It was a brave step, because before that no one had said it was possible to create a generator of optical range. Then it became a new, independent science – optics." The inventions were the first practical demonstration of quantum electronics.

Taking a stand

Prokhorov was regarded as a politically controversial figure by the west. In the 1950s, he established the Radio Spectroscopic Laboratory at the Nuclear Physics Research Institute of the Moscow State University, of which he became a full professor in 1957.

The research institute was also a key contributor to the Soviet program to build a counter system to US president Ronald Reagan's plans for a "Star Wars" defence system – so called because it would destroy ballistic missiles in space. But Prokhorov, who became a lifelong member of the Communist Party in 1950, was anything but a conformist in an era when expressing views that didn't align with the Party's was both rare and dangerous. Even under strict government controls, he was known for his independent streak and outspokenness.

As the editor-in-chief of the multi-volume reference book the *Great Soviet Encyclopedia*, a position he held from 1969, Prokhorov ignored orders to have the dissident physicist Andrei Sakharov excluded from the encyclopedia. Father of the Soviet hydrogen bomb, Sakharov later turned anti-war activist, winning the Nobel Peace Prize in 1975. He was also one of the regime's most courageous critics, a defender of human rights and democracy.

This act of defiance by Prokhorov might have sent him to exile in Siberia like his father before him, had he not been so highly respected.

But ironically, in 1983, he was one of four Soviet scientists to criticise Sakharov, who argued that the US might have to match the Soviet

Peter Whitson

The town of Peeramon has the dual 'honour' of being Prokhorov's birthplace and home of the oldest pub in Queensland.

Union in nuclear weapons before effective arms reduction negotiations could begin. In the previous year Prokhorov had been one of 97 Nobel Prize winners who called for a freeze on his country's development and deployment of nuclear weapons.

Yet in 1984, he lent his name to what was seen as a Soviet propaganda campaign. Prokhorov was one of four Soviet Nobel laureates to sign a letter to President Reagan on behalf of an imprisoned Native American political activist, Leonard Peltier. However the letter was perceived to be part of a Soviet ploy to deflect attention from Sakharov's hunger strike, staged on behalf of his wife who had been refused permission to travel to the West for treatment of a heart condition.

In the political sphere, his fame grew even more when he refused a government invitation to become a deputy in the Soviet parliament, famously declaring, "I am not a politician. I am a scientist".

Prokhorov continued to use his position to engage world leaders throughout his life. In 1999, he was one of several scientists to sign a letter to the Presidents of the USA and Russia urging them to work together to help Russia make the transition to a democratic country. That same year he was the chair of the Russian Committee for the Observance of the United Nations International Year of the Elder Person.

Although Prokhorov could have easily rested on his laurels after he won the Nobel Prize, he went on to create various types of lasers. In later years, his work focused on fibre and integrated optics and the creation of optical communication systems. He also supervised research in microelectronics, plasma physics, controlled

> His fame grew even more when he refused a government invitation to become a deputy in the Soviet Parliament, famously declaring, "I am not a politician. I am a scientist"

thermonuclear fusion and laser medicine. Between 1973 and 1981, he was chief of the Department of General Physics and Astronomy of the Institute of General Physics.

In 1981, he became director of the Natural Sciences Centre of the P.N. Lebedev Physical Institute and in 1983 he founded the General Physics Institute in Moscow, part of the Russian Academy of Sciences, and served as its director until 1988. He also served as editor-in-chief of *Laser Physics*, a bi-monthly international journal.

End of an era

Prokhorov was 85 when he died of pneumonia on January 8, 2002 in his home in Moscow. Scientists and politicians worldwide mourned his death, saying it marked the end of an era for Russian science. Russian President Vladimir Putin said in an issued statement: "His name is linked to outstanding discoveries that in many ways defined 20th century civilisation".

Yevgeny Dianov, head of the Fibre Optics Scientific Centre, told Russian television: "His independence allowed him to speak freely to the Soviet authorities." Meanwhile, the Russian winner of the 2000 Nobel Prize in Physics, Dr Zhores Alferov, said that his death was "a bitter loss for world scientists".

Outside of Russia, Art Guenther of the University of New Mexico's Centre for High Technology Materials in Albuquerque, New Mexico, spoke glowingly of Prokhorov as "a lifelong committed scientist, a consummate gentleman, a Russian patriot, and a global opticist. His contributions to quantum optics

Prokhorov's research contributed to the new study of optics.

Optical Society of America

CSIRO Tropical Forest

A plaque commemorating the birthplace of Prokhorov in the CSIRO grounds, Atherton, Queensland.

have formed the basis for much of today's photonics evolution and applications."

He added, he was always impressed with the "respect and adoration of the Russian scientific community for his leadership, contributions, and friendship. He will be sorely missed."

After his death, television footage showed Prokhorov's desk stacked high with files and paperwork. His glasses lay on top of one of the piles, as if he had only just taken them off. The scientist did not own a computer, which he half-joked would "interfere with his thinking". Instead he preferred to laboriously handwrite all his notes on hundreds of scraps of paper which could still be seen littering his office.

Prokhorov is buried in Moscow's Novodyevichy Cemetery, the final resting place of many prominent Soviet and Russian scientists, writers and composers. And lest we forget, on the other side of the world in Atherton, Queensland, there is a memorial plaque celebrating the achievements of Peeramon's most famous son. ■

Kathy Graham is a science writer based in Sydney.

Vital statistics

Name:	Aleksandr Mikhailovich Prokhorov
Born:	Peeramon, Queensland, Australia
Birthdate:	July 11, 1916
Died:	January 8, 2002 in Moscow
University:	Leningrad State University; Institute of Physics, Academy of Sciences; both in the USSR
Married:	Galina Alekseyevna Shelepina in 1941
Children:	A son, Krill
Lived:	Mainly in Russia

Awards and accolades

1948:	The Mandelstam Prize
1959:	Lenin Prize
1967:	Honorary professorship from Delhi University
1971:	Honorary professorship from Bucharest University
1969 & 1986:	Named a Gold Star Hero of Socialist Labour – the USSR's highest civilian award.
1988:	The Lomosonov Gold Medal for outstanding achievements in Physics

Prokhorov was also a member of many professional associations. He was elected a corresponding member of the Soviet Academy of Sciences in 1960, and became a full academician in 1966 and a member of its Presidium in 1970. He was an honorary member of both the American Academy of Arts and Sciences and the Optical Society of America. He also received: the Order of Lenin; Order of the Patriotic War, 1st class; and Medal for Valour for his bravery.

Why he was awarded the Nobel Prize

The Nobel Prize in Physics 1964 was awarded to Prokhorov with Nikolay Gennadiyevich Basov, also of Russia, and Charles Hard Townes of the United States for work in the field of quantum electronics.

The project that earned the three scientists their prize was a device that generated an intense beam of pure microwave radiation called a "maser". This stands for "microwave amplification by stimulated emission of radiation". The same principles of quantum electronics were later applied to the generation of an intense beam of pure light; hence the word "laser", which stands for "light amplification by stimulated emission of radiation". Between 1950 and 1955, Prokhorov and Basov devised a way of exciting electrons in ammonia molecules into higher energy states. The electrons were then relaxed back to their original lower state; the transition was accompanied by a burst of pure microwave radiation (a photon). The idea was then refined to generate shorter wavelengths of light. Applications of the research can be found in products such as compact discs and modern surgery.

In awarding the 1964 Nobel Prize in Physics, the Nobel committee recognised the contributions of all three physicists. One half was awarded to Prokhorov and Basov and the other half to Townes "for [their] fundamental work in the field of quantum electronics, which has led to the construction of oscillators and amplifiers based on the maser-laser principle."

References: A.D'A. and D.D'A Webling, *Nobel Prize Laureates and Australia* to 2000
Johnson's Russia List, Clara Ferreira-Marques, obituary, January 9, 2002
OSA's News, obituary, January 8, 2002
BBC News, obituary, January 8, 2002
Guardian Unlimited, Pearce Wright, obituary, January 11, 2002
Nobel Prize Winners Biographical Dictionary
The Nobel Museum
OE Magazine, obituary, Rich Donnelly, February 2002
Notable Twentieth Century Scientists
The Nobel Scientists: A Biographical Encyclopedia

The refugee who touched a nerve

By Julian Cribb

BERNARD KATZ
1911-2003
Nobel Prize in Physiology or Medicine 1970
Shared with Ulf von Euler and Julius Axelrod

The Nobel Foundation

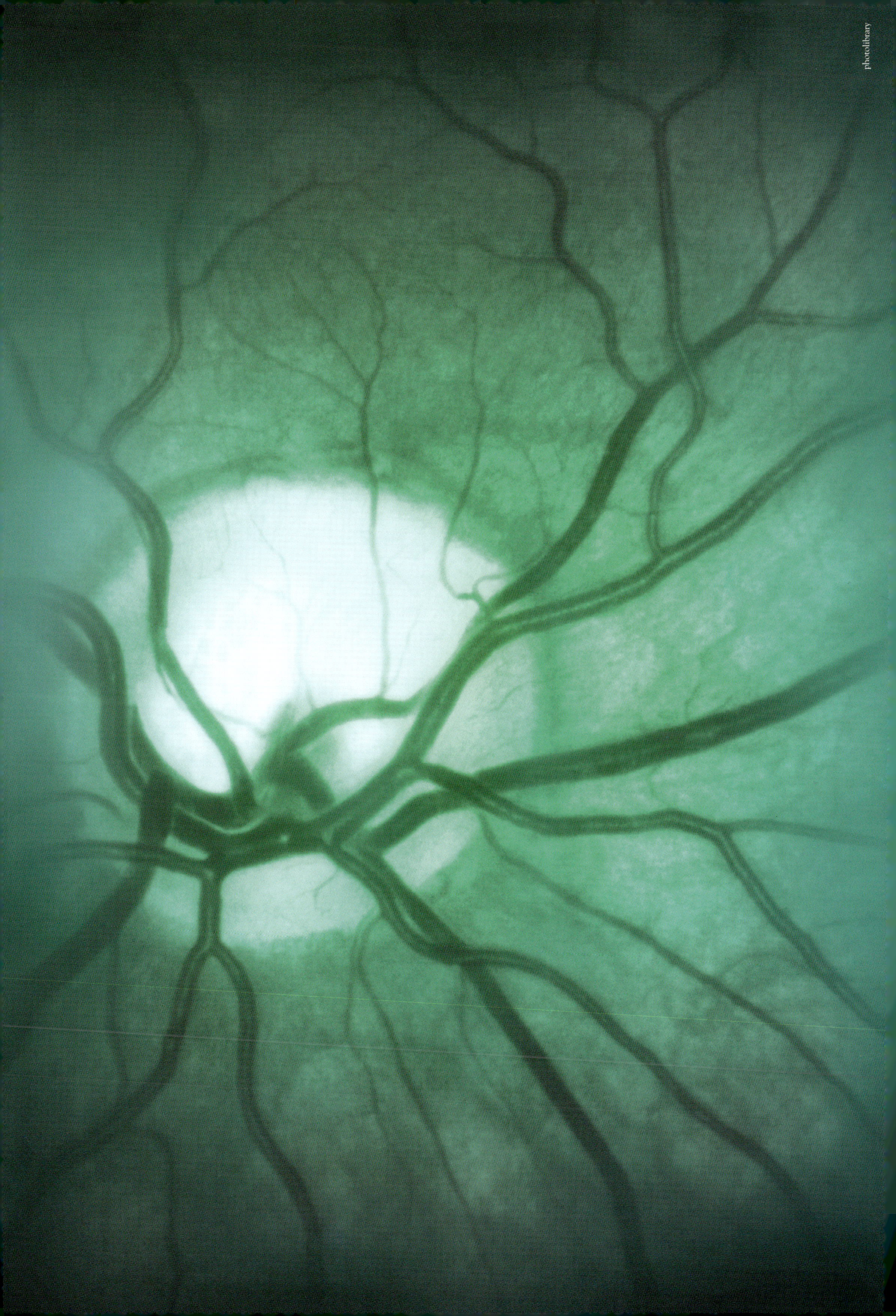
photolibrary

From fleeing persecution in Germany to contributing to the world's understanding of the human nervous system, Bernard Katz's life was anything but ordinary.

Bedraggled, penniless and feeling distinctly like Dickens' David Copperfield arriving on the doorstep of unknown relations, the young refugee nervously ascended the long, creaking flight of stairs to the top floor of University College, London.

It was a defining moment for 23-year-old Bernard Katz. Behind him lay the chasm of Nazi Germany that in February 1935 was already beginning to engulf both Jews and scientists like him. All his hopes were pinned on an outspoken British physiologist and Nobel laureate, Archibald Vivian Hill (A.V.) whom, a year earlier, had launched a scathing attack on the treatment of scientists under the Nazis, provoking the wrath of the Third Reich's scientific gauleiter, Nobel laureate Johannes Stark.

Katz was reasonably confident of a sympathetic hearing with Hill. However, he had no inkling that the daunting staircase was also taking him towards a series of brilliant advances in human knowledge, an Australian citizenship and a lifetime love-match with an Australian girl, to a British knighthood and to science's highest honour, the Nobel Prize.

Bernard Katz worked out how nerves communicate with the muscles and helped to lay the foundations for modern neuroscience and neuromedicine. He was born in Leipzig in 1911, the only child of a Russian father, Max Katz, and a Polish mother, Eugenie Rabinowitz.

However, although he was born in Germany, Katz never acquired German nationality. For his first six years he was a subject of the Tsar of Russia, as his fur merchant father had migrated to Leipzig in 1904 to escape the internal unrest that preceded the Russian Revolution. As a result, the family was regarded as "enemy aliens" during World War I. Then came the Revolution and any expatriate Russians who did not wish to become Soviet citizens found themselves without any citizenship.

Katz later said: "I did not foresee I was going to remain stateless and without a passport for the first 30 years of my life until I finally acquired this much coveted possession in Sydney, Australia."

At age nine, Katz had an early taste of what it meant to be an alien Jew when he was refused entry to the leading secondary school in Leipzig, even though he easily topped its entrance examination.

He was instead schooled at the Konig Albert Gymnasium, which had a strong reputation for the humanities and classics. Yet even in the school corridors, Katz was hardly protected from the anti-Semitic sentiment being encouraged by the fledgling Nazi Party. In 1922, at the tender age of 11, Katz heard about a fellow student who had "called the boys together and informed them of a marvelous plan that his father had discussed with him at home. The plan was that the Jewish population of Leipzig should be invited to assemble in the underground fair hall and, after closing the doors, should be killed off by filling the hall with poison gas".

Despite all of this, Katz prospered at school. He interspersed his studies with chess – for which he developed a lifelong passion – reading, walking and swimming.

The normal German high school curriculum took nine years. Katz forged ahead so strongly in his studies that he was encouraged to skip a year, and thus spent only eight years (1921-29) at the Gymnasium. His school reports were excellent, save for subjects such as gym and singing, in which he freely admits his performance was deplorable. He often tried to avoid those classes.

Katz was a quiet, bookish and solitary child. He said: "I was never what one would call a loner but I kept away from gatherings. I did have good friends, but usually only one at a time."

Among these was a young Jewish scholar Heinz Wydra, with whom Katz shared a desk and who introduced him to a way of thinking that later held significance for his approach to science. It was Wydra who taught Katz the game of chess.

In Katz's later years, his addiction to chess was replaced by a similarly obsessive addiction to physiological experiments. "The excitement produced by the occasional successful experiment was similar to the enjoyment of a good performance at chess," he said. "In both cases, the sudden flash of insight after a long struggle in the dark and the intuitive vision of a solution to a seemingly intractable problem are exhilarating experiences which one encounters in both kinds of activity."

During his last three school years, Katz had to choose between continuing his Latin and Greek and a more mathematical and scientific curriculum. Against his character, he opted for the classics because they left him more time to pursue his obsession with chess. He then "tended to drift off to one of the cafes in Leipzig which had been invaded by chess players", spending long hours over the chequered board nursing a single cup of coffee.

I did not foresee I was going to remain stateless and without a passport for the first 30 years of my life until I finally acquired this much coveted possession in Sydney, Australia

Katz later regretted this youthful indulgence, not so much because he missed out on the natural sciences, which he quickly caught up with when he undertook pre-university courses, but because he didn't pursue his maths.

Katz embarked on his medical studies at the University of Leipzig in April 1929. A lecture by Professor Victor von Weisacker on the social impact of medicine left a huge impression on Katz, as he realised that there could be great intellectual satisfaction in being a doctor.

Confronting prejudice

Throughout school, Katz had encountered occasional – and to him inexplicable – examples of prejudice on account of his racial background as a Russian Jew. As an Ost Juden, or Jew from Eastern Europe, he was surprised to find that even Jews of Germany and Western Europe, as well as non-Jews, looked down on him.

The World War I Marshal von Hindenburg, now Chancellor, was clearly favouring the Nazi Party and Adolf Hitler. "I finally decided there was little future for decent people and none at all for me in the Germany of those days," Katz said. "It was at that time, in the summer of 1932, that I began to make plans to emigrate as soon as I had completed my medical course."

Meanwhile, the challenges of university study were fuelling in Katz the prodigious focus and appetite for understanding that characterised his life's work. Katz had the benefit of an outstanding physics teacher in Dr Peter Debye, who a few years later gained the Nobel Prize in Chemistry.

The young student also spent time in the Institute for Medical History, gaining a broader understanding of the social context of his studies. Here he shared an enthusiasm for ideas with like-minded students and eminent philosophers, historians and writers on medicine.

(L-R) Kuffler, Eccles and Katz striding into the future, 1942.

Kuffler, Katz and Eccles 30 years on.

Katz was by now an active Zionist, running public appeals to raise funds to buy land for the new settlers in Palestine. He was also working part-time for two ear-nose-and-throat specialists, learning the basics of the medical trade and supporting himself financially. As his studies progressed, he homed in on what was to become his life's work – the study of how nerves function.

"I was attracted to the subject of neurophysiology at quite an early stage, from about 1930 onwards. In those days, the establishment of the laws of electric excitation of nerves and their precise mathematical formulation were regarded as a great thing. In retrospect it seems a somewhat naive approach."

He said that the exact fitting of so-called "strength-duration" curves to electric stimuli of various shapes and intensities was considered a wonderful achievement, even though it was only a formal exercise which shed little light on the physical mechanism of excitation.

"Nevertheless, I felt it was fascinating that one could make accurate and repeatable measurements of electric excitability on living tissues and express the results by a simple mathematical equation."

According to Katz, all one needed to do the experiments were some calibrated boxes of simple electrical gear, resistances, condensers and an isolated nerve-muscle preparation of a frog.

Escape

As he approached his final exams, Katz found the political atmosphere around him thickening. "I felt increasingly revolted by the behaviour of the majority of my fellow students who no longer bothered to conceal their Nazi sympathies and anti-Semitic vulgarities," he said. "I made use of every opportunity I could find to withdraw into the solitary atmosphere of my laboratory."

Many of Katz's most respected tutors were also starting to display support for Hitler. His isolation was further evidenced when, in order to win an academic prize, he had to submit an essay under a pseudonym. Katz chose that of Johannes Muller, the father of German physiology. He won the prize but was not officially allowed to receive it – although his professor later slipped him the money "under the counter". While still a student, he also had two research papers published. One of them caught the eye of one of A.V. Hill's colleagues, Ulf von Euler (with whom Katz was later to share the Nobel Prize).

"When the Nazis took over early in 1933, I still had some 20 months to go before completing my medical course," Katz recalls. "It was not at all clear whether I would be able to do so. For me, the only practical alternative was to join the exodus to Palestine and try to make myself useful in a kibbutz."

At the urging of his tutor, however, Katz decided to stick it out.

In 1934, he read in the English journal *Nature* of A.V.Hill's stinging critique of the policies of Nazi Germany towards scientists. Katz decided that, if it was possible, he would emigrate and go to work with Hill. His friends put him in touch with the famous Zionist leader Dr Chaim Weizmann. Meanwhile, British relatives promised to try to get him a visa – no easy task for a stateless person.

To meet Weizmann, Katz had to slip across the border into Czechoslovakia to Karlsbad. Weizmann promised to try to get some funds to help the young scientist. It turned out to be no more than £50 a year for two years, but the elated Katz hastened home to complete his degree and plan his escape.

"At the beginning of February 1935, I packed my bags and, equipped with travel tickets, a Nansen certificate with a temporary British visa, and the princely sum of £4, I took the train to Holland," Katz said. "I transferred to the Channel ferry at Flushing and arrived at Harwich the next day."

Weizmann promised to try to get some funds to help the young scientist. It turned out to be no more than £50 a year for two years, but the elated Katz hastened home to complete his degree and plan his escape

The following day he reported to Hill in London to begin what he later called "the most inspiring period of my life". Under Hill, whom Katz would later acknowledge as "the person from whom I have learned more than anyone else, about science and about human conduct", he studied the fundamental properties of synapses, the junctions across which nerve cells signal to each other and to other types of cells. Researchers had known since the days of Galvani (1737-98) that nerves were sensitive to electric discharges and then found that nerves actually generated minute amounts of electricity themselves. But much remained to be discovered.

Katz was accepted as a PhD student by Hill at University College, London (UCL), working in his laboratory until August 1939. He liked to joke that Hill only took him on "as an experiment".

Although Katz had arrived at UCL speaking little English, his writing style was simple, precise and unpretentious, something that he attributed to excellent teachers at the König Albert Gymnasium. Although the scientific papers that Katz wrote decades ago would have to be added to today, not one of his beautifully constructed sentences would need to be corrected.

He completed his doctoral studies in 1938, and was awarded the degree of Doctor of Science (London University). He also won a Beit Fellowship, then regarded as the premier award in the biophysical sciences.

In 1939, as a Carnegie Research Fellow, Katz was invited to study with John Carew Eccles, who was to win the Nobel Prize in 1963, at his laboratory at Sydney Hospital. Katz arrived in Australia a month before World War II broke out.

In Sydney, Katz teamed up with Stephen Kuffler, who had recently arrived from Vienna. For the next two years, Eccles, Katz and Kuffler worked together on transmission at the neuromuscular junction – the point where the nerves communicate with the muscles – producing landmark papers and forming lifetime friendships. This work laid the ground for the modern understanding of how one cell communicates with another.

At the time, a fierce dispute was raging among researchers about how the nerves communicated with the muscles – whether the electrical nerve impulse leapt across the nerve-muscle junction or whether synaptic transmission was mediated by the release of a chemical messenger (acetylcholine) as had been proposed by Henry Dale in the 1930s.

Although Dale's ideas were gaining acceptance, Eccles continued to argue strenuously in favour of purely electrical transmission. Katz admitted that he and Kuffler would have "the occasional stand-up fight" with Eccles over the question. Katz later liked to recall an incident in which he was helping Eccles mow his lawn, only to shear through the power lead to the electric mower, narrowly escaping a lethal shock. To avoid further accidents, Eccles hastily swapped the electric mower for a petrol-driven one. Katz would say that this was the precise moment he converted Eccles from "electrical" to "chemical" transmission between nerves and muscles.

Katz was naturalised as an Australian citizen in 1941. With the war clouds continuing to thicken, he felt a strong obligation to serve his adopted country and he joined the Royal Australian Air Force in 1942. He spent the remainder of the war in New Guinea and the Pacific as a radar officer.

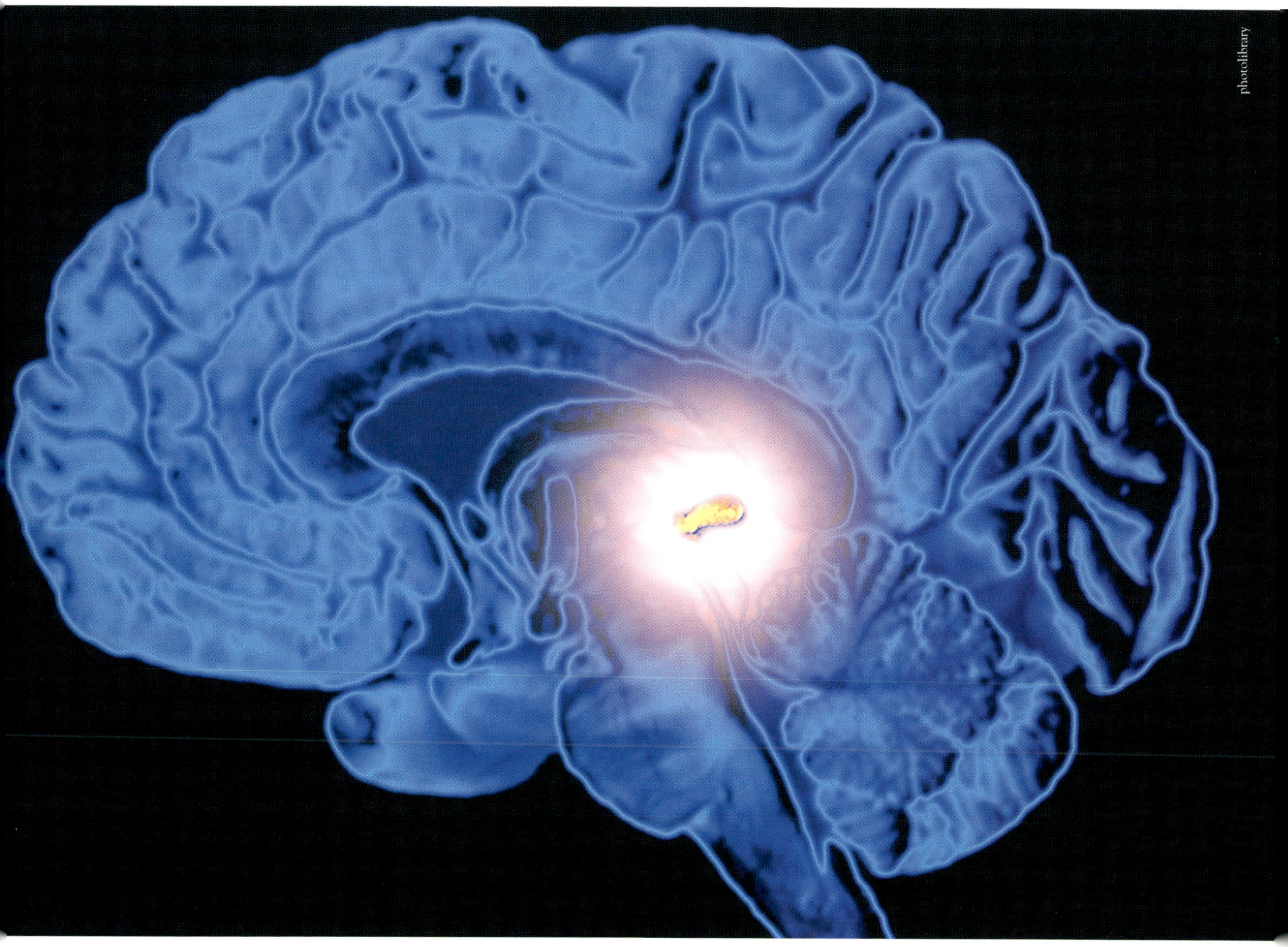
photolibrary

A pineal gland: Katz was particularly interested in its production of melatonin in response to light.

His department became a Mecca for postdoctoral students from all over the world. His influence on the training of a large number of the world's greatest scientists was huge

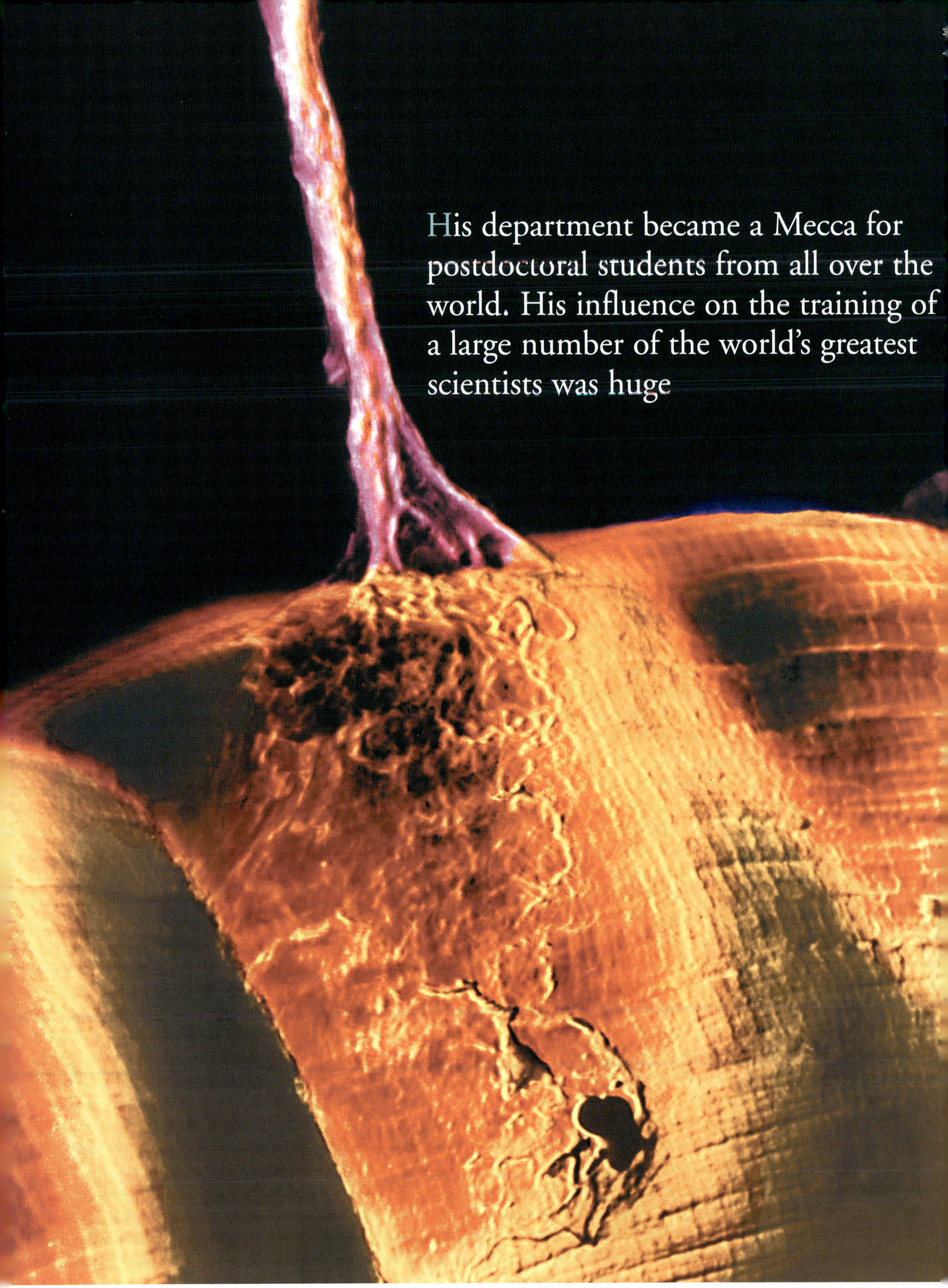

A muscle meets a nerve. Katz's experiments demonstrated that pores in the muscle cells were the means by which the nerves signal.

In Australia Katz also met a young, non-Jewish Sydney woman, Marguerite (Rita) Penly, who would become his life companion. They married shortly after World War II ended in 1945, and had two sons, David and Jonathan. For Katz, his jaunts with Rita to meet up with friends in Europe, Israel and America, were among his happiest times.

The newly married couple returned to England in 1946; there, Katz rejoined A.V. Hill's team at UCL and helped set up the Biophysics Department. There he embarked on some of his most important work. The new electrophysiological laboratory was soon in operation, using war-surplus electronic apparatus that the team had scrounged, rebuilt and recalibrated.

In 1952, when Hill decided to retire, Katz succeeded him as professor and head of department, leading it to become what colleagues later regarded as "one of the shining pinnacles of British biomedical science, attracting talented postdoctoral workers from around the world". In the same period, he also began a key collaboration with the Marine Biological Labs at Plymouth, studying the nerve currents in squid.

Much of Katz's work in these days focused on the biochemistry and action of acetylcholine, the transmitter molecule (or neurotransmitter) with which nerves activate muscles.

Groundbreaking discovery

In 1950, Katz and his colleague Paul Fatt, made an accidental but groundbreaking discovery. Even when it was supposed to be at rest, the end region of the muscle was fizzing with spontaneous electrical activity caused by the discharge of packets of molecules of acetylcholine from the nearby nerve ending.

They revealed that the acetylcholine opens minute aqueous (watery) pores in the muscle cells which allow the electrical current to flow through, causing the muscle to contract. Their experiments demonstrated these pores were the basis of chemical synaptic transmission – the means by which the nerves signal. The pores later came to be known as ion channels and are the basis for all nerve activity as well as many other cell functions.

The team then turned its attention to the second step in the transmission process, the mechanism by which acetylcholine activates its receptors.

In 1957, del Castillo and Katz proposed a two-step model by which the receptor of the nerve signal was activated. This model still forms a starting point for discussion of receptor mechanisms.

During the 1960s, Katz and Ricardo Miledi returned to the issue of transmitter release, undertaking a series of studies. Their work not only provided some of the first definitive evidence for the idea that calcium entry was essential to transmitter release but also allowed the input-output of a single synapse to be studied.

Further groundbreaking work followed in the early 1970s and their studies were so original that their first two *Nature* papers cited only four references to other scientific papers.

In 1967, Katz gained the Royal Society's prestigious Copley Medal and, in 1970, recognition of his achievements was crowned with the award of the Nobel Prize, which he shared with Ulf von Euler of Sweden and Julius Axelrod of the United States. The Prize was for: "their discoveries concerning the humoral transmitters in the nerve terminals and the mechanism for their storage, release and inactivation".

Katz's work had immediate influence on the study of organophosphates and organochlorines, the basis of new post-war study for nerve agents and pesticides, as he determined that the complex enzyme cycle was easily disrupted.

While Katz's research into nerve processes was of the most fundamental kind, it also came to have immediate significance amid the rising public concern about the effect of toxic chemicals. It was clear that the delicate electrochemical processes involved in nerve signaling could be easily disrupted. This provided other researchers with a powerful means for exploring the effects of chemicals on the human nervous system – including why they caused lethal paralysis – as well as opening new possibilities for the treatment of brain and nerve disorders.

Although Katz trained only five PhD students (including Liam Burke, an emeritus professor of physiology at Sydney University), his influence on his science was prodigious. A colleague recalls: "His department became a Mecca for postdoctoral students from all over the world. His influence on the training of a large number of the world's greatest scientists was huge."

And while B.K., as he was known to colleagues, could seem forbidding (there are some who recall traumatic experiences in presenting the first draft of a paper to him), there are stories of his jokes and light-hearted asides that cut through the pomposity of boring committee meetings.

In the latter part of his research career, Katz became interested in the biochemistry of the pineal gland, in particular, its production of melatonin in response to light. Although generally viewed as an Australian or British scientist, Katz's German origins and Zionist beliefs were recalled and celebrated in 1993 with the creation of the Bernard Katz Minerva Center for Cell Biophysics. This was a German-Israeli joint venture embracing the Hebrew University of Jerusalem, the Israel Institute of Technology in Haifa and the Max Planck Institute for Medical Research in Heidelberg.

After retiring as head of Biophysics at UCL in 1978, Katz, who had never entirely lost his German accent, remained active in research administration, working with the research council of the Royal Society as one of its two secretaries.

UCL colleague David Colquhoun remembers that even in retirement, Katz "continued to referee papers with an astonishing speed, often within a day or two, and he took a direct and lively interest in new developments for many more years. In the 1980s I remember him coming, almost running, down the stairs, asking to see David Ogden – at that time a post-doc in my lab – because he'd seen an abstract that David had submitted for a Physiological Society meeting and wanted to discuss it."

Katz was devastated in retirement when his wife Rita developed a prolonged illness. Even though he was in his late 80s and frail, Katz made the long journey to visit her twice a day. Katz, who shared his wife's love of literature, was widowed in 1999.

In April 2003, the stateless alien who had the nerve to change the world died at the age of 92. Of his life, *The Times* newspaper reported: "The work of Sir Bernard Katz constitutes an extraordinary contribution to our understanding of the nervous system. In his research he combined an uncanny instinct for separating the important from the trivial.

"Characteristically, he would approach a fundamental problem ... and by tackling the question in a new manner, opened up fields that had never been dreamt of." ■

Julian Cribb is a science writer and adjunct professor at UTS.

Vital statistics

Name:	Bernard Katz
Born:	Leipzig, the second largest city in the former East Germany
Birthdate:	March 26, 1911
Died:	April 23, 2003 in London
School:	Albert Gymnasium in Leipzig
University:	University of Leipzig; University College, London
Married:	Marguerite Penly in 1945 in Australia; widowed 1999
Children:	David, Jonathan
Lived:	Although naturalised as an Australian citizen in 1941, he returned to live in London in 1946 where, after his retirement in 1974, he remained active in research council and science administration.

Awards and accolades

1933:	Siegfried Garten Prize for physiological research
1939:	Beit Memorial Research Fellowship
1952:	Fellow of the Royal Society in 1952
1961:	Fellow of the University College, London
1965:	Feldberg Foundation Award
1967:	Baly Medal; Copley Medal
1968:	Fellow of the Royal College of Physicians
1969:	Foreign Member of the Royal Danish Academy of Sciences and Letters; Academia Nazionale Lincei (1969); American Academy of Arts and Sciences
1969:	Awarded a British knighthood
1970:	Nobel Prize in Physiology or Medicine

Katz's publications included *Electric Excitation of Nerve* (1939), the classic text *Nerve Muscle and Synapse* (1966) and *The Release of Neural Transmitter Substances* (1969), which is still widely read and admired.

Why he was awarded the Nobel Prize

Katz was renowned for his work on nerve biochemistry and the pineal gland. He uncovered fundamental properties of synapses, the junctions across which nerve cells signal to each other and to other types of cells. Katz was especially interested in the electrical events that occur when impulses in motor nerves induce muscle activity by acting at motor end-plates. These special structures in the muscle, with condenser-like properties, are charged by the nerve impulses and their discharge in turn activates the muscle. Through the discovery of the existence of "miniature end-plate potentials" Katz demonstrated that the messenger substance between the motor nerve and the muscle end-plate, acetylcholine, was released from the nerve terminals in small packages.

New substances for the treatment of high blood pressure and Parkinson's disease are the results of an increasing understanding of the mechanisms of chemical transmission. In the future, continued research will lead to new discoveries of importance for the understanding of the nature of mental disease and psychical disturbances.

References: The Nobel Museum, 2003
Les Prix Nobel, 1970
Katz, Bernard, *An Autobiographical Sketch*, Deutsches Museum Bonn
Nature Neuroscience, obituary, Stuart Cull-Candy & Donald Jenkinson, 2003
The Guardian, obituary, April 24, 2003
The Times, obituary, April 28, 2003
The Independent, David Colquoun, obituary, April 26, 2003
The Scientist, Susan Mayor, obituary, April 30, 2003
International Brain Research Organisation, Max Bennett, obituary, April 2003
Brain & Mind magazine, April-July 2003
The Sydney Morning Herald, John Carmody, obituary, May 7, 2003
Encyclopaedia Britannica

1971: Sir Bernard Katz at Melbourne University.

Wit, words and war

By Helen Verity Hewitt

The Nobel Foundation

PATRICK WHITE
1912-1990
Nobel Prize in Literature 1973

Getty Images

Self-doubt, staunch loyalty and violent rages are just some of the qualities that characterised Australia's only Nobel laureate for literature.

One of Patrick White's duties as a Royal Air Force (RAF) intelligence officer during the Second World War was to rifle through the pockets of the dead enemy for maps, letters, diaries, or any scraps of information. German pilots, Italian soldiers, "the yellow flesh melting like butter into the sand and saltbush the corpses washed up by the sea were the worst."*

Another of White's duties was to censor the letters the airmen wrote. He became obsessed with the role. He was a young writer plunged into the most extreme human circumstances, death at his elbow, privy to the most private longings and despairs of his fellows in their communications to their loved ones.

Literature and art were stays to him. During the inevitable periods of tedium and waiting, he read the Bible from cover to cover and all of Charles Dickens' novels. Of Dickens, he said: "As blood flowed, and coagulated in suppurating wounds, as aircraft were brought down in flames and corpses tipped into the lime-pits of Europe, I saw Dickens as the pulse, the intact jugular vein of a life which must continue, regardless of the destructive forces which Dickens himself recognised."

Goya, one of White's favourite artists, had recorded in his series "The Disasters of War" all the scenes White was now witnessing. Goya's passion, in which the cruellest human behaviour is depicted together with the most sublime, was also to be White's passion. At the same time, the ultra-sensitive White was noting: "The background of trivialities, tantrums, adulteries, service feuds and wranglings for posting and perquisites, against which a great war is fought."

His acute ear for the colloquial makes him a master of dialogue. As a novelist, White tried to capture the whole huge web of human life in its profuse complexity, contradictions and confusion.

The effects of war

Echoes of war sound through most of White's work. His first post-War novel, *The Aunt's Story*, is a turbulent portrait of Europe on the eve of war. The megalomaniac protagonist of his novel Voss had his source in Hitler, and had come into White's imagination while he was stationed in the Egyptian desert. During a year's posting in Palestine, White absorbed "something of the Hebrew archetype" which fed into his portrait of Himmelfarb, the refugee rabbi who is crucified in *Riders in the Chariot*. At the end of *The Twyborn Affair*, White's flamboyant alter ego Eadith Twyborn, is killed in London by a bomb during the Blitz. His two pre-War novels, while accomplished, are of their time; White's post-War novels exist on another level, for all time. What he called "Hitler's War" changed everything.

"In a green pork-pie hat and a black polo sweater", the young White had blossomed and revelled in 1930s London, entranced by the theatre, the art, and the deep cultural history of that great city. London meant intellectual, artistic, emotional and sexual liberation.

He had escaped from an extremely privileged but conservative world in Australia. The large White family had made a pastoral fortune during the 19th century. A saying in the rich Hunter Valley region used to go: "The Whites and the Wrights and the not-quites". (The Wrights being the family of the poet Judith Wright).

White, born in 1912, was expected to go on the land after completing his upper class education at an English boarding school and at Cambridge, where he read German and French. He did spend a year jackerooing in the Monaro between school and Cambridge, but felt like a stranger in Australia. In his autobiography, *Flaws in the Glass*, White repeatedly describes himself as a "cuckoo" in his family, a "changeling"; his mother accused him of being a "freak".

White later said he and his mother couldn't live in the same hemisphere, although he wrote dutifully every week. Explosive family tensions are a major theme in all of White's novels.

White never saw either of his parents working. His father had retired from the land at the age of 42 and agreed to fund White's ambition to be a writer in London. It was probably a relief

to everyone. "An artist in the family was almost like a sodomite; if you had one you kept him in the dark." London also meant escape from his mother's persistent attempts to match him up with some suitable girl. White had known from an early age that he was homosexual.

In 1936, White established himself in London, found marvellous mentors (and lovers), immersed himself in modernist art, tried his hand at writing for the stage and enjoyed the company and stimulation of committed actors, artists, critics and collectors. Lifelong friendships were forged at this time with the artists Roy de Maistre and Francis Bacon, among others.

White's first novel *Happy Valley* was published in 1939 and his second *The Living and the Dead* was in the offing when Hitler invaded Poland. White felt he had to enlist. "A sense of duty always sounds priggish when put into words." In *Flaws in the Glass* he repeatedly plays down his war service, but about one third of the autobiography is devoted to those five years: years in which he wrote nothing.

White accepted a commission in Air Force Intelligence and while waiting for a posting experienced the beginning of the Blitz.

"The incredible bombs had begun falling." His reality was being blown up. White's first posting was to the frontier between Eritrea and Sudan, then to Egypt, where he fell in love with the Mediterranean and Levantine melting pot of Alexandria.

He was involved with the campaign which culminated in the relief of Tobruk. The Corps HQ to which White was attached was cut off by German snipers. White, a very poor driver, had to steer a Dodge truck through the night, zigzagging through minefields reaching the ruins of Tobruk at dawn.

In 1941, in Alexandria, White met Manoly Lascaris, a Greek Alexandrian who was to become "the central mandala" in his life. Their loving partnership was to last until White's death in 1990, despite White's self-confessed jealousy and violent rages. Lascaris' unshakeable Greek Orthodox faith was a stable point which attracted White, himself always a restless spiritual seeker.

White spent a year on duty in Athens following its liberation and became a lifelong Grecophile. The country's mix of mythology, history and physical beauty and the fatalistic, complex psychology of its people spoke to his own intense temperament.

By 1948, White and Lascaris were living at Castle Hill, outside Sydney. White's mother was now living in London. At a time when so many Australian artists and writers were expatriating themselves, White's decision to live and work in Australia was of vital importance to the

Explosive family tensions and misunderstandings are a major theme in all of White's novels

White (right) with Manoly and their dogs at Castle Hill.

White and poet Judith Wright attend a nuclear arms meeting at the University of Melbourne.

Several years of despair following the War gave way to a revived faith in a creative divinity

development of Australian culture. Lascaris had wanted to leave Greece, and post-War London seemed like a graveyard to White.

The Australian landscape drew him back, along with the flourishing art world that he discovered in Sydney and the plentiful food, in contrast with rationed London. "Anyone who has experienced hunger will remember a destroyer of the spirit even greater than lust." Members of Lascaris' family had starved to death during the German occupation of Greece.

Several years of despair following the War gave way to a revived faith in a creative divinity. White's epic narrative, *The Tree of Man*, in which a lifetime of struggle ends with an affirmation of faith, was published in 1955. Deeply romantic by nature, White was a romantic modernist like many artists and writers whose work retained the mystic or transcendental strain of northern European romanticism, while fully absorbing the devastation of the World Wars.

Romantic modernism recognises the upheavals and horrors of the 20th century but clings to spiritual hope. *Voss* was published in 1957 to great acclaim; *Riders in the Chariot*, which coincided with and contributed to the high tide of visionary romantic modernism in Australia, followed. This huge metaphysical novel envisages an Australia in which Aboriginal, Christian and Jewish mysticism meet and recognise one another.

Published in 1961, *Riders in the Chariot* shows remarkable insight into the experience of Aboriginal children who were taken from their mothers. In the boyhood of his Aboriginal painter character Alf Dubbo, White evokes the double standard which turned a blind eye to white men fathering unacknowledged children with Aboriginal women.

White outlines the assumptions that those children would be better off in white society; the children's vulnerability to abuse in their foster homes; and the passive powerlessness of the children, with no one to trust or turn to.

Dubbo's adulthood is an account of homelessness, alienation, loss of cultural identity, mental anguish, repeated betrayals by white society, alcohol abuse and early death through disease. White also expresses his contempt for the corrupting, contaminating effect of commercialism on Aboriginal creativity in this novel.

White and Lascaris moved to Centennial Park in Sydney in 1964. In the big white house at Martin Road, White completed *The Solid Mandala*, and wrote *The Vivisector*, his portrait of a visual artist based on a number of artists he knew including Francis Bacon, Roy de Maistre and Sidney Nolan. Published in 1970, the novel was thought to be in the running for a Nobel Prize but it was apparently judged to be too bleak.

White succeeded in winning the 1973 Nobel Prize for his next novel, *The Eye of the Storm*, about old age and revelation. *A Fringe of Leaves*, based on the historical events of a 19th century shipwreck off the Queensland coast, followed. White's last novels were *The Twyborn Affair* and *Memoirs of Many in One*. During the decades at Martin Road, White became one of Sydney's legendary figures with his looming, charismatic presence, his unmistakable voice unlike any other, his piercing gaze, his uncompromising forthrightness, his rages, and the obvious intensity of his being.

Only a strictly disciplined work regime allowed White to satisfy his need for solitude and silence, and his equally strong needs for frequent trips back to Europe and for the creative stimulation and refreshing amusement to be found in the company of his fellow artists in any medium.

Falling out

Many of his friends still miss his tart wit, his love of gossip and his generosity. Everyone wanted to know White but psychological vivisection by that scalpel eye was a danger. Victims included such old friends as Barry Humphries, Geoffrey Dutton and Sidney Nolan.

Nolan and White shared a dynamic friendship for more than 20 years. They were both passionate about Rimbaud, French modern painting and Greece. Both reinterpreted the history and mythology of the European explorers of Australia, sharing a feeling for their vulnerability, courage and arrogance. Nolan's paintings provided the landscape for *Voss*, the armature of *A Fringe of Leaves*, and significantly influenced *The Vivisector* and *The Eye of the Storm*.

It was highly appropriate that Nolan accepted White's Nobel Prize on his behalf in Sweden in 1973. But White's beloved friend Cynthia, Nolan's wife, committed suicide in 1976 and the

National Library of Australia

Artist Sidney Nolan shared a dynamic friendship with White before they fell out.

friendship between White and Nolan ran aground. White attacked the painter in *Flaws in the Glass*; Nolan responded with a savage homophobic painting of White and Lascaris.

The ability to enter imaginatively into the psyches of his characters is White's forte. He himself felt fragmented, made up of many in one, able "to range through every variation of the human mind, to play so many roles in so many contradictory envelopes of flesh."

Stage-struck from youth and his days as a "stage-door Johnny" in London, White wrote plays as well as novels for the characters that inhabited his head. The best known include: *The Ham Funeral*; *The Season at Sarsaparilla*; *A Cheery Soul*; and *Big Toys*. Lascaris was always apprehensive about the writer's forays into the world of the theatre. White's temperament became particularly volatile in that exciting hothouse environment. He developed close working relationships and friendships with the theatre director Jim Sharman and actors including Kerry Walker and Kate Fitzpatrick.

Everything was grist to White's mill. His own childhood reappears in many guises in his novels, in all its asthmatic, solitary angst and aspiration.

The family drama is played over and over, with variations. The same four basic "sets" (two childhood homes, and the houses at Castle Hill and at Centennial Park) are rearranged to accommodate a multitude of scenarios. Overseas haunts, such as London and Greece, are also used. White asserted the characters in his books arose from his unconscious, all aspects of himself but also borrowed from neighbours, friend and foe, and whomever crossed his path.

For example, many of the supporting cast in *The Vivisector* are recognisable characters in the Sydney art world. Characters belonging to Lascaris' extended, complicated family add to the throng. White's huge imaginative life never ceased; every night he dreamt vividly, and those dreams flowed into his writing. In addition to the plays and novels, White also wrote a number of short stories, most of which are collected in *The Burnt Ones* and *The Cockatoos*.

He said he hated writing, but had no choice, that the novels forced their way out of him. In 1974, he opened the Henry Lawson Festival of Arts and took the opportunity to reiterate his romantic understanding of the creative spirit. Lawson, he said, was "a tortured manic depressive soul like many other artists. I know this is an unfashionably romantic view of the creative artist, but I think the fashionable opinion has been developed largely by intellectuals with little of the artist in them."

White once described himself as "a skeleton at the Australian feast", he raged against what he saw as Australian complacency, insularity, hypocrisy and materialism. "A pragmatic nation, we tend to confuse reality with surfaces. Perhaps this dedication to surface is why we are constantly fooled by the crooks who mostly govern us."

Passionate crusader

White devoted more and more time to political and environmental issues, although it took him away from his work. He was deeply involved with the battle to save Fraser Island, he supported Jack Mundey's Green Bans, he gave the keynote address at the inaugural meeting of People for Nuclear Disarmament and continued to be a frequent marcher and speaker at their rallies.

He gave much practical support to the Aboriginal Treaty Committee and to the Aboriginal Dance Theatre. In White's vision, there can be no true culture or "home" in Australia for non-indigenous Australians without reconciliation with the Aboriginal "spirit of the land" and the original inhabitants. He profoundly desired an Australian culture free of overwhelming British influence, in which there would be room for the voices of all comers – an expanding fusion and profusion of ever-growing complexity and richness. White did everything he could to foster this development.

He was a generous patron to young artists, and he gave some 250 works to the Art Gallery of New South Wales. Having won a number of literary prizes, he decided in 1967 to accept no more (in order to give other writers a go) and

National Library Australia

White was an active participant in anti-nuclear marches in the early 1980s.

He said he hated writing, but had no choice, that the novels forced their way out of him

White relaxing close to home in Centennial Park, 1984.

turned down the Miles Franklin and the $A10,000 Britannica Award in that year. Suspicious of honours, he turned down a knighthood and, later, an Order of Australia. He did accept the Nobel Prize in 1973 but then used all the money to establish an annual literary award for older Australian writers who had not received their due recognition.

White died in 1990 after a bronchial collapse following a lifetime of chronic asthma. In an epigraph to *Patrick White Speaks*, a 1989 collection of articles and public addresses by White, the poet and critic Dorothy Green expressed the feelings of many when she wrote: "What interests me most about this book is its consistency with the novels – the moral stance is firm from the beginning.

"As novelist and citizen, Patrick White is the voice of our country's conscience. He begs us to search our hearts."

Helen Verity Hewitt's book, Patrick White: Painter Manque, *was published by Melbourne University Press in 2002.*

Vital statistics

Name:	Patrick White
Born:	London, England
Birthdate:	May 28, 1912
Died:	30 September, 1990 in Sydney, Australia
School:	Tudor House, a private school in NSW, Australia; Cheltenham College, England
University:	Kings College, Cambridge
Partner:	In 1941 White met Manoly Lascaris, with whom he lived until his death
Children:	None
Lived:	Mainly in Sydney

Awards and accolades

1941:	Gold Medal of the Australian Literature Society for *Happy Valley*
1957:	W.H. Smith Award for *Voss*
1961:	Miles Franklin Award for *Riders In the Chariot*
1965:	Gold Medal of the Australian Literature Society for *Riders in the Chariot*
1973:	Nobel Prize in Literature
1974:	Australian of the Year

Novels:

1939:	Happy Valley
1941:	The Living and the Dead
1948:	The Aunt's Story
1955:	The Tree of Man
1957:	Voss
1961:	Riders in the Chariot
1966:	The Solid Mandala
1970:	The Vivisector
1973:	The Eye of the Storm
1976:	A Fringe of Leaves
1980:	The Twyborn Affair
1986:	Memoirs of Many in One

Short Stories

1964:	The Burnt Ones
1974:	The Cockatoos
1988:	Three Uneasy Pieces

Plays

1947:	Return to Abyssinia
1965:	Four Plays

Nonfiction

1981:	Flaws in the Glass: A Self-Portrait

Why he was awarded the Nobel Prize

Patrick White was awarded the Nobel Prize in Literature in 1973 for his novel, *The Eye of the Storm*. In the presentation speech, the Swedish Academy said he won "for an epic and psychological narrative art which has introduced a new continent into literature". *The Eye of the Storm* places an old, dying woman in the centre of a narrative which revolves round, and encloses, the whole of her environment, past and present, until we have come to share an entire life panorama, in which everyone is on a decisive dramatic footing with the old lady. Because of the Nobel Prize, Patrick White's literary fame has spread throughout the world.

*All quotations, unless otherwise indicated, from White's 1981 autobiography *Flaws in the Glass.*

Discovering the chemistry of life

By Julian Cribb

JOHN WARCUP CORNFORTH

1917-

Nobel Prize in Chemistry 1975

Shared with Vladimir Prelog

National Institute for Medical Research, London

A crippling ear disease that left him deprived of hearing did not stop John Cornforth in his quest to uncover the wonders of science.

Known to his friends as "Kappa", John Warcup Cornforth was born on September 7, 1917, in Sydney. His father was a British-born school teacher of English and classics and an Oxford graduate; his mother was a hospital nurse and descended from a German minister who had settled in New South Wales in 1832. John was the second of four children.

He started suffering the first signs of deafness, caused by the progressive disease otosclerosis, at age 10. By the time he was 20, Cornforth was unable to hear and had permanent tinnitus, or a ringing in his ears. However, he developed a profound insight into aspects of both nature and human nature that would eventually lead him to the pinnacle of scientific attainment: the Nobel Prize.

Cornforth's childhood was spent in Sydney and in the rural surrounds of Armidale, in the New England region of NSW, where his later fascination for understanding the natural world underwent a gradual awakening.

"Looking back at the time some 70 years ago when the love of science took hold of me, I think of no big event but of many small things that influenced me," he recalls.

He says as a child he read books and learned lessons but did not have much curiosity about the natural world. This began to change when he looked at the stars – in fact, he became known for his ability to tell time by the stars – and he says he entered science through astronomy.

At Sydney Boys High School, a young chemistry teacher, Leonard Basser, captured and inspired Cornforth's eager mind, encouraging him both to think and to experiment. His deafness was coming on gradually and chemistry appeared to be a profession he could pursue in spite of it.

A laundry laboratory

At 14, Cornforth constructed a small chemistry laboratory at home in his mother's laundry and was conducting his own experiments with a sense of growing absorption and fascination that never deserted him. "At the time one could buy small amounts of many common chemicals and I made a little laboratory at home, with improvised equipment to study chemical reactions," he said. "I soon discovered that the organic chemicals were the most interesting. With the help of a textbook on practical organic chemistry I made many preparations using cheap chemicals to prepare those that were too expensive to buy.

"This was more satisfying than astronomy: you could change things by your own effort. At that time I was rapidly losing my hearing so I suppose the work attracted me also for its impact on the other senses: the beauty of crystals and distilled liquids, the colours of dyes and smells – both good and bad."

Cornforth saw himself as a craftsman of the fundamental components of nature. "As a carpenter or carver learns to work with the grain of wood or bone, I learned that each substance has its own nature and can be easy or difficult to handle according to the procedure chosen," he said. "I began to see experiments as I see them now, not only as procedures to answer questions or to make compounds but as opportunities to observe what happens and to learn from mistakes."

John Cornforth was well ahead of his years at school and entered Sydney University at the age of 16. By this time he was unable to hear any lectures and became an accomplished lip reader. He was also blessed with an extraordinary visual memory and learned a great deal from handbooks and journals. Many of them were in German, which he did not know, but he found a German dictionary and looked up each word until he understood them all. Cornforth says that reading the original scientific literature helped him to become a scientist because it showed the evidence behind the things that he was being taught – and some of this evidence was wrong.

"The most liberating thing was the realisation that the literature wasn't entirely correct," he said. "It gave me quite a shock at first and then a thrill, because I can set this right. Always, and ever since, I've relied upon the primary literature exclusively. I don't believe a word I ever read in any textbook. I began to see science as a continuous process of discovery and correction and myself as a part of this process."

During his time at university there came a second moment of revelation, this time on a bushwalk in the Blue Mountains with friends. "One morning we were resting beside a river," he said. "I turned over so that my face was close to the grass and I began to count all the different kinds of plants that I could see. There were more than 20, all different, each beautiful in its own way. For me this was a kind of conversion because I had never looked at things in that way before. This was really the beginning of my curiosity about living things."

Cornforth says he brought back from this walk some fruits including wild grapes and some berries with a bitter taste. He took them into the laboratory and extracted pure compounds from them.

"This was not a very good way to study the chemistry of life but I began to be interested in the life sciences and to read biological text books," he said. "At that time they mostly described and classified things that nobody understood. But later, when I started to work with life scientists, I could understand their viewpoints and could use my chemistry to solve problems that interested us all. After that experience I started to look at what constituents of plants I could extract."

> I began to see science as a continuous process of discovery and correction and myself as a part of this process

Cornforth continued to be enthralled by the breadth and complexity of chemistry. "What you can do is to form in your mind a pattern of what is possible and what is not possible in chemistry," he said. "This helps you to make new compounds and to understand new reactions and structures. When the literature or one of your own experiments presents you with a new fact, you compare it with the pattern in your mind. Often the new fact fits into the pattern easily and reinforces it but sometimes the fact does not fit."

Cornforth says that when this happens, you check and sometimes you find that a mistake has been made. But if there is no mistake you must change your pattern to fit the new fact and you learn more about science on these occasions than at any other time.

Curiosity questions

Science for Cornforth begins with curiosity. "You ask questions, you read what other people have written and then you begin to find ways of answering your own questions," he said. "You never stop learning."

For Cornforth, knowing the fundamentals of how something was made or worked never detracted but rather magnified its wonder. "Keats once said that Newton's explanation of the rainbow killed the beauty of it for him. But for me, what I know about nature simply enhances the beauty of it. I am sorry for the people who look at a flower and don't understand anything at all about what is going on."

It isn't unusual for Cornforth to quote poets or poetry. He says: "I use poetry as a substitute for music. I have no particular vocabulary of music in my head, but I have thousands of lines of poetry that I can recall as I like. The curious

National Institute for Medical Research

John Cornforth.

University of New England

Armidale in 1924 where Cornforth spent his early childhood.

thing is that some pieces I have known for 65 years or more. And when I recall them, I find that I have sometimes improved them!"

The young Cornforth was also well known for his ability to produce limericks on demand.

In 1937, Cornforth graduated with first-class honours and the University medal. Along the way, he'd also acquired his nickname "Kappa", which came from a habit of scratching the Greek letter into his laboratory glassware to stop fellow students from "borrowing" it.

After a year of post-graduate research he was awarded an 1851 Exhibition scholarship to work at Oxford with Robert Robinson (who was later to nominate his colleague for the Nobel Prize).

Only two of these prized scholarships were given in Australia each year. The other went to fellow student Rita Harradence, also of Sydney and also an organic chemist.

In Sydney, both Cornforth and Harradence had been working in a laboratory designed by Robinson, who had become Sydney University's first professor of pure and applied organic chemistry after he arrived there from England in 1912.

Cornforth and Harradence met for the first time after she had a small accident with a valuable piece of laboratory equipment – a Claisen flask – which were difficult to come by in those days.

As a keen improviser of equipment, Cornforth had adapted an old Bunsen burner tube and taught himself to blow glass. By this stage he'd earned a reputation as a glass blower and equipped himself with a proper blowpipe for the job. One of Rita's friends suggested she get him to perform the repair, which he duly did.

It was the beginning of a lifetime partnership in which Rita became his co-researcher in the chemistry lab, his ears and often his interpreter in communicating with others.

Both also belonged to the Sydney University Bushwalkers Club, and walking remains one of Cornforth's abiding passions.

In a reminiscence for the club's archives, Cornforth recalls: "The longest day's march I ever did was when Dick Welch and I arrived early one Sunday morning at Jenolan and found that we couldn't get transport that day. We wanted to be in Sydney by Monday and we knew that a train to Sydney left Mt.Victoria around midnight and we decided to catch it; the walk was around 70 kilometres if I remember right. We made it, and fell asleep in our seats on the way back while our leg muscles stiffened. It was agony to get going again ... I remember saying to Dick around the middle of the walk: 'Tomorrow, we'll be laughing like hell at ourselves' and this was true. And I've never lost the taste for walking acquired in those days, and still walk around eight kilometres a day."

In 1941, Cornforth and Harradence married. They have three children and several grandchildren.

It was through Cornforth's grandchildren that he discovered he had an unusual talent, given his profound deafness.

"Sometimes my grandchildren have asked me to sing, and although I've explained to them that I can't sing they made me try," he said. "Apparently I can produce a melody quite accurately from something that I heard when I could hear. How that happens I don't know."

Everything for Cornforth always came back to Rita. The husband-and-wife team ended up collaborating on 41 scientific papers.

"Throughout my scientific career, my wife has

The young Cornforth at Sydney Boys High School.

Cornforth (second from left) with teachers and fellow students at Sydney Boys High School.

been my most constant collaborator," Cornforth said. "Her experimental skill made major contributions to the work. She has eased for me beyond measure the difficulties of communication that accompany deafness; her encouragement and fortitude have been my strongest supports."

War broke out as Cornforth and Rita journeyed to Oxford, his father's alma mater, and after completing their work on steroid synthesis for doctorates, they became part of the chemical effort on penicillin, which was the major chemical project in Robinson's laboratory during the war.

"We made contributions and I helped to write *The Chemistry of Penicillin* (Princeton University Press, 1949), the record of a great international effort," Cornforth said.

Among his most satisfying personal contributions to the chemistry of penicillin, Cornforth recalls the penicillamine story. At the time the research team was getting a pure penicillin preparation to be able to break up the molecule in various ways and penicillamine was an important fragment. Penicillamine is a chelating agent, that is, it attaches to other chemicals in the body and aids in their removal. Penicillamine is now used to remove excess copper associated with Wilson's disease. It is also used to treat severe rheumatoid arthritis.

"Analysis indicated the penicillin preparation only had five carbon atoms and Robinson immediately wrote a comprehensive table of the structures that it could be," Cornforth said. "But he was misled by one piece of analytical evidence, which was what used to be called the C-methyl determination. Using the Kuhn-Roth oxidation method you boil the substance you are analysing with chromic and sulphuric acid and then you distil over any acetic acid that has been formed from that oxidation.

"The amount of acetic acid was supposed to represent the number of methyl groups attached to a carbon atom. Because penicillamine gave hardly any acetic acid in that assay, they didn't like any formulae that had a methyl group. But that made it very difficult to think of a formula at all.

"Robinson saw what he thought was the least improbable formula and he asked Rita and me to synthesise it," Cornforth said. "I had to visit Manchester at about this time and on the way up, it occurred to me that penicillamine might have too many methyl groups, not too few." Subsequent experiments proved he was right.

After the war ended Cornforth returned to his earlier work on the synthesis of sterols. His close collaboration with Robinson continued after he and Rita joined the scientific staff of the Medical Research Council and worked at its National Institute, first at Hampstead and then at Mill Hill. By 1951, Cornforth and Robinson were able to complete, simultaneously with Woodward at Harvard, the first total synthesis of the non-aromatic steroids. Steroids are a class of naturally occurring organic substances and their derivatives that are of great importance in biology, medicine, and chemistry.

However, Cornforth and Robinson were always more than mere collaborators in the laboratory; they shared a close friendship that lasted until Robinson's death in 1975.

"The nature of our friendship was a continuous sequence of differences of opinion," Cornforth said in 1985. "He was an argumentative person and I have known the same thing to be said of myself."

At the National Institute for Medical

The Blue Mountains near Sydney where Cornforth had a revelation that led to his curiosity about living things.

Cornforth at Sydney University.

I know to what extent scientific advances are the product of an ambience created by many people, not just the few who tend to have the best ideas

Research, London, Cornforth came into contact with biological scientists and formed collaborative projects with several of them. In particular, he shared an interest in cholesterol with George Popják. They began to collaborate in studies of its synthesis in living organisms.

This was the beginning of what Cornforth recalls as the richest and most fulfilling period of his career. "I spent 16 of the best years of my life in an extraordinary place, the National Institute for Medical Research, and I know to what extent scientific advances are the product of an ambience created by many people, not just the few who tend to have the best ideas.

"At this time Konrad Bloch was beginning his work on the biosynthesis of the sterols and Popják and I began to conduct experiments in which the disciplines of chemistry and biochemistry could be applied."

In 1962, Cornforth and Popják left the Medical Research Council and became co-directors of the Milstead Laboratory of Chemical Enzymology set up by Shell Research. Lord Rothschild helped establish the laboratory and Cornforth worked under him until Lord Rothschild left Shell in 1970. At Milstead, a project was developed that looked at the study of the stereochemistry of enzymic reactions.

Stereochemistry is the study of how the properties of a chemical compound are affected by the spatial arrangement of atoms in molecules and complexes. Not only can a compound have more than one geometric form, but chemical reactions can also have specificity in their stereochemistry, thereby forming a product with a particular three-dimensional arrangement of the atoms. This is especially true of reactions in living organisms. Cornforth mainly studied enzyme-catalysed reactions. Stereochemically specific reactions have great practical importance, as many drugs, for example, are active only in one particular geometric form.

Cornforth's stereochemistry work continued after 1968 when Popják left Milstead to go to the University of California at Los Angeles. In 1975, the same year that Rita chose to retire, Cornforth left Shell to take up a position as Royal Society Research Professor at the University of Sussex.

Cornforth's work at Milstead led directly to the 1975 Nobel Prize in Chemistry for clarifying the stereochemical control of enzymic reactions. Some substrates for enzymes contain carbon atoms that become chiral – left- or right-handed in structure – when one of the attached hydrogen atoms is replaced by deuterium (an isotope of hydrogen that has twice the mass of ordinary hydrogen).

This labels the compound, and examination of the label in the product can throw important light on the reaction mechanism.

One problem was how to measure the tiny optical activity generated by replacing one hydrogen with deuterium in otherwise symmetrical molecules. (Later it became possible to make methyl groups chiral by using all three hydrogen isotopes and to measure the chirality without using optical rotation).

Optical assistance

"It was necessary to measure the optical activity and looking at all the methods that were available, I could see that it was going to be nearly impossible," Cornforth said. "I wrote to a friend in Australia who was an expert in optics and I asked him whether there was anything in the pipeline for measuring very small optical activity.

"He told me to go to the National Physical Laboratory as they were evolving a prototype. It was a marvellous instrument but it was a lash-up of all kinds of components. We did our bio-chemistry and chemistry and got two specimens of mono-deuterio succinic acid in which the molecules are mirror images of each other.

Analysing chemical properties led to Cornforth's Nobel Prize. His special interest was enzyme-catalysed reactions.

As a scientist, Cornforth was happiest in the laboratory.

As for the ceremony, I couldn't hear a word of what was said. I amused myself by looking around at the audience

We had a few milligrams of each of these and we also had a third specimen which we had much more of, which was made from a product in which we were sure of the stereochemistry.

"We took these three specimens to the National Physical Laboratory at Teddington and they did the optical rotations. They came out beautifully. The two dispersion curves of the two compounds were mirror images of each other, one going up like this, the other going down like that, a perfect dispersion. I think that's the day I remember with the most pleasure in my experimental life."

There came another memorable day when, shortly after Cornforth had moved from Shell to the University of Sussex in 1975, his wife Rita telephoned to tell him she had just heard it announced on the BBC news that he had been awarded a Nobel Prize for his work.

"I was quite surprised," he said. "I had estimated my chances at about one in three. As for the ceremony, I couldn't hear a word of what was said. I amused myself by looking around at the audience. It was in this sports stadium, an enormous place, because the town hall was being refurbished but I could see in the darkness of the auditorium these flashes of bright light. They kept on like this and I couldn't make out what they were. And finally I realised all the women were wearing their jewels, and that was what was causing the flashes of light. That was the thing I remember most of all from the ceremony."

As well as a gold medal and prize money, Nobel Prize recipients receive a personal diploma depicting their achievement. Cornforth's diploma had a drawing of two molecules, which were supposed to be the mirror-image forms of acetic acid. Associate Professor Damon Ridley, who spent study leave in 1978-79 with Cornforth, said: "But the cartoonist drew two identical structures, so the certificate has a scientific flaw. This appeals to Kappa's sense of humour."

In 1977, Cornforth's Nobel Prize was followed by a British knighthood, which came after wide professional recognition of his skills as a scientist. Memorable for others was Cornforth's brief yet powerful speech of acknowledgement on behalf of himself, Vladimir Prelog and their colleagues: "That our work has been considered worthy of such distinction is a great satisfaction to us both but I think that we derive equal satisfaction from the sense of being two in the great company of those who approach the truth.

"In a world where it is so easy to neglect, deny, corrupt and suppress the truth, the scientist may find his discipline severe. For him, truth is so seldom the sudden light that shows new order and beauty; more often, truth is the uncharted rock that sinks his ship in the dark. He respects all the more those who can accept that condition; and in returning thanks tonight we are saluting all those who make our load lighter by sharing it." ■

Vital statistics

Name:	John Warcup Cornforth
Born:	Sydney, Australia
Birthdate:	September 7, 1917
School:	Sydney Boys High School
University:	Sydney University; Oxford University
Married:	Rita Harriet Harradence in 1941
Children:	Three children
Lives:	In England, where he is the Royal Society Research professor at the University of Sussex in Brighton

Awards and accolades

1953:	Elected to the Royal Society
1953:	Chemical Society's Corday Morgan medal
1965:	Flintoff medal
1965:	Biochemical Society's Ciba medal, awarded jointly awarded with George Popjak
1967:	Stouffer prize
1968:	Royal Society's Davy Medal
1968:	American Chemical Society's Ernest Guenther award
1972:	Prix Roussel
1975:	Nobel Prize in Chemistry
1975:	Australian of the Year
1977:	Awarded a British knighthood
1982:	Royal Society's Copley Medal

Why he was awarded the Nobel Prize

In 1975, John Cornforth and Vladimir Prelog were awarded the Nobel Prize in Chemistry. Cornforth was awarded it for his work on the stereochemistry of enzyme-catalysed reactions. This subject is quite complex and relates to geometry in three dimensions. It is concerned with the delicate mechanism of important reactions in biological systems, where a group of atoms takes the place of a certain hydrogen atom among two or three, which may appear to be equivalent. The problem is to decide which of the hydrogen atoms is replaced and if nearby groups retain their positions or if they are rearranged in some way. The enzyme leads the process in a quite uniform way. Without this guidance, chaos would break out in the biological system.

Cornforth "marks" the hydrogen atoms by making use of hydrogen's three isotopes; ordinary hydrogen (mass 1), heavy hydrogen or deuterium (mass 2) and radioactive hydrogen or tritium (mass 3). He then makes use of their differing reaction speeds (the lightest reacts the quickest). Only trace quantities of tritium can be used; no more than a millionth or so of the molecules involved in the reaction are marked. The planning of this research is an outstanding intellectual achievement. Among the biologically significant reactions the stereochemistry of which has been clarified is the synthesis of steroids from mevalonic acid via squalene and the condensations of acetate with glyoxylic acid to malic acid and with oxalylacetic acid to citric acid. Since Cornforth showed the way, similar research has been initiated in other quarters.

References: Cornforth JW, "Becoming a scientist"
Cornforth JW, autobiographical notes
Cornforth JW, "Scientists as Citizens", Address to Royal Australian Chemical Institute, 1992
Nobel Museum, biography and Nobel speech
Thomas, R, Interview with Sir John Cornforth, Vega Science Trust 2001
Encyclopaedia Britannica
http://www.usyd.edu.au/about/publication/gazette/features/nobel.shtml

Survival of the cells

By Brad Collis

University of Melbourne Archives

PETER CHARLES DOHERTY
1940-
Nobel Prize in Physiology or Medicine 1996
Shared with Rolf Martin Zinkernagel

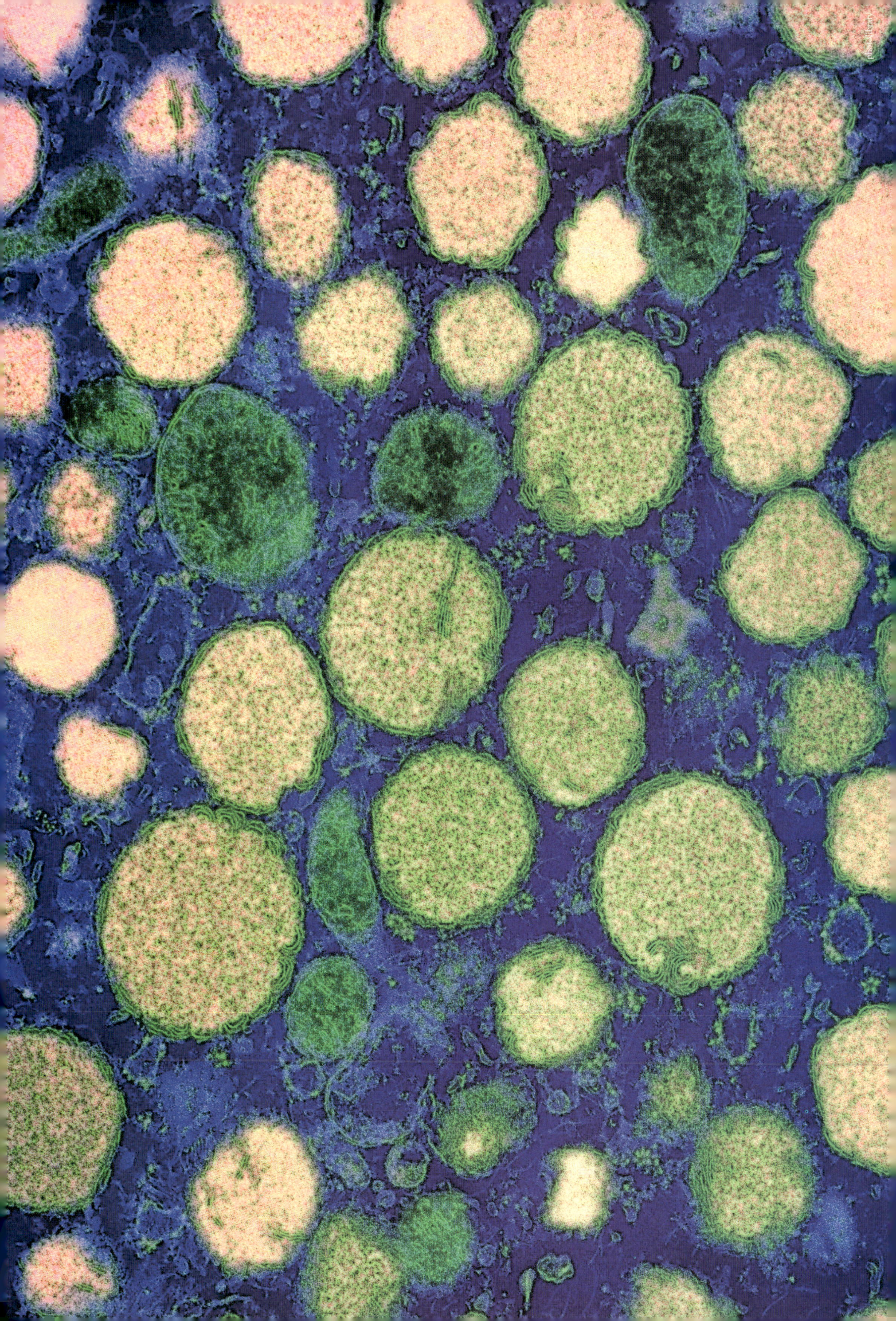
photolibrary

Research into certain diseases was heading down the wrong track until Peter Doherty's work on immune responses was released.

Peter Doherty has always regarded himself as somewhat of a misfit; invariably "out-of-sync" with his circumstances. He has always loved the icons of Australian outdoor life, the sun and surf, yet as a boy he was invariably forced indoors by an Irish-inherited complexion that rapidly turned him lobster red in the sun.

On leaving university he was a veterinarian when he wanted to be a researcher and, 40 years later, a Nobel Prize-winning scientist with no time to be the novelist that he had aspired to become. His academic interests at school had been literature and history, but he also excelled in physics and chemistry and he felt these were more likely to find him a good job. He was tempted by marine biology, but a visit to an open day at the University of Queensland's Veterinary School settled the matter.

From an early age, this tendency to walk adjacent to, rather than on, the routine path occupied by his peers shaped him into an avid, liberal reader and observer to which was added a dash of Irish rebel – all handy ingredients for someone drawn to the allure of bold ideas and scientific discovery.

It also equipped him well for another unique aspect of his life at this point in time. Peter Doherty is the only Australian accorded science's highest honour to still be working and he is quite willing to use the clout that a Nobel Prize bestows to push the cause of science and education in the political arena and in the community generally.

Doherty's story still has some distance to run. However, in the meantime, he is an intriguing living history; a person whose experiences are still touchable by aspiring young scientists looking for a role model. He didn't follow the ordered career paths of predecessors like John Eccles and Frank Macfarlane Burnet, but he has triumphed as an individual who refused to be subdued by the jealousies of university politics.

Peter Doherty was born on 15 October 1940 in Brisbane as the eldest son of Eric Doherty, a telephone mechanic and administrator, and Linda Byford, a piano teacher. The family lived in a tough outer suburb of Brisbane where he went to the Indooroopilly State School. Near his home a cement works caked everyone and everything in grey dust, providing a reason to stay inside to read and soak up the sounds of his mother's passion for classical piano.

Early influences

He recalls: "There was always music in the house including Chopin, Debussy, and Beethoven, and books, so my early influences were classical composers, reading, and the wireless. When I look back, the ABC had a very strong influence on me. I still remember being absolutely moved, listening to a program one day on Icelandic sagas. I think that's the key to education – being able to reach and intrigue a child."

At school Doherty says he was good at science, bored by maths, and passionate about history and ideas. These ideas came from books, people, and exploring art galleries. His interest in artistic expression saw him buy a second-hand Leica camera and even build his own darkroom and enlarger.

"I was interested in learning and creativity because I think I realised even then that education was a way up, and a way out."

Even so, school life was tough. It was authoritarian and Doherty, then and still, has never been one to acquiesce to authority or dogma, which made his entry into university absolutely liberating. He says: "It was a transformation for me, a real freedom to learn and explore like never before."

He was influenced by authors as diverse as Aldous Huxley, Jean Paul Sartre and Ernest Hemingway – and it was reading Hemingway that resolved him to be "the man of action rather than the philosopher", which is why he decided to study veterinary science.

"I was also going though a religious phase and captivated by the whole 'feeding the world' ideal, and veterinary science seemed one way to play a part. Of course, I was just 17 years old and would probably have made a very different decision had I been more mature."

Doherty graduated with his BVSc in 1962 from the University of Queensland and because he had been supported by a bonded government scholarship, he was required to spend several years working as a rural veterinary officer. He was keen to do laboratory-based research and made the mistake of saying so. His employer, the Queensland Department of Agriculture and

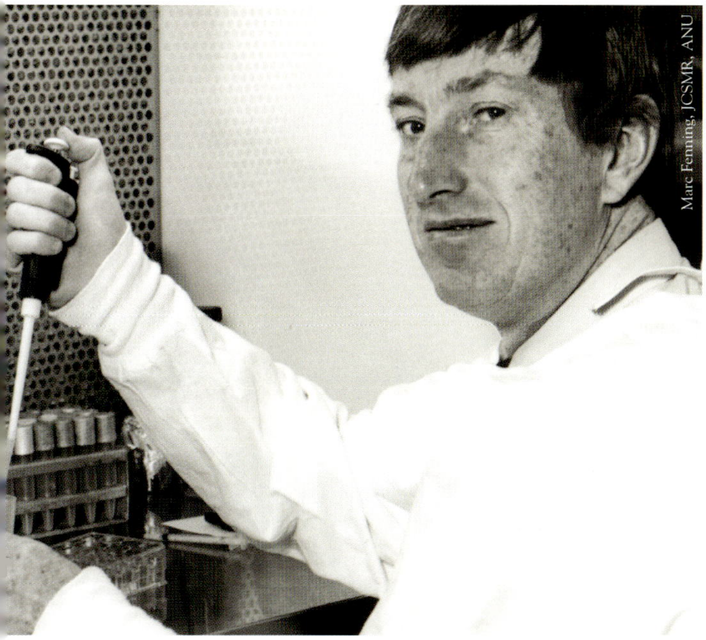

Immunology experiments at the John Curtin School of Medical Research.

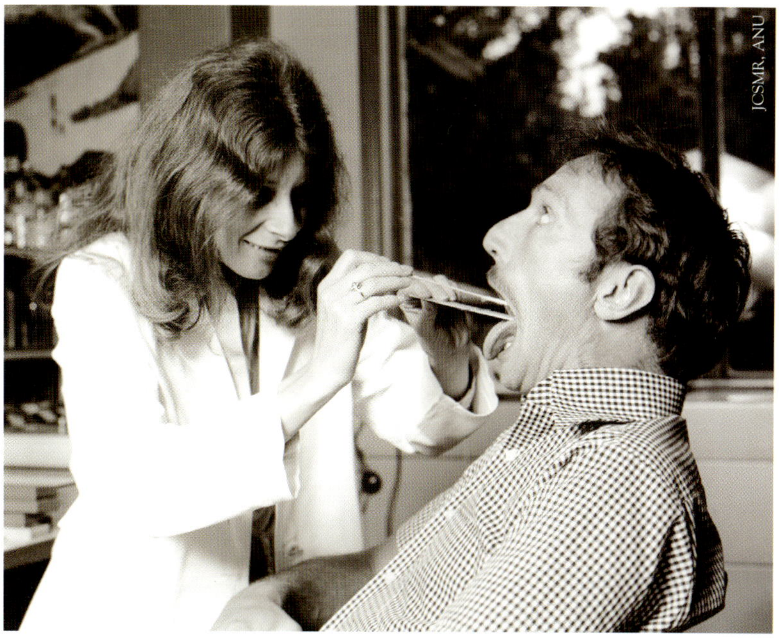

Rolf Zinkernagel, who shared the Nobel Prize with Doherty, gets tongue tied.

It was reading Hemingway that resolved him to be "the man of action rather than the philosopher"

Stock, promptly sent him to the country as a rural veterinary officer.

"I spent some months driving large distances to perform post-mortems on cattle and pigs that had died of unknown causes and to survey cattle for various venereal diseases. This resulted in the diagnosis of Trichomoniasis in an area where it was thought that complete eradication had been achieved. Realising that I was a danger to their regulatory effort, the Department quickly brought me back to the state veterinary laboratory, the Animal Research Institute (ARI) at Yeerongpilly."

At the ARI he undertook an epidemiological study of bovine leptospirosis and a thesis on this was later the basis for his masters degree. At the ARI was a young microbiologist – Penelope Stephens – who had been contracted to develop a viral diagnostic service. They graduated about the same time and she was his first real girlfriend. They married in 1965.

Still determined to be a research experimentalist, Doherty responded one day to an advertisement in *Nature* magazine for a training post in experimental pathology at the Moredun Research Institute in Edinburgh. He got the job, which also allowed him to enrol as a part-time PhD student with the University of Edinburgh.

While there, he helped to run a diagnostic neuropathology program for the Scottish Veterinary Investigation Service and undertook a research project on the tickborne, encephalomyelitis virus. The thesis earned him his Edinburgh PhD in 1970 – just three years before his historic work back in Australia that would lead to the Nobel Prize.

While in Scotland, Penny worked at the Institute of Animal Genetics until they decided to start a family and, in a few years, they were the parents of two boys. Doherty and his family thoroughly enjoyed life in Scotland. Doherty recalls: "For the first time, I could spend the whole day outside without the penalty of sunburn. Our long vacations were used for camping holidays in Europe, including our first trip to Scandinavia and Stockholm with a young child in the back of a Volkswagen van."

A lucky sidetrack

The family considered staying in Britain permanently but Doherty was keen to learn more about the developing field of immunology and the John Curtin School of Medical Research (JCSMR) in Canberra was the place to do it.

"I was actually supposed to come back to work with the CSIRO at Geelong (Australian Animal Health Laboratory) but decided to go to Canberra first to learn more about immunology,

Their findings demonstrated conclusively the need for the cellular immune system to recognise simultaneously both "foreign" molecules and "self" molecules

and then join the CSIRO. Well of course we made our great discovery and got stuck, otherwise I probably would have spent the rest of my career happily in Geelong.

Instead, Doherty became the most recent Australian to be awarded the Nobel Prize when in 1996 he and a former colleague, Rolf M. Zinkernagel, were recognised in the category of Physiology or Medicine. The Prize was awarded for ground-breaking research, carried out in 1973-74, that fundamentally changed medical science's understanding of immune response.

"It was a classic science story," Doherty says. "Both Rolf and I were interested in trying to see exactly how an infecting micro-organism actually causes damage and disease, which meant finding out a way to quantitate the cellular invasion.

"By chance, an American researcher Richard Carp was visiting as part of his research into scrapie (a sheep disease) and we got chatting. He explained how he collected cerebral spinal fluid (from mice) where it accumulates at the back of the brain. This provided us with a source of cells in which a viral invasion could be quantitated."

Doherty and Zinkernagel discovered that white blood cells (lymphocytes) must recognise both an invading virus and certain "self" molecules – the so-called major histocompatibility (MHC) antigens – in order to kill the virus-infected cells. This concept of simultaneous recognition of both "self" and "foreign" molecules now forms the basis for a new understanding of the cellular immune system.

At the time they began their research, it had already been discerned how antibodies, the circulating defense molecules, recognise and kill targets such as bacteria. It was far less well understood, however, how the white blood cells recognise and kill virus-infected cells without destroying the normal uninfected cells.

Another mystery was the reason behind nature's creation of immunological uniqueness, the small, but important differences that exist between molecules called transplantation antigens which white blood cells recognise as "self" or "foreign". These differences were a major obstacle to organ transplants.

In their research, the two scientists were able to use the mice to study how the immune system, and particularly T-lymphocytes (a sub-set of white blood cells), could protect animals against infection from a virus able to cause meningitis. Doherty calls T-lymphocytes the "hit men" of the immune system. The infected mice developed these killer T-lymphocytes, which in a test-tube could kill virus-infected cells.

Receiving an honorary doctorate from the chancellor of the University of Technology, Sydney, Sir Gerard Brennan, 2003.

But there was an unexpected discovery: the T-lymphocytes, even though they were reactive against that particular virus, were not able to kill virus-infected cells from another strain of mice.

What decided whether or not a cell was eliminated by these killer lymphocytes was not only if they were infected with the virus but also if they carried the "correct" variant of histocompatibility antigens – those of the infected mouse itself.

Zinkernagel's and Doherty's findings, which were published in *Nature* in 1974 demonstrated conclusively the need for the cellular immune system to recognise simultaneously both "foreign" molecules and "self" molecules.

It became possible for the first time to understand that the true function of transplantation antigens was not to provide an obstacle to transplantation. Instead, their function is to bind and present molecules from viruses and other micro-organisms to white blood cells in such a way that the white blood cells understand whether they should become aggressive or stay calm. As a consequence it became obvious how each individual, thanks to his/her unique set of transplantation antigens, also carries his/her unique immune system.

It also became possible to understand why evolution has created these large immunological differences between individuals within a species. Immunological diversity is advantageous for the individual and the species. Thus, there will always be some individuals who survive even the most severe epidemics. In return, individuals carrying a certain variant of transplantation antigens have an increased susceptibility to autoimmune diseases such as rheumatoid arthritis or multiple sclerosis and this is possibly the price they pay for their forefathers having survived a severe epidemic.

Doherty and Zinkernagel's explanation of the self/non-self selectivity of the immune system advanced research in many clinical areas. It led to advances in therapeutic efforts to strengthen the immune response against invading microorganisms and certain forms of cancer, and to efforts to diminish the effects of autoimmune reactions in inflammatory diseases, such as rheumatic conditions, multiple sclerosis and diabetes. It has also proved to be a great advance in the field of organ transplants.

In an interview with science writer Roger Beckmann for the Australian Academy of Science, Doherty provides his own explanation of his research: "It shows us that the immune system can recognise a third state – an altered self – as well as self and non-self. When a virus has infected a cell and the cell is displaying viral antigens in addition to its own, it has become altered self. That's what's recognised and dealt with rather than the viral antigens per se. The point is that the body treats altered self in much the same way as non-self. A virally modified cell is destroyed in the same way that a transplanted cell from another individual would be.

"People were wondering why the body should have a system for combating transplanted tissue when this state clearly never arises in nature.

"We suggested that the recognition of alloantigens, that is MHC antigens differing from your own, was there not to frustrate transplant

Prime Minister of Australia John Howard with Doherty.

With Professor Suzanne Cory of The Walter and Eliza Hall Institute of Medical Research.

surgeons but to help the body 'see' altered self." The essential importance of the work is that it changed the entire direction of research into diseases such as cancer, multiple sclerosis, diabetes, and even AIDS – research that was going down the wrong path.

It opened up crucial new fields of study that now occupy thousands of medical scientists around the world who are consequently getting much closer to understanding and combating some of humanity's most devastating diseases.

The research gave medicine a crucial step forward, although it's not just the scientific advance that has made Doherty and Zinkernagel's collaboration folkloric in the annals of Australian science. The real reason they are said to have teamed up in the beginning was that Zinkernagel liked to sing arias from *The Marriage of Figaro* and Doherty, an aficionado of classical music, was said to be the only other person happy to share such an operatic workspace. Then, despite the significance of their research being quickly recognised around the world, the two scientists, Doherty in particular, found themselves unwanted in Australia.

No vacancies

Doherty's position at the university was only temporary and when he applied for a permanent job he was told quite bluntly that there wasn't one. Both men subsequently went to the United States – Zinkernagel because he had intended to move there anyway, but Doherty because he needed a job. For him, it was a move forced by institutional bloody mindedness and jealously.

In a 2002 interview with journalist Luke Slattery for *The Australian* Magazine, Doherty recalled: "We made this big discovery. It was very exciting. And of course with a lot of very bright people who were a bit prima donna-ish really, there was a lot of resentment too. The difference between that institution and an American institution is that if you'd made a really big discovery like that they would have done everything possible to keep you. These guys were really glad to see us walk out the door."

Doherty was 34 with a young family, and so he accepted an offer to work in Philadelphia as associate professor at the Wistar Institute in Philadelphia. He quickly became involved with the highly regarded Immunology Graduate Group that worked closely with researchers at the University of Pennsylvania.

After the disappointment of his rejection at home, the time in Philadelphia was fulfilling and rewarding: "The Wistar/Penn axis was a highly interactive, and very open, intellectual environment. I collaborated extensively … and was part of a large, campus-wide multiple sclerosis research effort. Penny went back to school, and developed a new career in the area of drug information. I wrote grants, was a member of the immunology circuit, worked with outstanding graduate students and became an established scientist and academic."

Ostracised

Buoyed by the direction his career had now taken, Doherty's confidence was up again, so much so that, "I made the major mistake of accepting an offer to return to the JCSMR as head of the Department of Experimental Pathology. However, my decision was made on emotional grounds rather than on the basis of what was actually being offered."

Basically there wasn't enough money to go around and what funding there was, according to Doherty, was not distributed according to the quality of people's work, as it would have been in an American university, but to "local heroes" and favourites.

Surprisingly he was invited to join a group that was asked to review funding arrangements and the group recommended scrapping the existing tenure and funding system. The recommendation (which some time later was eventually adopted) was rejected vehemently by the university's hierarchy and Doherty and his associates were quickly ostracised. For the second time, he decided he was wasting his time in Australia and left.

It was an emotionally tough period with Doherty, one of the most brilliant researchers in his field in the world, unable to work in the country that he loved with a passion.

"I was very angry for a long time, vowing never to return to Australia or go anywhere near a university. I was only interested from then on in working at a specialist research institution."

The opportunity to rebuild his research career came with the resources offered to him by the St. Jude Children's Research Hospital (SJCRH) in Memphis. He was appointed as professor of Biomedical Research and chair of the Immunology Department where he devoted himself to studying cancer-causing viruses, in particular the Epstein-Barr virus (EBV) thought to be responsible for a number of diseases including nasopharyngeal carcinoma – a tumour of the nasal passages and throat. EBV has also been cited as a possible link to Hodgkin's disease, a cancer affecting cells of lymph nodes.

Doherty remained at St Jude's (where Penny worked as a hospital volunteer) until their final return to Australia more than a decade later in 2001. By this time Doherty was an acclaimed Nobel laureate and had been named Australian of the Year in 1997.

"I had gradually reconciled myself because I did come into contact with Australians who were

It changed the entire direction of research into diseases such as cancer, multiple sclerosis, diabetes, and even AIDS – research that was going down the wrong path

working hard to try and make a difference; to make the country work. Also Australia is small enough to be able to do things differently if it chooses, and it occasionally does this very well."

There were also family ties. His youngest son was by now a Melbourne barrister, although the eldest had settled in Seattle as a neurologist.

In reflecting on his career, Doherty felt his characteristics as a scientist stemmed from a non-conformist upbringing, a sense of being something of an outsider, and looking for different perceptions in everything from novels, art and experimental results.

In his interview with Beckmann, he said he loved immunology because he loved puzzling out complex, intricate systems. He also made the case for scientists to be free to be able to pursue conceptually-driven research, rather than be

Doherty's findings changed the research direction of a number of scientists.

We never set out to make our discovery – we weren't aiming in that direction at all. But when we found something unexpected we followed it

Doherty in his laboratory at the St Jude Children's Hospital in Memphis, USA.

locked solely into end-use driven research. This is an issue that has been dividing scientists and science policy-makers in Australia in recent years.

Keeping curious

"Conceptually-driven research is what is likely to yield some of the biggest benefits. But with this stuff you can't be sure where it will end up. Real curiosity-led work cannot be confined by a short time-horizon and it doesn't guarantee an outcome. Plenty of research leads up blind alleys. But you have to know that those blind alleys are there in order to find the right pathway.

"Of course, that doesn't mean you don't need applied research – it's essential – but you need to get the balance right between the two. Many governments, with their short time horizons, tend to favour the applied side too much."

When asked about the qualities needed to be a successful researcher, Doherty listed persistence and the need to be totally absorbed in what you do. "You also need to have an open mind, and be prepared to drop one line of inquiry and follow another if it looks interesting."

He adds: "We never set out to make our discovery – we weren't aiming in that direction at all. But when we found something unexpected we followed it."

In summarising his own approach he said he liked complexity, and was always delighted by the unexpected: "Ideas interest me. Intellectually, I march to the beat of my own drum and have little interest in competing in 'races'. There are too few people working in the area of viral pathogenesis and immunity, too little funding, too many problems and too little time."

Vital statistics

Name:	Peter Charles Doherty
Born:	15 October, 1940 in Brisbane, Queensland
School:	Indooroopilly State School
University:	University of Queensland; University of Edinburgh
Married:	Penelope Stephens in 1965
Children:	Two sons, James and Michael
Lives:	Melbourne

Awards and accolades

1983:	Paul Ehrlich Prize for Immunology; Fellow, Australian Academy of Science
1986:	Gairdner Prize
1987:	Fellow, Royal Society
1993:	Alumnus of the Year, University of Queensland Alumni Association
1995:	Albert Lasker Basic Medical Research Award
1997:	Australian of the Year
2001:	Melbourne University's Laureate Professor

Why he was awarded the Nobel Prize

Peter Doherty was awarded the Nobel Prize for his work on immune responses. He and his colleague Rolf Zinkernagel, with whom he shared the prize, were trying to see how an infecting micro-organism causes damage and disease. The two scientists discovered that white blood cells must recognise both an invading virus and certain "self" molecules in order to kill the virus-infected cells. This concept of simultaneous recognition of both "self" and "foreign" molecules now forms the basis for a new understanding of the cellular immune system. Doherty and Zinkernagel's explanation of the self/non-self selectivity of the immune system advanced research in many clinical areas. It led to advances in therapeutic efforts to strengthen the immune response against invading micro-organisms and certain forms of cancer, and to efforts to diminish the effects of autoimmune reactions in inflammatory diseases, such as rheumatic conditions, multiple sclerosis and diabetes. It has also proved to be a great advance in the field of organ transplants.

References: The Nobel Foundation, Press Release: The 1996 Nobel Prize in Physiology or Medicine, 7 October 1996
Sean Henahan, Access Excellence, Science Updates, http://www.accessexcellence.org/WN/SUA08/nobmed10.html
A.D'A. and D.D'A Webling, *Nobel Prize Laureates and Australia to 2000*, private monograph
Professor Lars Klareskog, presentation speech, The Nobel Prize in Physiology or Medicine 1996, http://www.nobel.se/medicine/laureates/1996/presentation-speech.html

I beseech you to take interest in these sacred domains so expressively called laboratories. Ask that there be more and that they be adorned, for these are the temples of the future, wealth and wellbeing. It is here that humanity will grow, strengthen and improve.

Here, humanity will learn to read progress and individual harmony in the works of nature, while humanity's own works are all too often those of barbarism, fanaticism and destruction.

LOUIS PASTEUR

ABIE

Adventures in Innovation

Australians have always taken risks and embraced new ideas. Our young country has achieved worldwide renown across the public, private and education sectors. ABIE is proud to present these stories that – like our Nobel laureates – celebrate achievement, showcase innovation and inspire success.

By Dr Alan J Jones

A vision for research

→ Australia is renowned for being a clever country but needs to look beyond its borders to really achieve success with its innovations.

For me, the word "innovation" immediately triggers an image of the industrial revolution, from canals, waterpower, the steam engine, railways, steamships, synthetic dyes and even "the dark satanic mills".

The negative impacts of rapid industrialisation in Europe have largely been cleaned up and as human beings we have learned more about the deleterious effects mankind can introduce to the environment. The waves of technological change into the 21st century have been much more rapid and penetrating and more complex in a technological sense. They range from petrochemicals to plastics, light metals such as aluminium and magnesium, microelectronics, computers, space, telecommunications, software, the internet and applications of biotechnology.

However, innovation is not just about technological change or the introduction of radical changes in technology. It is a much more subtle process of turning ideas into wealth, a process of adopting change to obtain improved outcomes. Central to the innovation process is the contribution of firms.

In the context of the firm, innovation is about the linkages and interactions with other organisations and institutions, including linkages with other firms – suppliers, customers and competitors alike and linkages with the knowledge base. (see Figure 1)

Figure 1: Collaborating for innovation

Adapted from the Conference Board of Canada, Innovation Report, 2003.

Knowledge can originate from any source – local and international – and increasingly no one firm can afford to stand alone. The ideas and the changes involved in the innovation process can also be subtle and incremental. They can include changes in organisational approaches within business, marketing and financial strategies, as well as the adoption and/or adaptation of technologies.

It is difficult to go past economist Paul Romer's view that innovation is about "the million little things" that improve the operation of firms and institutions.

Figure 2: Innovation creates wealth

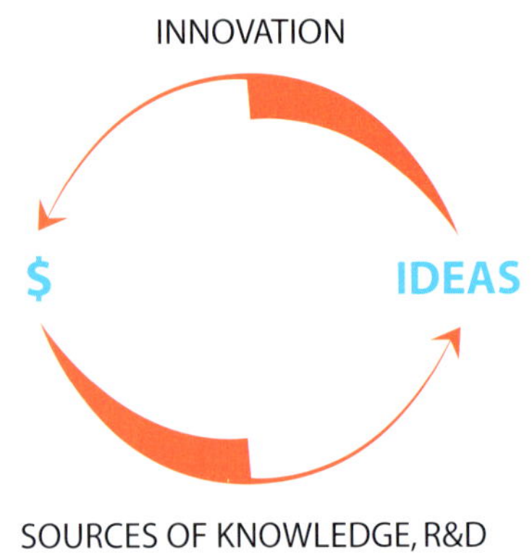

Figure 2 shows that innovation creates wealth but that the generation of knowledge, for example research and development for innovation, demands some re-investment of that wealth.

In speaking with firms about their capabilities and capacity to innovate, I am greeted by a plethora of examples that could be used to illustrate the factors that lead to successful outcomes in the innovation process.

Management matters

The first and most critical parameter for innovation in a firm is associated with management. They must exude the qualities of leadership and have realistic visions about the firm's direction.

Above all, these people are characteristically receptive to ideas and to change. In the case of the self-employed, the motivation is for personal success. For public companies the motivation is for management to be able to say that they have added to the bottom line and subsequently to shareholder value of which they may well share a part.

Managers are essentially good communicators and good listeners but are also resolute decision makers when it comes to the "crunch" on pulling the plug on non-returning projects.

In this context I need to mention a couple of "innovation heroes", namely Denis Hanley, the founder of Memtec, the membrane filtration company, which is now French-owned and the late Paul Trainor, the founder of the former Telectronics Pacing Systems, which captured substantial world markets in heart pacemakers.

Exporting for expansion

The second part of the vision for Australian-based innovative firms is recognition that the Australian market is generally too small to sustain a long-standing business. Businesses that are technically based must be export-oriented but also need to understand their overseas markets and the considerable need for technical support in those markets. Many choose to work with reliable local market operators.

Dr John Pulsford, R&D manager of Varian Australia, says that in the event an Australian firm has some significant intellectual property in the scientific instruments business, it is likely to find it too expensive to penetrate and maintain overseas markets.

Some of the more innovative subsidiaries of multinationals based in Australia can see the advantages of exporting, including automotive products and components. This contrasts sharply

Getty Images

The race for nnovation: if Australia wants to compete in the world market, it needs to be receptive to all opportunities that arise.

with the view of simply being present in Australia to serve the local market.

Know thyself

Third, the successful firms are not "one-product firms". They continuously seek product, process, service and quality improvements. They know their customers and suppliers. They identify their own weaknesses and seek to strengthen their position through collaboration or alliances or, where their business strategy permits, they capitalise on the strengths of others by conducting or commissioning research and development from expert sources such as research institutions and universities.

Larger Australian companies such as

BHP Billiton and Orica frequently seek out new ideas to take the "brain products to the market", but so do many of the multinationals with Australian operations, especially in the biotechnology and pharmaceuticals sectors.

> Ten more of these medical device companies and our balance of payments problems would be addressed

Where an internal or external research and development strategy does not provide adequate or rapid solutions to particular problems, an acquisition strategy has been adopted. While we remember the US-based Cisco Systems acquiring "our" local company Radiata, Toll Logistics, Australia's largest transport company, has become a master of this strategy. So has the Western Australian Nautronix Group, when in 1990 it acquired a division of Honeywell.

> We need to be receptive to all opportunities to be more competitive in this globalised world

Some innovative firms have grown by capitalising on the excellence of Australia's research and development system. They include the medical device companies Cochlear and ResMed. To paraphrase Australia's chief scientist, Dr Robin Batterham: "Ten more of these and our balance of payments problems would be addressed".

Other innovation comes from the biopharmaceutical research houses of Biota Holdings, AMRAD and Starpharma, with the latter receiving permission from the US Food and Drug Administration to test its dendrimer carried drugs, developed at the former Biomolecular Research Institute in Melbourne. There are also innovative companies such as Lochard, which specialises in the design, manufacture and support of airport management systems using digital analysis of sound (originating from the University of Melbourne). Another example is Softlaw, which offers a specialised system of "legislative rule-based technology" used by the Department of Veterans Affairs to pay out compensation entitlements. This technology originated from the research work of the founders at the Australian National University (ANU) and University of Canberra.

Some of the less heralded research and development that has led to innovation is found in the food company Burns Philp. It commercialised the developments in the genetics of yeast strains from the ANU. The CSIRO developed the application of carbon-fibre technology to numerous aircraft components by the Boeing Corporation. The management of VIPAC Engineers and Scientists once stated to me that they placed high value on "the intellectual property between people's ears". This serves to move the discussion to other modes of innovation.

Quiet achievers

In-house innovation is probably more common than that originating from institutional research and development. Bishop Engineering, for example, has a long history of developing advanced steering systems for the automotive industry and its market penetration has gone relatively unnoticed, with about one in four cars now incorporating its system.

> Part of the vision for Australian-based innovative firms is recognition that the Australian market is generally too small to sustain a long-standing business

A joint venture with Daimler Chrysler has led to the Bishop Group becoming a lead innovation centre for the automotive giant. Meanwhile, the company Austal builds boats but its innovation comes from understanding the technology of aluminium construction, originally picked up by investigating the aerospace sector. Austal now designs and builds high performance 30- to 90-metre passenger and vehicle ferries and acquired Oceanfast and Image Marine as part of its diversification into smaller luxury vessels.

Finally let me mention a common approach to innovation in Australia where we take the best ideas that are available regardless of source and integrate them into a system to meet customer needs. Consider the developments of Australia's latest passport system. Alphawest has combined the optical character recognition technology of Microsystems Technology of the United States and Optika's (USA) Acorde system for imaging and workflow in working with Passports Australia to develop a highly innovative system.

I apologise to the numerous Australian-owned and multinational companies that have provided me with a true feeling of how innovative this country of ours really is. But we need to be patient. New high-growth innovative companies cannot be expected to hit their straps for 10 years and, in the biotechnology/pharmaceuticals area, a more realistic time frame is probably 20 years. The tradition of innovation in Australia must continue, be it through technology adoption or adaptation, or through better ways of doing things through organisational, financial and marketing improvements. We need to be receptive to all opportunities to be more competitive in this globalised world.

The role of government

Today governments, industry and commentators around the world recognise that innovation and the knowledge that underpins it are critical drivers of productivity and economic growth in the new millennium. However, most of the elements of the innovation system are private and most innovative activity goes on outside government.

A country's wealth is primarily generated by the activities that go on in the private sector. Thus, the role of government is to create the best environment in which the system can function and to create cohesion between all the players in the system. This is achieved by working in a fully coordinated and cooperative manner. ■

Dr Alan J Jones is the director of Innovation Analysis in the Industry Policy Division of the Department of Industry, Tourism and Resources.

The important thing is
not to stop **questioning.**

ALBERT EINSTEIN

By Professor Alan Pettigrew

Australia's medical miracles

→ A snapshot of the work of Australia's medical research community reveals many world-leading innovations.

Twenty years ago, a book was published in Australia that captured the lives of ordinary Australians at work and play over a 24-hour period. The photographs in the book – *One Day in Australia* – were taken at various locations around the country and featured some spectacular images. The book opened many eyes to the scope, grandeur and individual makeup of our country.

I believe it would also lift the veil from many eyes if, on one particular day, we could capture the work being done by the small army of Australian health and medical researchers as they battle against the mysteries of illness and disease and work towards the goal of better health for all Australians.

Research breakthroughs

What might we see on one such day? In fact, some wonderful things. For instance, the delight on the faces of the team at the University of Queensland's Centre for Immunology and Cancer Research which, after 18 years work, developed a vaccine that works by protecting women against infection by the human papillomavirus (HPV). This development has been rightly hailed as a major breakthrough in the fight against cervical cancer.

> A visit to the University of New South Wales will find Associate Professor Levon Khachigian firing "magic bullets" to seek out and destroy harmful genes involved in blood vessel diseases

We might also drop into the laboratories of National Health and Medical Research Council (NHMRC) Industry Fellow Dr Geoff Dandie, whose team is working with bioactive molecules extracted from cow's milk to see if we can prevent or treat the inflammatory bowel disease suffered by an estimated 60,000 Australians.

A visit to the University of Melbourne to see what our latest Nobel laureate, Professor Peter Doherty, is up to in his area of immunology would definitely be worthwhile. Doherty won his prize for discovering how the immune system recognises virus-infected cells, a breakthrough of immense importance to HIV sufferers.

In Canberra at the Australian National University, renowned epidemiologist Professor Tony McMichael is busy strengthening the centre's research into environment and population health. He returned from London to take up the post of director of the National Centre for Epidemiology and Population Health.

A visit to the University of New South Wales would find associate Professor Levon Khachigian firing "magic bullets" to seek out and destroy harmful genes involved in blood vessel diseases. As one Australian dies from cardiovascular disease every 10 minutes, the potential new therapeutic agents based on this

research could have enormous health and economic benefits.

A roving observer would be interested in the work of Dr Bronwyn Kingwell and her team at the Alfred Baker Medical Institute and Monash University. The group has made a discovery that may give rise to a new way of treating Type 2 Diabetes by mimicking the effects of exercise.

Not far away, at the Alfred Psychiatry Research Centre, schizophrenia sufferers are participating in clinical trials and responding well to a world-first treatment using estrogen. Professor Jayashi Kulkarni and her team have been testing the hypothesis that estrogen may be a useful new treatment for the devastating illness.

The inquisitiveness of Australian researchers is deep and wide. There are teams looking at bone disease, Aboriginal and Torres Strait Islander health and the use of milk in protecting teeth.

There are collaborative efforts in diabetes (with the Juvenile Diabetes Research Foundation); a tripartite agreement on

Australia's scientists are quietly achieving great medical outcomes here and globally.

indigenous health research (with Canada and New Zealand); and a push to improve the health of developing countries in conjunction with the Wellcome Trust (UK) and New Zealand.

Australian innovations have benefited communities here and abroad. The cochlear implant, for example, has meant thousands of people are now able to hear. Thousands more are back at work after heart attacks, and many live active, healthy and rewarding lives after being bombarded with deadly rays to kill off their cancer.

Estrogen may be a useful treatment for the devastating illness of schizophrenia

Many of the initiatives above are currently funded by NHMRC grants or have benefited previously from the council's support. In 2003, the NHMRC allocated about $A300 million in government funds for health and medical research in Australia. A proportion will be for fundamental research, but much will be directed towards particular problems. All applications will have satisfied the NHMRC's peer review process for allocating funds.

Do we get value for money? Access Economics released a report for the Australian Society of Medical Research which showed that for every $1 spent on research, the dividend for Australia was $5.

The report also stated that if research enabled a 20 per cent reduction in the incidence of cancer, the benefit to the nation would be $184 billion. Another statistic stated a reduction of just 15 per cent in cardiovascular problems would save $34 billion – more than the Federal Government's annual health budget. The benefits of research are clearly dramatic.

The report, entitled "Exceptional Returns: the value of investing in health R&D in Australia" called for our governments to commit an additional $1 billion annually to the country's research budget for each of the next five years. This would raise our commitment from its current 0.25 per cent of GDP closer to that of the United States (0.7 per cent) and the United Kingdom (0.8 per cent).

If we want to find a cure for cancer, eliminate the scourge of heart disease, overcome HIV/AIDS, and lessen the debilitating effects of mental illness, if we want our asthmatic children to breathe easier or if we just want a better quality of life in old age, then our commitment to research on the national stage and as a major international player must not just be undiminished but be enhanced.

I am looking for an outcome where the NHMRC will be able to allocate more than double the funds it has available to it now. It will be well worth it, not just in terms of a national health outcome, but also in terms of increased economic wealth for the country. These long-term aims require sustained commitment by individuals, industries and governments. And who knows, maybe we might also be able to produce a few more Nobel laureates. ■

Professor Alan Pettigrew is chief executive officer of the NHMRC.

Nobel laureates and unsung heroes

There is probably no brighter international spotlight than the one that lights up the truly outstanding researchers who win Nobel Prizes. Some of the winners will tell you that a Nobel Prize changes their life – they become public figures who are in demand, their views are sought and their presence is requested at functions.

There are many signposts that point to the excellence of Australian research and researchers. For instance, we publish 12.6 research papers in the field of "health" per million of population. In the UK, the figure is 11.3; in the US it is 9.0. There are other indicators too, including international prizes, the granting of patents, the establishment of start-up companies and, most importantly, the improving health of the nation.

The nation's success in research comes from understanding our national limitations. Funds are limited and competition is intense – thus, the way forward is often through cooperation at home and abroad, stretching the dollar by tapping into excellence wherever it exists.

Sharing wisdom

As we address our own health research priorities and our own health agenda, we must also contribute to the worldwide knowledge base and learn from the problems experienced elsewhere by others.

By their nature, most researchers are outward looking, despite appearances to the contrary as they sit hunched over laboratory instruments. Nobel laureates are very much like beacons – they tell our researchers that geography doesn't have to limit them and that clear thinking and excellent work will have its rewards. The reality is, however, that Nobel Prizes are very hard to come by and that, for most, lesser rewards will have to suffice.

Apart from our Nobel laureates, we have other heroes: Fiona Stanley, Don Metcalf, Barry Marshall, Graeme Clark and Gustav Nossal are all well known.

Death of diseases

Australians owe their researchers a huge debt. Less than 70 years ago, as the NHMRC held its first meeting in Hobart, the infant mortality figure in Australia was 39.78 per 1000 births. In the mid-1990s, it was less than six. In the first half of the 1900s, diphtheria, tetanus and even leprosy were illnesses with which Australians had to contend. Today, they are all but wiped out. Most babies born today will live well beyond 80 and the quality of life for most will be good.

However, the work is not over. Right now, in university and medical research institute laboratories across the country, dedicated men and women are setting about trying to further unravel the mysteries of asthma, cancer, heart disease, mental illness and arthritis, among others. They are all heroes – even if unsung.

By Dr Ian Cooke

A vision for public health

→ Developing vaccines from plants is just one initiative being undertaken by the Macfarlane Burnet Institute in Melbourne.

The Melbourne-based Macfarlane Burnet Institute for Medical Research and Public Health (Burnet Institute) continues to practise the same commitment as its founder, Nobel laureate Sir Frank Macfarlane Burnet, in the fields of virology and infectious diseases. One of its main focuses is assisting countries such as Nepal, Indonesia and Vietnam, which have fewer resources to devote to public health issues.

The Burnet Institute is spearheading a range of initiatives aimed at reducing the impact of infectious diseases in Australia and around the world. Some of these initiatives include improving the diagnoses of HIV cases and developing a measles vaccine within genetically modified food crops.

Director of the Institute, Professor Steve Wesselingh, believes that to achieve the Institute's vision of delivering benefits to the health of individuals and entire populations, basic research needs to be integrated with public health action.

"Most of the Burnet Institute's achievements have involved the development of appropriate and sustainable technologies to prevent or control infectious diseases in resource-poor countries," he says.

Global fight against HIV

One project involves providing clinical support for the management of Victoria's HIV-infected patients. It also provides support for some high-caseload general practices and hospitals based in Sydney and within Tasmania.

Another project which has had tremendous success is the introduction of diagnostic kits to resource-poor countries. The kits enable practitioners in these countries to perform diagnostic treatments at low cost. They perform as well as the diagnostic equipment used in more sophisticated environments. The Institute is also involved in training the personnel.

The service has been so successful that a number of staff have taught Indonesian scientists how to perform the low-cost CD4 analyses in Bali and now similar collaborative projects are being undertaken in the Indian cities of Chennai and Mumbai. CD4 is a type of molecule that is found on the surface of one type of T-cell (white blood cells that play an important role in the body's immune system).

"We are not just carrying out research," Wesselingh explains. "We are demonstrating how the research can be applied in resource-poor settings. It is in settings such as these that interventions are most needed."

Vaccine alternatives

Another project underway at the Institute's Tattersall's Children's Vaccines Research Laboratory is the possible manufacture of vaccines for measles and malaria within genetically modified food crops. Wesselingh says plant-based vaccines, such as the measles vaccine, have the potential to overcome many of the limitations of live vaccines, including the need for refrigeration and injection.

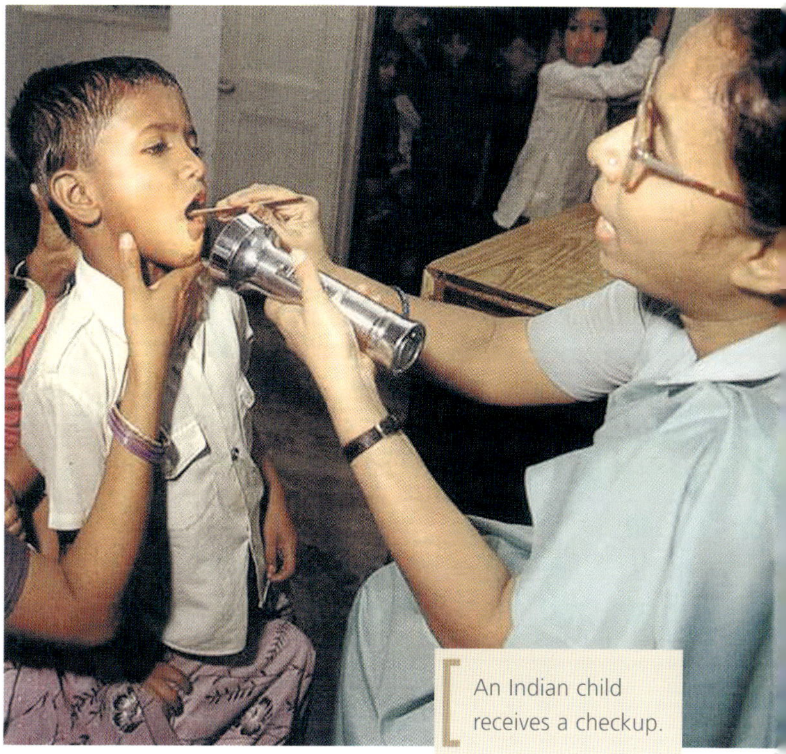

An Indian child receives a checkup.

"Traditional vaccines need to be kept in cold temperatures and this is almost impossible in resource-poor countries," he says. "If this research succeeds, it means vaccines can be put into genetically modified food crops that are grown locally, such as rice or millet, so people then ingest the vaccine."

Working with vaccines has also been pioneered in Lombok in Indonesia with the introduction of a routine hepatitis B immunisation program. This was the first demonstration of the usefulness of hepatitis B vaccination in a developing country's immunisation program.

It has contributed to the models used now by the Global Alliance for Vaccines and Immunisation and the Gates Foundation to support universal hepatitis B immunisation in the 70 poorest countries of the world.

The Institute is also at the forefront of advocating safe drug-using projects locally. It also wants to establish these projects internationally in countries such as Nepal, Indonesia, China and Vietnam. Dr Nick Crofts, deputy director of the Institute, says its Centre for International Health is also leading the way with its range of health development activities.

"We are leading the way in educating and training local people to maintain important community health projects such as refugee health, HIV/AIDS and communicable diseases," he says. ■

Dr Ian Cooke is associate director, research and development at the Burnet Institute.

By Gayle Bryant

Medical missions

→ From its early days of studying snake and spider venoms, The Walter and Eliza Hall Institute is now researching cures for malaria.

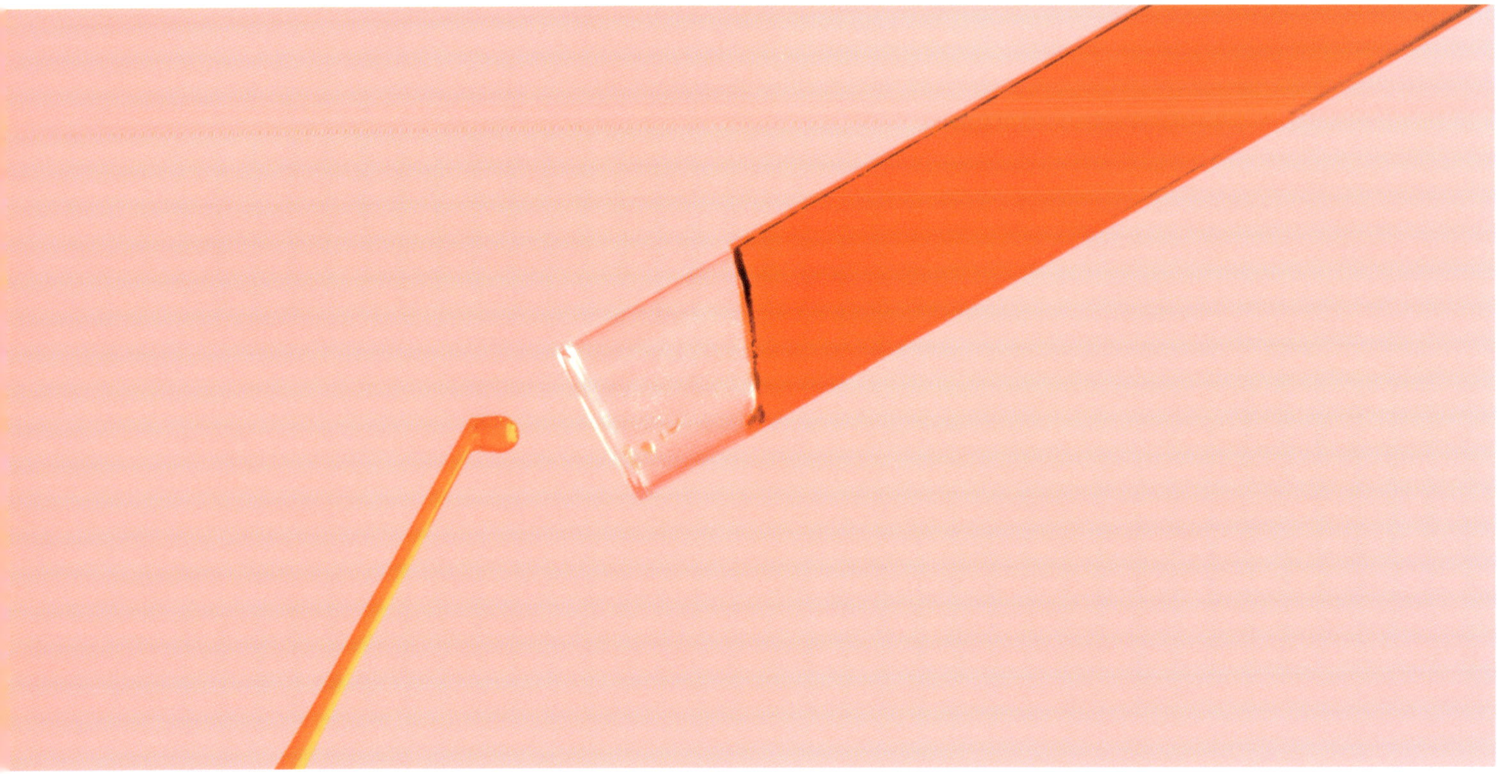

Melbourne's Walter and Eliza Hall Institute (WEHI), founded in 1915, attracts scientists from all around the world to its dynamic research programs on cancer, immunology, autoimmune disorders and infectious diseases. The goal is to develop a deeper understanding of the causes of these diseases, including any underlying genetic predisposition, and to use this knowledge to develop better diagnostics, treatment and prevention strategies.

Combating cancer

Curing cancer may seem like an impossible dream but research at WEHI has already greatly enhanced cancer treatment. More than 4 million patients undergoing cancer therapy have benefited from treatment with hormones known as CSFs (colony stimulating factors) that accelerate the recovery of their white blood cell production and so protect them from lethal opportunistic infection. This treatment stems from decades of painstaking research carried out at WEHI by Don Metcalf and his team, starting in the 1960s. The early clinical trials, performed with colleagues at the Royal Melbourne Hospital and the Ludwig Institute, also led to bone marrow transplantation being replaced by blood stem cell therapy. The CSFs are now being trialled for new applications, including Crohn's disease.

Director of WEHI Professor Suzanne Cory says cancers are caused by an accumulation of genetic mistakes. "Research at WEHI showed for the first time that mutations, which inhibit the natural process of cell death, contribute to cancer development," she says. "Paradoxically, cell death is a critical part of the life process. In every second of every person's life, close to a million old and damaged cells self-destruct on cue, to be replaced by about the same number of new cells."

Cory says there is a need to know in intimate detail how the cell death program (apoptosis) works. "Our studies should enable us to develop novel therapeutics that can activate apoptosis of cancer cells more effectively than current treatments," she says.

WEHI's quest for more effective therapeutics for cancer and other diseases has been greatly boosted by the recent recruitment of a team of structural biologists and chemists led by Professor Peter Colman, whose groundbreaking research resulted in the development of the anti-influenza drug Relenza.

Curing cancer may seem like an impossible dream but research at WEHI has already greatly enhanced cancer treatment

WEHI has also purchased a state-of-the-art x-ray crystallography machine following a $A1 million donation. X-ray crystallography allows photographs to be taken of proteins that are magnified

100 million times their size. When such a detailed picture of the protein is available, it is possible to understand how that protein performs its complex chemical tasks in the body.

Infection influences

Understanding the complex circuitry of the immune system is central to developing better strategies for combating autoimmune disorders and infectious diseases. WEHI's reputation in this arena began with the pioneering studies of Frank Macfarlane Burnet on animal viruses in the 1930s. Burnet became director of WEHI in 1944 and for more than a decade WEHI made major contributions to influenza, herpes, psittacosis, Murray Valley encephalitis and poliomyelitis. Indeed, the method Burnet developed to manufacture flu vaccines remains in use at Commonwealth Serum Laboratories to this day.

In 1957, to better understand the body's response to pathogens, Burnet abruptly switched the attention of the institute to immunology. He reasoned that early exposure to foreign antigens might "trick" the developing immune system into accepting foreign material as "self". This bold hypothesis, known as immunological tolerance, was vindicated experimentally by Peter Medawar in London and the two shared the Nobel Prize for Medicine in 1960. Burnet's most influential legacy to immunology, however, has been his explanation of how the body can make such a prodigious array of antibodies – the "clonal selection theory".

When Burnet retired in 1965, he was succeeded by his brilliant former student, Gustav Nossal, who expanded the institute greatly in size and scope. Immunology remained a central focus, with major contributions to clonal selection, tolerance, the genetic basis for antibody diversity, and thymus development. Of note, building on his earlier discovery of the vital immunological function of the thymus, an organ until then thought to have been but an evolutionary relic, Jacques Miller showed with Graham Mitchell that immunity depended on two distinct types of lymphocytes, later called B cells and T cells. Today, WEHI immunologists remain at the forefront in deciphering precisely how these lymphocytes interact with each other and with remarkable scavenger cells known as dendritic cells to generate protective immunity.

Unfortunately, the immune system sometimes turns rogue and attacks the body, provoking a range of autoimmune disorders which afflict millions of people worldwide. Starting under Burnet, Ian Mackay developed the concept and key features of autoimmunity and pioneered the treatment of autoimmune (lupoid) hepatitis with immunosuppressive drugs. This legacy continues at WEHI.

The juvenile diabetes program has yielded reliable early indicators of disease onset and a promising preventative vaccination regime. Other programs are tackling rheumatoid arthritis, coeliac disease and multiple sclerosis.

Looking back on the proud history of WEHI, Cory reflects how far the organisation has travelled since it was founded in 1915 with funds from the charitable trust established by Eliza Hall using the fortune amassed by her entrepreneurial husband, Walter.

"The small group studying venoms of snakes and spiders in a laboratory within the Melbourne Hospital in Lonsdale Street would marvel at the technology available to the 600 scientists, students and support staff housed in our current ultra-modern institute in Parkville," she says. "They would be amazed at how far medical research has progressed. But they would clearly recognise the mission of The Walter and Eliza Hall Institute as the 'mastery of disease through discovery'." ■

Malaria parasites invade a red blood cell. The body's immune response causes the high fever, typical of malaria, that contributes to the death of about 2 million people every year.

Diverse research activities

Research at The Walter and Eliza Hall Institute of Medical Research (WEHI) is organised into seven key divisions and groups. They are:

- Cancer and Haematology
- Molecular Genetics of Cancer
- Immunology
- Infection and Immunity
- Autoimmunity and Transplantation
- Structural Biology
- Genetics and Bioinformatics

Vanquishing infectious diseases still remains a priority, the principal targets now being malaria and leishmaniasis. A perpetual scourge of humanity, malaria kills at least 2 million people every year and debilitates hundreds of millions more.

WEHI's 20-year commitment to developing an effective malaria vaccine has yielded several promising lead molecules. WEHI researchers have also tracked down how the parasite acquires drug resistance and identified the malaria genes responsible for the lethal stickiness of infected red blood cells.

Meanwhile, leishmaniasis, transmitted by sandflies, affects at least 12 million people each year and about 300 million people are at risk.

To date there is no vaccine and resistance is developing against the commonly-used antimony drugs. The WEHI's Infection and Immunity division is working hard to develop a vaccine against leishmaniasis and to identify parasite targets for the design of new and specific drugs.

By Nic Svenson

Bringing proteins to life

→ Understanding and identifying how proteins and their bioactivities work can provide opportunities for solving life sciences problems.

For most of us, the word "protein" conjures up images of meat and fish, but protein is much more than this. Hair, nails, blood, cells, hormones, plants and even bones are all made up of millions of different proteins. Identifying how proteins do specific things within cells is an easy way of telling what a cell is up to. For example, if certain cells overproduce a growth-promoting protein, it may be an indication that the cells are on their way to becoming cancerous.

Detecting, identifying and analysing proteins in a high-throughput way is called proteomics, a term coined by Australian scientists at the Australian Proteome Analysis Facility (APAF). The set of proteins produced by a cell or an organism at any time under any circumstance is a highly dynamic set of proteins that is called the proteome, in the same way that all genes collectively are called the genome.

Studying proteins might seem a bit old-hat in these days of decoding DNA, but proteomics is the fastest-growing area of life sciences. Professor Mark Baker, chief executive officer of APAF, says this is because a road map (the genome) is just the instructions on how to get somewhere, while the streets, the car and the traffic lights are the different proteins – they actually do the work in the cell. He adds that while animals such as butterflies and caterpillars have identical genomes, their proteomes are quite different.

Much of the technology used in proteomics was developed in Australia and is now distributed globally

Scientists at APAF spend most of their time providing proteomic services to the research community, but they also teach students, train other scientists, provide advice on setting up and running proteomics laboratories and conduct ER&D, which stands for new technology evaluation, research and development.

"When APAF started in 1995, we were the first dedicated proteomics centre in the world," Baker says. "Much of the technology used in proteomics was developed here in Australia and is now distributed globally."

Technology incubator

A separate company established by the original APAF team is Proteome Systems (PSL). It is dedicated to the development of proteomics technology, to the discovery of biomarkers and drug targets, and to proteomic bioinformatics.

Baker says that APAF is proud of the long relationships the organisation has with its commercial partners and he is keen to build new opportunities across Australia and internationally. Indeed, Baker seems to have expansion on his mind. APAF is one of the major national research facilities and he wants it to grow beyond the existing nodes – at Macquarie University, University of Sydney, University of NSW and TGR Biosciences in Adelaide.

Getty Images

Caterpillars and butterflies: same genomes, different proteomes.

TGR Biosciences has a range of rapid, high-throughput functional proteomic bioactivity assays. This means purifying proteins by what they do, unlike traditional methods which separate proteins on the basis of physical characteristics such as size or charge. This system can also be used in the search for, and to improve, new drugs. Baker describes the relationship between APAF's nodes as "synergistic and complementary".

"With new companies being established, enthusiastic ER&D partners and a growing client base, it is clearly a universe that is expanding," he says. ■

Customised flour

Agricultural scientists are using proteomics and genomics to develop better flour. Some wheat varieties produce flour ideal for making biscuits, while some types are better for bread. In each case, different proteins are involved. By identifying which proteins lead to each trait, scientists can unravel their genetic code and food producers can make more informed decisions about the type of wheat they wish to grow.

By Professor Judith Whitworth

Raising the Curtin

→ Development of HIV vaccines and anti-cancer drugs have taken this institution to the forefront of medical research in Australia.

An HIV vaccine is the latest notable development to come out of the John Curtin School of Medical Research at the Australian National University. This vaccine is the result of research from the laboratory of Professor Ian Ramshaw. Ramshaw's work follows the tradition of the school's focus on immunology. This includes Peter Doherty and Rolf Zinkernagel's Nobel Prize in 1996 for their work on immunology carried out in the 1970s, along with Frank Fenner's Japan Prize for his share in the eradication of smallpox, an effort continuing over 50 years.

Ramshaw's research has resulted in the commencement of clinical trials of the HIV vaccine conducted by an Australian consortium that includes the University of New South Wales, the University of Melbourne and the CSIRO. The trials are funded by a major grant from the US National Institute of Health.

Meanwhile, a new concept in vaccine development within the Division of Immunology and Genetics at the John Curtin School is the scrambled antigen vaccines or "savines". The technology, which has been patented, will allow the development of more effective and safe vaccines using a totally synthetic approach to molecularly re-engineer whole pathogens or large numbers of antigens simultaneously. This avoids the problems of using the whole virus which would risk introducing the infection itself.

Vaccines for viruses

Ramshaw says an effective vaccine must be able to protect a genetically diverse human population against a wide range of viral isolates. "We are currently developing the next generation of HIV vaccines to address this issue," he explains.

At the school, research was once characterised by dedicated individuals working long into the night or small groups combining their collective genuis toward a groundbreaking discovery. These days, successful research is often the result of large teams and collaborations. The school currently boasts more than 200 external collaborations, as well as a range of intra-campus and internal partnerships.

One initiative is the result of more than two decades of work on sulfated polysaccharides by Professor Chris Parish and his colleagues. It is paying off with the production of a number of compounds that have promising clinical applications that include uses as anti-inflammatories and anti-cancer drugs.

Sulfated polysaccharides are simple molecules made up of chains of sugars (saccharides) with sulfate groups, joined in long chains.

The most advanced of these compounds is PI-88, an anti-cancer drug that has a double action in inhibiting the development of tumour blood supply and in preventing metastasis by inhibiting the enzyme heparanase. The cloning of this enzyme by Dr Mark Hulett and Dr Craig Freeman of the John Curtin School was a significant achievement in advancing the development of the compound.

PI-88 is already in phase II clinical trials and has shown extremely encouraging results in treating cases of advanced melanoma. A cancer patient on the trial said recently: "The drug I am on now (PI-88) allows me to function and have a life."

Marc Fenning, Multimedia, JCSMR, ANU

Getting cultured: viruses are being targeted with new vaccines.

The school has big plans for the future, including focusing on the core areas of functional genomics, proteomics and bioinformatics. The construction of a new building will provide state-of-the-art laboratories for these disciplines. Already, a new building to house the Australian Phenomics Facility next to the school is well advanced. This facility, funded by the Federal Government, will be headed by Professor Chris Goodnow of the John Curtin School's Medical Genome Centre – a research unit of the John Curtin School. ■

Professor Whitworth is director of the John Curtin School of Medical Research.

Heart of the matter

An important consideration in the process of innovation is to assess the scope of the issue or problem. After all, innovation generally seeks to bring about positive change – and this can only be measured if the baseline is understood.

The cost of healthcare is an important political issue around the world, causing pressure on the Medicare and Pharmaceutical Benefits Scheme budgets and prompting alarming projections for the future. Medical research – discovering the causes of diseases and preventing them – offers the only remedy that is both better and cheaper.

Heart and vascular disease and stroke are due to atherosclerosis and are the largest causes of death, disability and health expenditure in the world today. Because the problem is so large the benefits of improvement are great.

It is not surprising, therefore, that a recent study showed that investment in medical research pays off in general – about $A5 return for every $1 spent by the public. However, the benefit in cardiovascular disease research is even greater – $8 for every $1 spent. This is because cardiovascular research has led to treatments that actually work. -

However, much more remains to be done as there are increasing rates of diabetes, obesity and heart failure in our community. Good research and innovation is a people business. It requires creative and skilled individuals. New technology is also important. Context matters here, too. The Baker Heart Research Institute has a name in clinical research brought about by its close association with clinical services and the Alfred Baker Medical Unit. It brings clinicians in apposition with basic scientists. Problems and solutions go back and forth and the results eventually impact on the health of our community.

Our approach is to form multi-disciplinary teams comprised of people with complementary skills to focus on the prevention and eradication of heart disease and strokes. This collaboration between mind and spirit is a natural breeding ground for rich ideas. Ideas are, of course, the genesis of innovation. Therefore we are confident that the Baker can deliver a future full of innovation to the community.

PROFESSOR GARRY JENNINGS
DIRECTOR
BAKER HEART RESEARCH INSTITUTE

leaders SPEAK

The *Australian Business and Investment Explorer (ABIE)* is an annual journal, this book being Volume 3. *ABIE* consistently seeks and edits relevant commentary from well-known as well as unknown thinkers and achievers. See other entries within this volume.

Good research and innovation is a people business

Australia requires an intense focus on technology-based wealth creation, preferably with global application

Technology the turbocharger

No society can be a truly just one without wealth creation. A fundamental truth is that the economic pie must be grown before it becomes divided and shared. In this way, in relative terms, we all win. And the basis, or enabler, of wealth creation is technology. If we apply technology to that which we already know, it is called productivity; if we apply technology to something completely new, and we deliver the concept to the marketplace, then we have innovation. In short, technology is the turbocharger of wealth creation.

We all require receptors of understanding – if we don't already have them – about technology's value-adding capabilities and its concomitant application.

Since Australia has such a huge and deteriorating current account deficit, we require an intense focus on technology-based wealth creation, preferably with global application. The way forward is to back people who are into the game early in potentially very large global markets.

In order to do this, we require venture capital to be applied to the most relevant concepts being developed by technology-savvy people who have unquestioned integrity. The money would be freed up and, one could argue, must be freed up from the massive superannuation funds that are being accumulated in this country.

At a minimum, 1 per cent of these funds should be targeted at potentially exciting wealth-creating opportunities. The money would be administered by people who have the appropriate experience, knowledge and awareness required in making circumspect equity placements.

We must understand that 80 to 90 per cent of these investments will most likely fail. However, these failures are important to the whole process. The people involved will become more aware and better educated in what to do and what to expect as they move on to other ventures. This is part of the learning process.

In most cases, gains from the winners will far outweigh losses from the losers. The time for us to start to invest in our own future is now. We have enough examples to know that we can do it; we just need more of the same. It is this approach that will eventually work.

PETER FARRELL
CHAIRMAN
RESMED

ABIE

By Rosemary Lynch

Gearing up for the future

→ Expanding its capabilities to meet new regulatory requirements is a priority at the Queensland Institute of Medical Research.

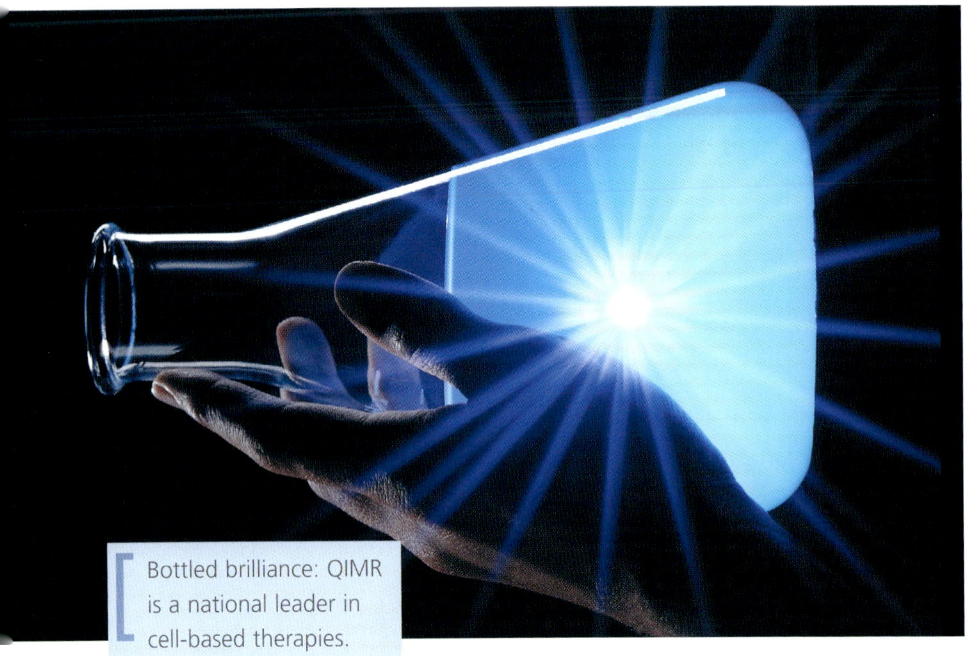

Bottled brilliance: QIMR is a national leader in cell-based therapies.

Images of Australia are often characterised by constant sunshine in the great outdoors. However, this exposure to the elements is also accompanied by the fact that Australians have the highest incidence of skin cancers – especially melanoma – in the world.

The Queensland Institute of Medical Research (QIMR) has a strong culture of investigating disease problems from a variety of perspectives and finding effective treatments for diseases such as skin cancer is a key focus. Increasingly, this has meant the use of more sophisticated equipment and facilities to conduct investigations on a broad scale.

"We are looking at melanoma and non-malignant skin cancers starting with the biomolecular mechanisms occurring in a cancer cell, through to understanding the prevalence of skin cancers in populations, and the manufacture and testing of anti-cancer therapies," says QIMR director Professor Michael Good. "We collaborate with industry and academic partners and also conduct contract research in this area."

Bench to bedside

Making the innovative move to create a facility for the manufacture of potential therapeutics at the same time as inviting a clinical trials company to co-locate within the institute makes QIMR unique as a "bench to bedside" medical research centre in Australia.

Successful research findings can be taken further along the "drug-to-market" pipeline by translation of those findings into fully GMP (Good Manufacturing Practice) compliant processes suitable to take into clinical trial. These products, potential anti-cancer and other therapeutic candidates, can then be tested in clinical trials at the co-located clinical trials company.

As a national leader in the area of cell-based therapies and vaccine delivery, QIMR made the strategic decision in 2002 to build a cell and molecular therapy manufacturing facility, called Q-Gen, that would meet the current and anticipated regulatory requirements and increasing standards of manufacture being demanded by the Australian Government Therapeutic Goods Administration and by the public. In taking this financial step, QIMR wanted to address not only its own requirements but to build something that would service Australian and international requirements for GMP manufacture.

Also located within the institute, and co-located within the grounds of the Royal Brisbane Hospital, is the clinical trials company Q-Pharm. Q-Pharm is partly owned by QIMR and is a joint initiative between QIMR and the University of Queensland. It specialises in early-phase clinical trials suitable for small-scale testing of various therapies before the large-scale trials that are only conducted by overseas "big pharma" companies with large budgets. Q-Pharm has conducted a range of trials and bioequivalence studies for local and international clients, and in its second year of operation is seeking to form alliances with pharmaceutical and biotechnology companies worldwide.

These added capabilities build on QIMR's substantial record in medical research and open up a new vista for future biomedical achievement.

"Critical in the pathway from investigation of a disease problem to putting a cure on the shelves is the generation of intellectual property and preservation of its value," explains Tracey Mynott, QIMR's business development executive. "Successful development and testing of new therapeutics to clinical trial stage maximises the potential value of a product within Australia and strengthens its position as a biotechnology leader." ■

Rosemary Lynch is the communications officer at the QIMR.

QIMR has extensive research programs in the following areas:

- Cancer
- Population studies
- Infectious diseases
- Disorders associated with organ transplantation
- Immune system disorders
- Human genetics
- Indigenous health
- Therapeutic development

By Nic Svenson

It's all in the genes

→ A leading research facility is working in the complex world of genomes to better understand diseases and their treatment.

Getty Images

Piecing together genetic information is all in a day's work at the AGRF.

Understanding our genetic make-up is important for a number of reasons such as identifying diseases and knowing how to treat them. One institution undertaking this task is the Australian Genome Research Facility (AGRF), a leading facility for the collection of molecular genetic information. Genes are contained in DNA and encode a set of instructions for every living thing in every single cell.

Knowing which genes a person or plant has helps scientists determine whether they are active or dormant, and even how those genetic instructions are written. This process is called DNA sequencing and is one of AGRF's technical strengths.

Extracting and analysing DNA is not cheap and the AGRF devotes 20 per cent of its time to developing more accurate, faster and less expensive ways of carrying it out (the rest of the time is spent processing samples from clients – mainly academic researchers).

When it comes to research, the AGRF has a policy of linking up with local and overseas manufacturers. For example, it is working with Victorian company Varian to replace manual techniques for separating what are often only subtly different strands of DNA with a high-throughput, automatic system. This is important because small changes in a plant or animal's genes can mean big differences in where they'll grow or what diseases they may develop.

"We're very keen on tweaking existing technologies to see if we can make them even better," says Dr Sue Forrest, director of the AGRF.

"Companies don't tend to mind if my team plays around with their equipment and procedures because, when it comes down to it, feedback benefits everyone."

The AGRF is also working with organisations such as nanoVic, miniFAB, CSIRO, the Nanostructural Analysis Network Organisation and the Cooperative Research Centre for MicroTechnology to make buzz phrases such as "lab on a chip" a reality.

These groups focus on refining procedures that currently use large pieces of equipment and expensive reagents so they can be more easily automated, making the processes less expensive.

For example, pocket medical devices are designed to extract the DNA from a patient's blood, analyse it and tailor medication to suit the variations in their genes. AGRF can partner this technology development and adapt it to the processes it uses in its services.

Although such processes won't happen overnight, Forrest is determined the AGRF will be heavily involved. "We need to be involved with companies that are pushing the boundaries," she says. ■

Querying the kangaroo

A project known as the Kangaroo Genome Sequencing Project provides a significant opportunity for Australia. The US Government, through the National Human Genome Research Institute (NHGRI), has offered to match Australia's efforts on sequencing the genome of a small kangaroo species, the tammar wallaby. Marsupials offer key insights into the genetic programming and evolution of mammals and life in general because they are so distantly related to humans.

The Australian Genome Research Facility (AGRF) needs $A6 million funding to participate in this project.

Director of the AGRF, Dr Sue Forrest, says: "This as an opportunity not only to further research an iconic Australian animal, but also to prove the value of a major national research facility program which provided the infrastructure enabling us to participate in projects of such global significance."

By Lisa Tait

Hear now. And always.

→ Hearing-impaired people across the world are receiving the gift of sound thanks to the Australian-designed cochlear implant.

The gift of sound: Jayden is one of 50,000 people who have received an implant.

Five-year-old Jayden Vella plays with other children, communicates with his parents and joins in with games in the playground. It is difficult to tell that he has a Nucleus cochlear implant system, a device that is designed to help him hear.

Before Jayden received his cochlear implant in May 2001, he lived in a world of silence. Jayden's mother, Annette, says the implant changed their lives. "We wanted Jayden to have as much of a normal hearing life as possible," she says. "Now, he can respond to different sounds, to his name and to instructions. He can interact with other children and join in with their games.

Most importantly, he is happier and less frustrated. I would say that the feeling you get seeing your child smile to the sound of your voice is like winning the lottery. It has been a new beginning for all of us."

Inspired by his close relationship with his father, who was deaf, Professor Clark's goal was to find a way to improve hearing and the quality of life for people who are deaf

The implant is an Australian innovation that has helped 50,000 people in 120 countries worldwide. The research behind the cochlear implant began with Professor Graeme Clark's work in the 1960s and is still continuing today. In 1967, Professor Clark, of the University of Melbourne, began researching the possibilities of an electronic implantable hearing device.

His endeavour would eventually lead to the creation of what we regard today as the Nucleus cochlear implant system.

Inspired innovation

Inspired by his close relationship with his father, who was deaf, Professor Clark's goal was to find a way to improve hearing and the quality of life for people who are deaf.

In 1978, Professor Clark led a team that saw Rod Saunders become the first person in the world to receive a multi-channel cochlear implant. Four years later, Graham Carrick received the world's first commercial, 22-channel cochlear implant. These were groundbreaking procedures that were the first in what has become a medical success story. It is also a story about how innovation has been the driving force behind a successful commercial entity.

In 1979 Nucleus, a group of medical equipment manufacturers, became interested in the commercial potential of Professor Clark's work. Two years later, the University of Melbourne, the Australian Government and Nucleus set out to develop a commercially viable cochlear implant and bring it to the market through a clinical trial.

Cochlear Ltd was established in 1982 as a corporate entity to continue commercial operations for the multi-channel cochlear implant system. Cochlear listed on the Australian Stock Exchange in 1995 and it now employs more than 800 people worldwide. It is consistently listed as one of Australia's top 100 companies by market capitalisation. Each year, about 15 per cent of Cochlear's revenue goes directly towards the research and development of new technology.

Commercial collaborations

Cochlear has continued to set the industry standard in innovation, safety and reliability, and its devices are used in more than 1000 clinics worldwide. Partnering more than 3000 cochlear implant professionals, the company takes a collaborative approach to hearing science and its translation into real patient benefits.

Today the company participates in more than 90 collaborative research programs with university and research hospitals in more than 35 countries. Cochlear's innovation is a result of direct collaboration with cochlear implant professionals and recipients throughout the world. Its research efforts are closely linked to the Bionic Ear Institute and the Cooperative Research Centre, in Melbourne, Australia. Cochlear exports 95 per cent of what it produces and is committed to growing the international market for cochlear implant technology. It is the only publicly-held company in the cochlear implant industry.

It is a company that is paying dividends through profit and the personal lives of hearing-impaired implant recipients. According to a study in Hanover in Germany, more than 80 per cent of children who receive a cochlear implant under the age of two, and who receive appropriate rehabilitation, attend mainstream schools.

The international market leader in cochlear implant technology, the company takes a long-term view to reinvesting into its key markets in order to support and underpin sustainable growth. It has regional offices in the US, UK, Switzerland, Germany, China and Japan. Cochlear's Nucleus product range continues to set the benchmark in product safety and reliability, as well as innovation and technological advancement.

The cochlear implant is not just changing developmental progress for young children. It can also assist hearing-impaired adults. Twenty-two-year-old Xanthe McLean had progressive hearing loss from the age of four, a condition which deteriorated markedly in 2001. "Family and friends had to repeat things at least twice, and it was much harder in social situations, as well as understanding lectures at university," says McLean.

It is the only publicly-held company in the cochlear implant industry

Without her Nucleus cochlear implant system, she would have almost certainly been forced to quit university because her poor hearing meant she would not have been able to function properly during her work placements in hospitals. McLean feared she would not be able to work in a hospital setting because she could not hear vital instructions.

Since the implant, McLean has had more confidence about tackling life's challenges. "I have a greater sense of independence in going about my daily life," she says. "I am always conscious of how much the implant has helped me and I am grateful that I was able to receive one and benefit from the gift of sound once more. It has given me a new lease of life. I am much more optimistic about my future and my ability to cope with life's challenges." ■

Lisa Tait is the media and promotions manager at Cochlear Ltd.

Global success

- Cochlear Ltd, established in 1982, is the first company in the world to bring multi-channel cochlear implants to the market.
- It was listed on the Australian Stock Exchange in 1995 and is consistently listed as one of Australia's top 100 companies by market capitalisation.
- The company employs more than 800 people worldwide.
- Devices are used in more than 1000 clinics worldwide and the company partners with more than 3000 cochlear implant professionals.

By Tracey Evans

A family affair

→ An umbrella approach to its diverse companies is the secret to Johnson & Johnson's success.

While the concept of adding wheels to a chair to increase the mobility of those unable to walk was a creative way of improving people's lives, it still had its limitations. People in wheelchairs still need assistance to get up and down stairs and curbs and reach other inaccessible places.

More sophisticated wheelchairs have followed, both power-driven and hand-driven, but stairs and some surfaces have remained a problem.

Enter inventor and entrepreneur, Dean Kamen, who's known as a determined advocate for science and technology. Kamen has been inventing marketable and successful products since his days as an undergraduate and now holds more than 150 patents for his inventions. His focus on innovative medical devices brought him to the attention of Independence Innovation, a Johnson & Johnson company, which partnered with him to develop a device to increase the independence of people in wheelchairs. As a result, the iBOT was born.

The iBOT, more formally known as the INDEPENDENCE iBOT 3000 Mobility System, is to wheelchairs what mobile phones are to the telegraph. It allows users to move around at eye level and to reach high places, it can climb up and down stairs without assistance, climb curbs and travel over grass, gravel, through sand and over uneven terrain, and it can be operated remotely to allow an assistant to drive it into a van. The chair uses a combination of electronics, sensors and software components which automatically adjust wheel position and seat orientation to stabilise the user and mimic the principles of human balance.

The iBOT's contribution to Johnson & Johnson's more than $US36 billion ($A48.9 billion) in sales last year is thanks to the company's exceptional approach to capturing and marketing innovation.

Johnson & Johnson constantly searches the world for innovative products and companies, adding to its umbrella of more than 200 operating companies. When a new company is identified as one that would fit in with Johnson & Johnson's mission, it is merged into Johnson & Johnson but remains independent. This decentralised and diversified structure helps capture and create innovation.

"Every year as many as 30 companies may be added to the umbrella, while about 10 may leave," says Nicholas Campbell, head of Corporate and Public Affairs, Johnson & Johnson Group. "Because this umbrella is always increasing and the overall company is growing, Johnson & Johnson has become one of the world's most comprehensive healthcare companies, continually growing and outstripping the market."

Actively seeking innovation

The process of seeking out innovation around the globe is facilitated by the company's Corporate Office of Science and Technology (COSAT), a team of people with years of experience and distinguished careers in science as well as research and development. They are located throughout the world and visit universities and companies to investigate advances in science and technology. In Australia, COSAT is represented by Johnson & Johnson Research Pty Limited (JJR), which is located at the Australian Technology Park in Sydney.

Johnson & Johnson constantly searches the world for innovative products and companies, adding to its umbrella of more than 200 operating companies

Johnson & Johnson's present-day profile represents massive progress since it began operations with 14 employees in 1886, manufacturing a simple surgical dressing. While some of its first products can still be found on supermarket and pharmacy shelves, many are now high-tech versions of the original such as the ubiquitous BAND-AID, first introduced in 1921. In the latest advance, the simple sticky dressing has become a "liquid bandage" which creates a clear, flexible, breathable seal to keep out water, dirt and germs to help prevent infection.

Meanwhile, the company has diversified far and beyond its early beginnings in sophisticated medical devices and pharmaceuticals. Two of the most recent releases from Johnson & Johnson's pharmaceutical businesses are a treatment for children with Attention Deficit Hyperactivity Disorder (ADHD), CONCERTA, and the atypical antipsychotic drug RISPERDAL CONSTA. Both medications have delivered significant benefits to patients' quality of life by reducing the frequency with which they need to be taken.

CONCERTA is an extended-release medication for ADHD treatment, taken in one morning dose and designed to last 12 hours, eliminating the need for children to take a second dose of their medication during the day at school. The tablet is designed to release the medication in a controlled pattern throughout the day.

RISPERDAL CONSTA is a groundbreaking new way of treating schizophrenia and bipolar mania. It is a long-acting, injectable medication that provides more certainty and safety for patients by ensuring they receive regular and accurate doses. "It's something that's never been achieved before and probably the greatest advance in psychiatry in 10 years," says Campbell.

As well as improving patients' lives, the new drug is proving a cost-saver for public health programs. Schizophrenia fills more hospital beds than almost any other mental illness, but a Swedish study has found that patients treated with RISPERDAL CONSTA are significantly less likely to be hospitalised due to relapse in their symptoms and those who were hospitalised had much shorter

From simple surgical dressings in 1886 to sophisticated medical devices and pharmaceuticals today.

hospital stays. The study found that there were substantial cost savings to the Swedish healthcare system as a result.

Enduring company values

Much of Johnson & Johnson's progress can be put down to the company's values, first set down by Robert Wood Johnson in 1943 in a document called "Our Credo" and still espoused today. The credo identifies Johnson & Johnson's customers, "the doctors, nurses and patients … mothers and fathers", as its first responsibility.

Employees are listed as the company's next responsibility with a promise to maintain "fair and adequate" compensation and "clean, orderly and safe" working conditions. No doubt radical for its time, the credo also upholds equal opportunity in employment and urges managers to "be mindful of ways to help our employees fulfil their family responsibilities". The third responsibility is to the community and the environment.

"The final responsibility is to shareholders," says Campbell, "based on the idea that shareholders will receive the best long-term return if customers, employees and the community are well looked after."

Johnson & Johnson was founded in 1886 and is now the world's most comprehensive manufacturer of healthcare products. It achieved record global sales of $US36.3 billion in 2002.

The Johnson & Johnson umbrella of worldwide operations consists of more than 200 operating companies in 54 countries and about 112,000 employees. These companies market healthcare products in more than 175 countries.

In Australia, Johnson & Johnson has been established for more than 50 years. Today there are more than 1000 employees in Australia and New Zealand and annual sales of just around $A1 billion.

There are eight operating companies in the region, each with their own managing director, that help form Johnson & Johnson's umbrella of worldwide operations: Johnson & Johnson Pacific; Johnson & Johnson Medical; Janssen-Cilag Australia; Johnson & Johnson Research; Johnson & Johnson Vision Care (Australia) – which manufactures contact lenses; Ortho-Clinical Diagnostics – which produces professional diagnostic products for hospital and clinical laboratories; DuPuy Australia – making treatments for musculo-skeletal disorders; and Tasmanian Alkaloids – which extracts morphine and thebaine from poppies for use in pain relievers.

Establishing innovative R&D

Politicians and academics have been lamenting Australia's brain drain for years but one scientist, with support from a major global corporation, has done his bit to reverse the trend.

Denis Wade had spent more than 30 years in academia, including stints at Oxford and various American universities, and finally as Professor of Medicine and Clinical Pharmacology at the University of New South Wales. In between, he'd been "in and out of the pharmaceutical industry on various continents", acting as a consultant. But he was nearing a self-imposed deadline for a career change. "I decided that nobody should be a university professor after the age of 50 and I got out with two days to spare," he says.

Wade was approached by Johnson & Johnson to establish a research and development facility in Australia and he jumped at the chance. "It gave me an opportunity to pursue my passion to try to commercialise Australian science and technology," he says.

That was 16 years ago, when biotechnology was booming in the United States and Europe. But at the time, most companies didn't seriously consider Australia as a player in the science and technology fields, says Wade. "There were some home-grown efforts in the embryonic stages then, but the overall level of activity here was pretty low.

"The thing that characterised the American venture capitalists involved in the early stages in biotechnology, compared with those in Europe, was that they were very hands-on," he says.

"They didn't just write cheques and wait to see what happened, they took an active part in the development of companies and worked out that the value lay in the intellectual property. I wanted to see that sort of thing happen in Australia."

As a result, Johnson & Johnson Research was established in 1987, as a subsidiary of Johnson & Johnson US, to focus on innovative Australian research that could provide new product opportunities in the area of human health. It is now a significant non-US based centre in the J&J Group, covering a wide spectrum of healthcare, with a major laboratory research and development focus on molecular genetics.

Its successes have included cutting-edge work in gene therapy. "We were one of only two or three groups around the world that pioneered human gene therapy," says Wade. "We're now leading the world with a revolutionary product made in Sydney, that's in testing in Australia and the US.

> It gave me an opportunity to pursue my passion to try to commercialise Australian science and technology

"Other things followed from that – we got involved in other areas of genomics and genetic diagnosis, and that's now been licensed to a major, global diagnostic and chemical supply company. That's another big coup for Australia."

Wade's reputation and the early breakthroughs made by Johnson & Johnson Research helped to reverse the trend for some of Australia's finest scientists to head overseas in search of opportunities. The company has also become a desirable target for scientists from around the world. "We had our pick of the very best people and that's unquestionably one of the reasons we've done so well," he says. Wade retired from Johnson & Johnson Research in 2003. For the last 10 years, he has been a member of J&J's US-based Corporate Office of Science & Technology and its Business Development Council. In 1999, he was made a Member of the Order of Australia for service to clinical pharmacology.

Since Wade's retirement, Dr Susan Pond now leads the team at Johnson & Johnson Research.

Tall poppies shine

With a mission to achieve growth through innovation, Tasmanian Alkaloids does not only have business expansion in mind. A more literal interpretation has seen the company, a wholly-owned subsidiary of Johnson & Johnson, grow a better poppy capable of producing more of the precious substance that forms the basis of many pain relieving drugs.

"There's a considerable need for pain medication around the world," says managing director Brian Hartnett, "And our inspiration is to help to relieve the pain and suffering of people around the world. That's how we see our role."

Hartnett's supplies of raw product are growing in up to 12000 colourful hectares around Tasmania. While poppies naturally produce morphine, they also contain trace amounts of thebaine, a far more versatile alkaloid.

The pharmaceutical industry is more interested in codeine than morphine and, as a result, manufacturers use a simple, one-step reaction to convert the morphine to codeine, says Hartnett.

While thebaine is not useful in its own right as a pharmaceutical, it is used to produce many other products. "You can make these products out of morphine," says Hartnett, "But it's much more difficult and expensive."

Doing the impossible

So when one of the scientists at Johnson & Johnson Research in Sydney encouraged a technique that might increase the amount of thebaine in poppies, Tasmanian Alkaloids set to work even though at least one international expert said it was impossible.

"At the time we were a very small company with very limited resources and nowhere to do the work," says Hartnett. On a shoestring budget, the company hired a mobile hut to use as a

Tall poppy: Tasmanian Alkaloids is meeting the huge demand for pain-relieving thebaine.

laboratory, employed a technician and scraped together the money to buy the necessary equipment. "The technician didn't like visitors because if you walked across the floor of the mobile hut, the instruments wobbled," recalls Hartnett.

"We developed techniques to grow lots of little poppies and test a drop of sap from each one without killing it and screened hundreds of thousands of plants. We had all that linked up through the instruments to the computers, scanning for changes in alkaloid profile. Then, one day we found a little plant that didn't have a morphine peak but there was a big peak for thebaine."

The plant was grown to maturity, its seeds carefully collected and more were grown. Each new plant recorded the same characteristics – no morphine, just thebaine. "Sometimes scientists find things which don't breed consistently; you get an aberration from one generation to the next. But this is a true mutant, it breeds true to every generation," says Hartnett.

"From that we've built a huge part of our industry. Over half our production is thebaine from other generations of plants of this type and it's been very successful because there's a huge demand for thebaine and we've been able to meet that demand."

The company's success was recognised in 2003 at the Australian Business Excellence Awards, organised by Standards Australia. Tasmanian Alkaloids took out the Bronze Award for its "use of the principles and practices of excellence to develop strategic direction and a product portfolio that has positioned them as leaders in the marketplace".

"There's a very good collaboration between the company, the Government, the farmers and the community," says Hartnett. "That makes it something quite exceptional." ■

By David Salt

Getting the good medicine

→ Australia's world-class reputation for the quality of its pharmacies is being reinforced with the introduction of a new range of programs.

Why is it that Australian pharmacies are the envy of other pharmacies the world over? Because they command community respect, they're able to maintain the highest levels of quality and they are an active partner in government and community initiatives. Central to all these achievements is one institution – the Pharmacy Guild of Australia – and the key to its success is its innovative approach to business.

To protect the standing of community pharmacies we have to be creative

Innovation is all about finding creative answers to meet emerging challenges. It is about looking forward, identifying issues before they become problems and then actively developing imaginative solutions. It is also about turning challenges into opportunities.

Is there a pharmacist in the house?

Many people in the community experience difficulty in taking the right medicine in the appropriate way. This might be because a number of medicines have been prescribed at separate times and some might have adverse effects when taken together. This sometimes happens if prescription medicines are taken together with other health-care products. Another difficulty may be that the person taking the medicine is old or infirm, and has trouble remembering what should be taken and when.

Whatever the reason, the inappropriate use of many medicines can have a serious impact on a person's health. In response to this problem, the Pharmacy Guild has organised a world-first scheme for the provision of special help to be available. The Home Medicines Review scheme was set up in partnership with the Federal Government and is a service provided by both doctors and pharmacists. If a doctor believes a patient would benefit from a Home Medicines Review, he or she can refer the patient to the community pharmacy of their choice and provide their medical details to the pharmacist.

The pharmacist then visits the patient at their home and looks at both their prescribed and non-prescribed medicines. The pharmacist will then discuss with the patient how the medications are being taken and any uncertainties they might have. Following this visit, the pharmacist reports back to the doctor who will then discuss with the patient any changes that might be needed.

The challenge for Australian pharmacies was to maintain their reputation for quality and relevance in a rapidly changing world. Medicine and health information is increasingly available from multiple outlets and everyone claims to be an expert. How do you distinguish yourself in such a rapidly evolving marketplace?

"We've long realised that if the Guild is to maintain the high level of standing of community pharmacies then we have to be creative and actively ensure the quality of their professional services," says John Bronger, national president of the Guild. "We simply cannot wait till government does it for us."

Established in 1928, the Guild comprises qualified, registered pharmacists who own their own pharmacies. Its members are owners of some 4500 pharmacies throughout Australia.

"Our members are a special breed," says Bronger. "In addition to being dynamic small-business owners, they're also highly qualified and committed health professionals who have spent a minimum of five years at university earning their qualifications.

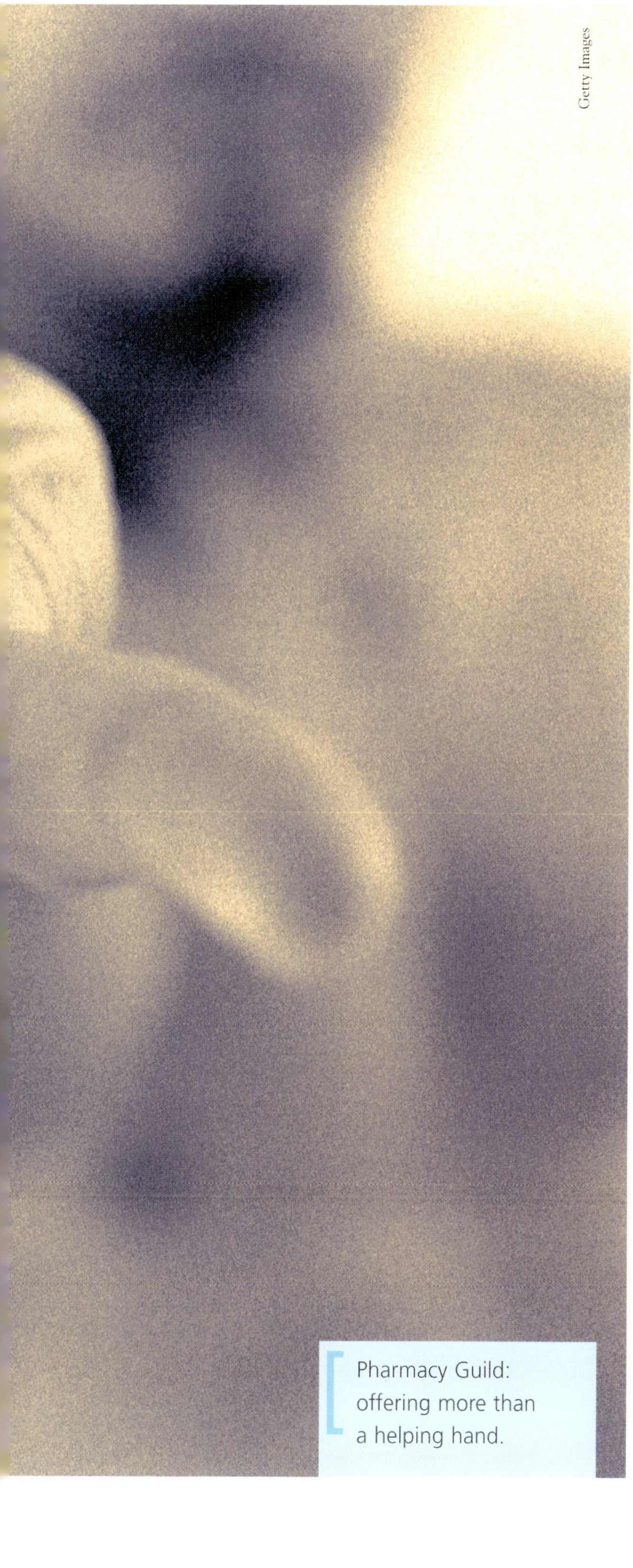

Getty Images

Pharmacy Guild: offering more than a helping hand.

Supporting rural health

A rural pharmacy practice provides a number of challenges and rewards that a city practice simply doesn't offer. With the right support, the potential is there for rural pharmacists to experience high levels of satisfaction from the wide range of work they perform. Rural pharmacists are respected members of rural communities and are fundamental to the local community structure.

The Rural and Remote Pharmacy Workforce Development Program was established specifically to strengthen and support the rural pharmacy workforce in Australia. It aims to improve the training and satisfaction of pharmacists already operating in rural areas while encouraging more city-based pharmacists to consider the benefits of moving to the bush.

The program, one of the most comprehensive rural pharmacy incentive schemes in the world, is funded by the Department of Health and Ageing and is managed by the Pharmacy Guild of Australia.

It offers a number of features, including education scholarships for rural pharmacists, an emergency locum placement service, scholarships for students from rural and remote areas wishing to study pharmacy, internship scholarships for undergraduate students, specific scholarships for indigenous students, rural and remote pharmacy infrastructure and support grants, placement of pharmacist academics in rural areas, a national rural pharmacy promotion campaign and a rural pharmacy newsletter.

Our members are a special breed

The Guild seeks to protect the interests of its members, and we need to be proactive about it."

For some time the Guild has sought to establish quality benchmarks for the way pharmacies conduct their business.

"We have to be an active partner with both government and the community to ensure that local pharmacies are relevant to the community's needs," he says.

Quality not quantity

To assist with this aim, the Guild developed the Quality Care Pharmacy Program to raise the standards of service provided to the public, improve the quality of professional practice and achieve better health outcomes for all Australians.

"Our Quality Care Pharmacy Program is a big winner for the

consumer and the pharmacy," says Bronger. "Pharmacies which have gained accreditation have a number of advantages. The performance of the pharmacy team is improved and service to customers is better. Staff feel empowered, productivity increases and customers are more satisfied. We have found that guaranteeing quality pays real dividends."

Through our innovative approach in demonstrating the value of community pharmacies, we've turned this challenge into a real opportunity for our industry

The Guild has also entered into three long-term Community Pharmacy Agreements with the Federal Government which have led to the creation of a range of schemes to ensure the quality and relevance of pharmacies around Australia. Examples of these schemes include the Rural and Remote Pharmacy Workforce Development Program, the Home Medicines Review project and an extensive program of research and development grants.

"These schemes are demonstrating how valuable community pharmacies are to the health and wellbeing of all Australians," says Bronger. "Take, for example, the Rural and Remote Pharmacy Workforce Development Program. The community pharmacy is often the only place some people have to go to receive professional health advice. This program, which the Guild manages, gives rural pharmacists the opportunity to get the training they need, and provides an emergency replacement service where required."

Bronger adds that while modern medicine is becoming increasingly effective, it is also much more complicated to self-administer. The Guild has designed another innovative scheme – the Home Medicines Review project – under which a patient can request a home visit from their local pharmacist at no cost.

"The Research and Development Grants Program administered by the Guild involves a total of $A15 million being distributed to a variety of pharmacy-related projects between 2000 and 2005," Bronger says. "Some of these projects have immediate and recognisable relevance for a great many Australians. They include an investigation into the value of pharmacists' roles in smoking cessation, the effectiveness of mentor support in the training of pharmacists and improved community pharmacy participation in the treatment of diabetes and asthma."

The programs that the Guild is running are envied the world over and are ensuring that pharmacies around Australia will remain relevant and valued by the wider community, according to Bronger. "Through our innovative approach in demonstrating the value of community pharmacies, we've turned this challenge into a real opportunity for our industry."

The programs that the Guild is running are envied the world over and ensuring that pharmacies around Australia will remain relevant and valued by the wider community

He adds that Australia is lucky to have a thriving and independent retail pharmacy network. "Independent pharmacies in many countries are finding it increasingly difficult to be commercially viable as the market becomes dominated by large multinationals that are more interested in selling product than providing the highest quality service and independent advice," he says. "The independent pharmacist is still alive and well in Australia. As a consequence, Australians have access to the some of the best-value medicine and advice you'll find in any country." ■

Rising standards

On average, each person visits a community pharmacy 14 times a year. That's more than 200 million visits a year in which pharmacists are able to provide professional advice and service. To make these visits more efficient, the Guild and other industry stakeholders have developed the Quality Care Pharmacy Program (QCPP), an industry-developed quality assurance program. It is an integrated system of performance standards, tools and processes based on business and professional standards that have been developed by both the Guild and the Pharmaceutical Society of Australia. Other major industry stakeholders that have contributed include pharmaceutical wholesalers, pharmacy marketing groups, pharmacy boards and pharmacy schools.

QCPP raises the standard of customer service in individual pharmacies across Australia – providing an industry-wide guarantee of retail service quality and professional practice. It is supported by a comprehensive range of materials, including a manual of 113 pharmacy standards that focus on professional services, retail skills and business management.

The Federal Government has recognised the value of the QCPP in delivering higher quality services to the public and has provided $50 million in financial incentives to encourage pharmacies to gain accreditation under the scheme. By 2003, 3500 pharmacies had been accredited.

By David Salt

Shooting for the stars

→ An Australian biotechnology company is leading the world in applying nanotechnology to the treatment of selected viruses.

Starpharma is on target with its use of nanotechnology.

The origins of the company Starpharma involved a single novel idea: could large, star-like molecules known as dendrimers become drugs that could be targeted to sites within the human body? Such was the potential of the idea that the Australian company was established in the mid-1990s to explore possible applications. The original idea has since blossomed into one of the world's most exciting and rewarding fields of technology – dendrimer nanotechnology.

Among a list of achievements, Starpharma – a Pooled Development Fund – has gained approval from the US Food and Drug Administration to start human trials of a dendrimer-based gel called VivaGel, designed to prevent diseases being transmitted during sex. The aim is to block a range of viral diseases, including HIV.

"It's the first time a defined nanostructure has been cleared for clinical testing," says Dr John Raff, Starpharma's chief executive. "In the US alone, sexually transmitted diseases (STDs) affect more than 13 million people annually."

He adds that throughout the world, more than 42 million people are living with HIV. VivaGel could be a solution to this growing problem as it blocks several STDs using one easily applied gel.

The secret of VivaGel's efficacy, and Starpharma's success, lies in the nature of its core science – the dendrimer. Dendrimers are synthetic, nanoscale structures that can be tailored for various applications. They are constructed into a series of controlled steps that begin with a central core molecule. Onto this core, numbers of small branching molecules are added (dendri means tree, mer means branch).

Modified molecules

These branching molecules are grown out, giving the dendrimer a star-like quality, which is the basis of the Starpharma name. The final generation of molecules added to the growing structure makes up the surface of the dendrimer and these are tailored to enable the dendrimer to bind to multiple targets – such as human cells, bacteria or viruses.

Each component of the dendrimer, which includes core, branching and surface molecules, are purposely selected to give the nanostructure desired properties and functions to enable it to carry a drug or a molecular structure of choice, and to bind to a target site on or around the body.

"This is a form of nanotechnology that is beginning to take off," explains Raff. "We have the chemistry to build these amazing structures, and the biology to understand what we need to build into them to make them into effective drugs and disease-targeting systems. Starpharma is coupling this cutting-edge science with excellent business and commercialisation skills."

This isn't blue sky research with applications still decades away. We are already manufacturing products

Raff says Starpharma aims to design and build dendrimers that are simultaneously active against a wide variety of diseases. They can be built into a desired level of bioavailability, making sure they are safe to administer at therapeutic doses.

"This isn't blue sky research with applications still decades away," Raff says. "We are already manufacturing products. VivaGel, for example, has already been shown to work in trials on monkeys where it blocked a humanised version of a monkey immunodeficiency virus. It is also active against HIV, chlamydia and herpes. We are now embarking on human trials – a world first – which means that the transition from the research to the real world is well under way."

Starpharma is also working on other areas such as treatments for respiratory viruses (such as the flu), cancers and a range of tropical diseases. Its technology is also being used by the US Government to improve the detection and countering of bioterrorist attacks. ■

By Bob Beale

Sleeping soundly

→ The lives of thousands of people around the world are being saved with the invention of a device that treats sleep apnea.

The story of ResMed has its origins in 1981 following a shrewd clinical insight by a Sydney medical research team. The team gave it form by cobbling together a length of hose, a glued-on facemask and an air blower that worked like a vacuum cleaner in reverse. The contraption made its wearers look like Darth Vader and it roared like a freight train.

However, although crude, that first device led to one of Australia's most outstanding success stories in medical technology: the first non-invasive treatment for the most serious form of sleep-disordered breathing (SDB) – obstructive sleep apnea (OSA) – and a thriving global business.

ResMed is the company that grew out of that simple beginning and today is a leader in the design and manufacture of devices to treat and diagnose sleep-disordered breathing. It has almost 1500 employees and a commercial presence in more than 60 countries.

The company has made the *Forbes 200 Best Small Companies in America* list for seven consecutive years and is one of only four medical device companies to do so. Among a string of other awards, it was named Australian Exporter of the Year in 2002.

ResMed's financial performance helps to explain why: it has had a positive cashflow and grown solidly for 12 straight years. In 2003, its revenues grew by 34 per cent. It now holds either first or second place in all the major markets of a fast-growing global industry and has an enviable record for business acumen, technical innovation and bringing better health to thousands of people across the world.

How did a small Australian start-up company, with what was then just one obscure product, come to be such a commercial force?

Positive pressure

ResMed's origins go back to that first realisation by Professor Colin Sullivan and colleagues at the Royal Prince Alfred Hospital that people who suffered from OSA – a common breathing disorder that can disrupt sleep hundreds of times a night – could benefit greatly from having positive air pressure applied to their airways while they slept.

Sullivan's pioneering work on breathing during sleep has helped to alert the medical world to the hidden epidemic of untreated sleep apnea and its serious consequences. Symptoms of OSA can include heavy snoring, sleeplessness, heart palpitations, morning headaches, depression, memory lapses and sexual dysfunction.

Sufferers are often so tired during the day that they are liable to suddenly doze off while at work, driving the car, reading or watching television. Recent medical research has strongly linked OSA to a number of serious, life-threatening comorbidities including hypertension, congestive heart failure, stroke, coronary artery disease, obesity and diabetes.

Sullivan's original device proved highly effective: the blower maintained positive air pressure – fractionally above normal atmospheric pressure – and delivered it through the mask. This kept the sleeper's airway open and unobstructed, enabling them to sleep safely and well.

However, it took another man with exceptional vision, Peter Farrell, a former academic at the University of New South Wales, to take the device into the commercial realm. Farrell was the founding director of the Baxter Centre for Medical Research (BCMR) when he first heard of this treatment for "snoring sickness". Despite initial scepticism, he did his homework and listened to Sullivan.

What clinched it for Farrell was watching film of an affected man who woke more than 400 times a night after his breathing stopped, his heart repeatedly racing and blood pressure soaring dangerously. With Sullivan's device in place, the man slept soundly.

When Baxter sold its respiratory home care division, Farrell founded ResMed in 1989 to manufacture and market the device. He raised $A1.2 million from investment by staff and private investors to buy the intellectual property developed by BCMR as well as Sullivan's original patents, which BCMR had previously acquired from the University of Sydney.

The company's vice-president, innovation, Dr Bob Frater, says Farrell is an inspirational leader who deserves a large share of credit for the company's success.

"He had the international contacts, the corporate experience and was able to get very skilled and able people on board," Frater says. "He has been a very strong, single-minded leader with a very clear idea of where he was going."

Frater, who left a senior post with the CSIRO to join ResMed in 1999, is in no doubt about the value of ResMed's technology since it was his own experience with sleep apnea that led to him becoming involved. "My wife was telling me that I snored very badly and I was always tired during the day," he says. "By the afternoons I'd be very sleepy and my staff told me later that my attention span was getting very short."

What clinched it for Farrell was watching film of an affected man who woke more than 400 times a night after his breathing stopped, his heart repeatedly racing and blood pressure soaring dangerously

A chance meeting with Farrell put him on the right track for treatment. "It made an incredible difference to my life, so much so that with the exception of international flights I haven't slept a night without the mask in the 12 years since," he says. "Without it I would have ended up with serious hypertension. I'm still around when the statistics indicated I would have been dead five years ago."

Getty Images

Silent slumber: ResMed helps sleep apnea sufferers get a good night's (or day's) sleep.

OSA affects millions worldwide and is about as prevalent as asthma or diabetes but only a fraction of sufferers are diagnosed and treated. Part of ResMed's secret to success has been a strong commitment to innovation. It spends up to 8 per cent of its net revenues on research and product development.

Its "think global, act local" business strategy has given it an international focus from the start. Farrell sets great store in hiring quality employees to meet local market needs, listening closely to feedback from patients and practitioners.

Patent protection

However, the company also puts a heavy emphasis on protecting its innovations and its market share through intellectual property. By mid-2003, it had 631 patents issued or pending and 244 designs registered or pending for a range of technologies.

ResMed has maintained its focus on the SDB market. While this market has continued to grow steadily, the strong links between SDB and a range of chronic diseases has created a huge potential market – potentially much bigger than asthma or diabetes.

ResMed's product range still includes air flow generator devices and air delivery systems such as masks, headgear and tubing to connect the system to the patient, as well as humidifiers and accessories to improve patient comfort, convenience and compliance with therapy. Its main manufacturing facility remains based in Sydney.

One change is that it has branched out into a growing range of sleep laboratory products and has forged strategic alliances with other companies involved in the diagnosis, treatment and monitoring of breathing and sleep. However, it retains a strong sense of teamwork and an ongoing commitment to quality, innovation, protecting intellectual property and growth that ResMed believes will keep it at the forefront of its market niche – that and a belief in improving the quality of life for people with sleep-disordered breathing.

According to Frater, when considering a new product Farrell asks, "Does it have the Alpha factor, do we really love it?"

"But careful financial analysis is vital too," he says. "The innovation needs to be exciting and intriguing, but it also has to pay the rent." ■

By Gayle Bryant

Diagnostic differences

→ Research being undertaken by a medical research institute is providing quick and less painful solutions for patients.

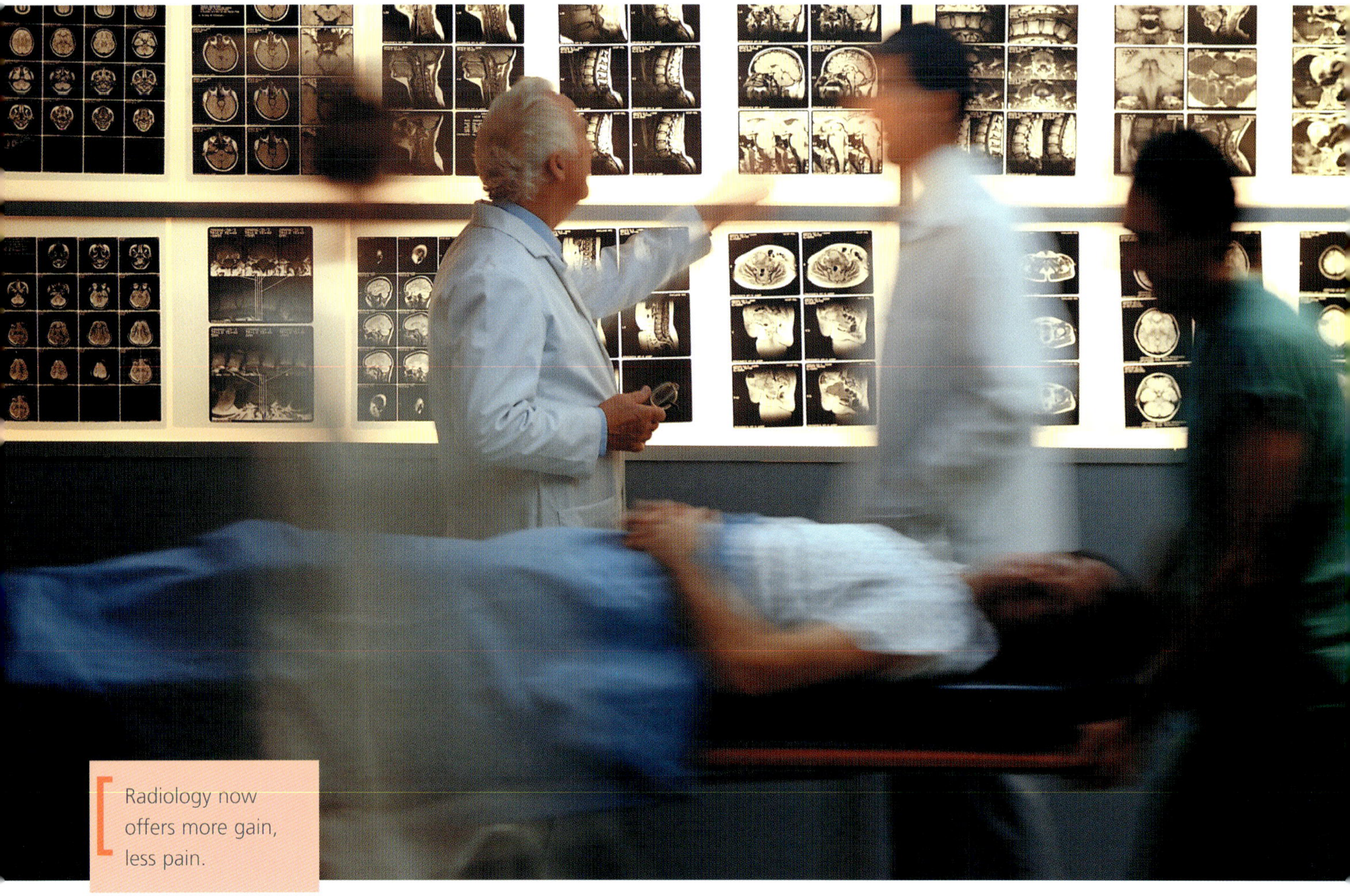

Radiology now offers more gain, less pain.

Research and innovation is crucial for a better understanding of how medical specialists can treat patients. Mayne Health Diagnostic Imaging is one institution which is at the forefront of medical research in Australia and has been the pioneer of a number of innovative procedures.

Dr William Clarke, managing radiologist at St George Private Hospital, is involved with a number of leading-edge practices, especially in vertebroplasty techniques. Vertebroplasty literally means changing your vertebra.

"Vertebroplasty is for people that have osteoporosis which weakens the bone," Clarke explains. "When the bones become weak they no longer have the ability to withstand the pressures of everyday life so it is easy to break bones from small injuries."

Cementing a solution

Vertebroplasty involves inserting a needle through the skin into the vertebral bone and injecting bone cement. "This changes the vertebral bone from being a soft squelchy broken bone back to its original state of being a compact part of the body," Clarke says. "It takes the pain out of fractures and the one-off injection can be carried out in day surgery."

Clarke says another interesting procedure his team carries out is fibroid embolisation, which is aimed at women who have fibroids in their uterus.

This procedure used to be done surgically but is now carried out on an out-patient basis

"About 50 per cent of women over 30 have fibroids," he says "They are usually small and don't cause problems. However, if they become large, they cause heavy bleeding and painful periods."

The traditional approach to remove fibroids was surgery which involved taking out individual fibroids or in some cases the entire uterus. The interventional radiology alternative is to put a small plastic tube in the leg into the artery that is supplying blood to the uterus and injecting small particles which temporarily blocks the arterial blood supply to the uterus.

"Because of the size of these particles they don't injure the normal muscle in the uterus but tend to cause an injury to the fibroid," Clarke explains. "It causes them to shrink and in about 90 per cent of cases it abolishes the symptoms that are bothering the patient."

Dr Lourens Bester, is managing radiologist at Westmead Private Hospital. One innovative procedure he performs involves administering radiotherapy and chemotherapy directly to the liver. "We call it targeted arterial chemo embolisation," he says.

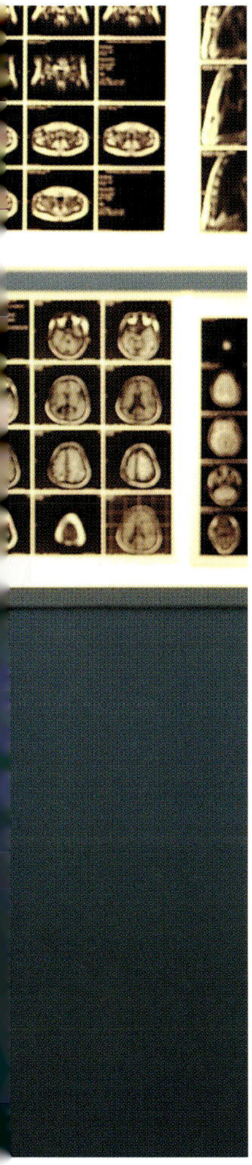

"We insert a catheter in the right groin into the artery that supplies the liver with blood. Then we infuse radioactive particles into the liver. This causes local radiation in the liver and the patient is either cured or their life is prolonged."

Bester says another leading procedure with chemotherapy patients is the use of venous ports to administer chemotherapy in the radiology department. The procedure takes 35 minutes and patients leave the department after an hour. "We implant the ports underneath the skin and attach a catheter to them," he says. "This is then put into a vein and chemotherapy can be carried out through it."

Bester says this procedure used to be done surgically but is now carried out on an out-patient basis. "The patient walks into the oncology department where we can directly puncture through the skin and draw blood for analysis or give chemotherapy. They are out within an hour," he says.

Brains behind new centre

The ability to research and study such innovative procedures has been enhanced since the establishment of the Mayne Clinical Research Imaging Centre in May 2003.

The Centre is based at the Prince of Wales Medical Research Institute, one of Australia's leading research centres. It will initially focus on research and clinical practice in brain, nervous system, joint and prostate diseases and will assist doctors, researchers and radiologists in mapping parts of the body and assisting in the diagnosis and treatment of a variety of illnesses.

The Centre was installed with a Philips 3 Tesla Magnetic Resonance Imaging (MRI) machine – only the third of such strength in Australia, and the first of its kind in New South Wales. The machine is designed to keep Australia at the forefront of brain and nervous system research.

Working with Philips allows the Institute and Mayne to be part of a larger international collaborative research team. Radiologist Dr Ron Shnier, Mayne's national director of diagnostic imaging and director of St George MRI, says the arrangement between the parties is a multi-faceted one that brings significant benefits to the research and medical community.

"We have an open door policy here that allows anyone that is doing research in this area to have access, as long as the research is of merit," he explains. "Our radiologists will be ideally placed to help the researchers, and train other radiologists. Patients will greatly benefit from this as we will be able to learn more about the brain and how particular treatments affect it."

Other areas being studied at the Centre include rheumatoid arthritis, in particular examining how much damage there is and how to treat it, and functional brain imaging.

"I think we will see more alliances between research and commercial organisations in the future," Shnier says. "These partnerships can hasten the technology available for standard radiology work and enable better treatment for patients. What we all want for the patients is better diagnosis and surgery that is less invasive." ■

Influencial imaging

Mayne Health Diagnostic Imaging (MHDI) provides the full range of radiology services performed by some of Australia's leading radiologists. Services include general x-ray, ultrasound, multi-slice computer tomography, nuclear medicine, mammography, extracorporeal shock wave therapy, bone densitometry, and state-of-the-art magnetic resonance imaging (MRI).

The group also supports ongoing research to drive advances in diagnostic imaging, primarily in the areas of interventional radiology and neurology.

MHDI comprises a national network of 130 public and private community and hospital-based diagnostic imaging sites. It employs about 160 imaging specialists nationally and provides radiology services to 30 hospitals.

The imaging capabilities of the 3 Tesla MRI system, installed in the Mayne Clinical Research Imaging Centre, will assist with procedures such as nerve fibre tracking, functional imaging and brain perfusion.

Fibre Tracking

Diffusion imaging of the brain yields a detailed image of nerve fibre tracks. With the high resolution and signal of the 3 Tesla MRI system, fibre tracks can be defined, segmented and clinically evaluated. Fibre tracking examinations can provide extensive information regarding tumour growth and its effect on the nerve pathways, neuro-anatomy and surgical planning.

Functional Imaging

Functional imaging is based upon imaging the changes of blood oxygen concentration levels and blood flow in the brain during physical or mental activity. It is a significant tool for both clinical and research investigation having the ability to depict areas of brain activation to quantify the amount of brain activation.

Perfusion Imaging

Perfusion imaging is similar to functional imaging and is based on contrast changes within the brain. The images can be colour mapped to qualify and quantify blood volume and blood transmit times.

By Nic Svenson

Cleaning up

→ An innovative Australian sterilisation company is expanding by finding new ways to improve old methods.

Steritech decided that to improve its business, it needed to work smarter with what it had. This family-owned business, established in Melbourne in 1971, is involved in the sterilisation business – an area that is receiving more and more attention. There is a greater awareness of contamination as an increasing number of goods need to undergo sterilisation to remove microbes or insects before they can be used.

"The medical products sector accounts for 40 per cent of our business," explains Steritech chief executive, George West. "Typical products that require sterilisation include boxes of dressings or gloves, individually tailored packages of instruments, and drugs that surgeons take into theatre."

Steritech carries out a lot of work for the Australian Quarantine and Inspection Service (AQIS), in particular eradicating pests or disease-causing organisms that could be attached to products entering the country. Exports are also treated to ensure they meet the requirements of other states or countries.

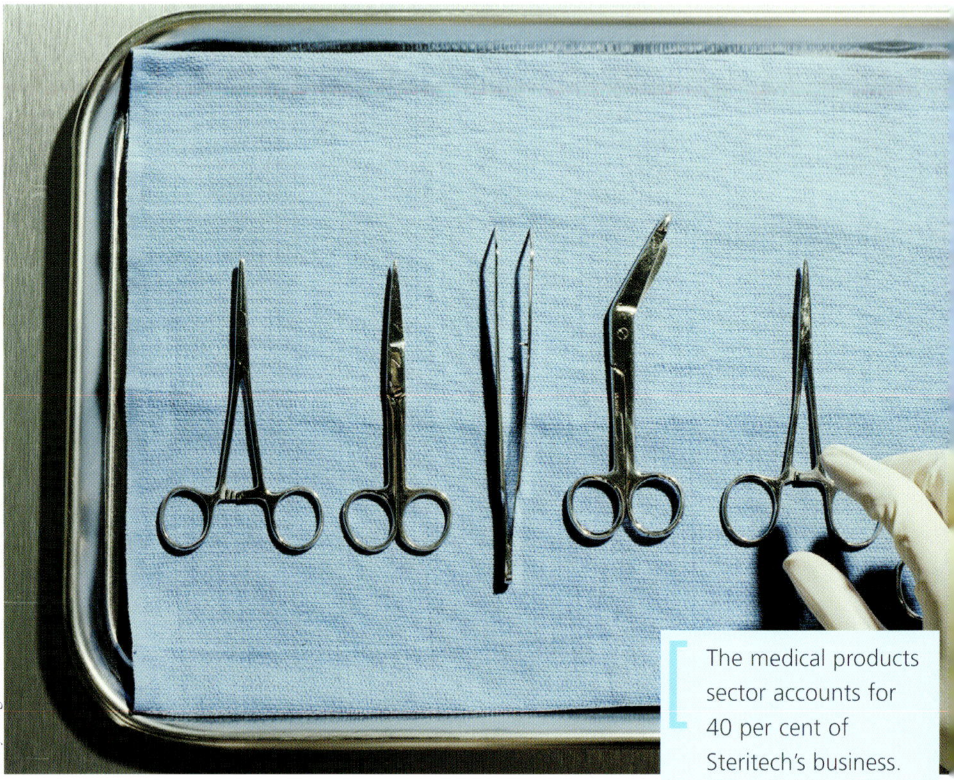

Getty Images

[The medical products sector accounts for 40 per cent of Steritech's business.

Pesky products

Nearly 80 per cent of all single-use medical items are sterilised by irradiation. This process involves placing products on a conveyer belt, which travels through a heavily-shielded chamber where they are exposed to just enough radiation to kill any pests.

However, this technique doesn't suit all products. For example, glass turns black, while teflon breaks down if it undergoes an irradiation treatment. But Steritech has a solution for these types of products – fumigation with ethylene oxide (EtO). Steritech has EtO chambers at its Sydney and Melbourne plants.

West says a number of medical companies approached Steritech after a company that was involved in EtO fumigation decided not to continue. "We were asked to consider entering this area, so we did," he says.

Steritech promotes its business at conferences and industry events. "We go to gatherings, such as custom brokers and freight forwarders conferences, and set up a display," West says. "Or we invite them to hold their meetings at our plant and give them a tour."

Steritech's strengths lie in finding novel uses for tried-and-true technology and tailoring its services to meet the needs of each client. As West explains: "We're cleverer at using the technology and we cover a far wider range of products than most other companies. For example, if someone wants to destroy seeds in wheat, we find out what those seeds are and test samples of them to determine the optimum level of radiation that is required to knock them out."

This focus on efficiency has a direct hip-pocket benefit for Steritech's clients because the stronger the dose and the longer the exposure, the higher the cost. ■

Nic Svenson is a science writer based in Sydney.

■ In the early 1970s, Steritech was the first company in the world irradiate wine corks to eliminate mould.
■ Each year thousands of beehives are irradiated to get rid of the microbe that causes American Foul Brood, which usually affects th larvae. Steritech and CSIRO pioneered a technique to kill the microbe in the early 1980s.
■ Steritech does not irradiate food destined for human consumptic but many containers and packaging materials are treated.
■ AQIS can order importers to treat items that might harbour exc pests or disease-causing organisms. These include wood carvings, drums made with animal skins and even stuffed toys.
■ Imported grain, such as stock feed and birdseed, are treated to prevent germination of unwanted seeds.

Every great advance in science
has issued from a new audacity of imagination.

JOHN DEWEY

By Dr Geoff Garrett

Facing the future

→ Like most organisations in today's environment, Australia's national science agency faces some complex challenges in order to stay relevant.

Australia faces some serious challenges over the coming 25 years as it seeks to be a society that is globally competitive, ecologically sustainable and socially enlightened.

Our Nobel laureates remind us that passion, commitment, excellence and partnership were the road to success in the 20th century. This is even more the case in the 21st century.

At an international level, we confront economic and trade pressures of globalisation, changing and increasingly volatile geopolitics, security threats, growing environmental pressures and a global revolution in science and technology. At home we are experiencing slower population growth, an ageing population with rising health costs, degradation of natural resources – particularly water and biodiversity – the recurrent threats of fire, drought, flood and climatic instability.

However, the need to sustain economic growth through increased productivity, the creation of new industries and jobs and the development of new energy sources is as strong as ever.

To equip Australia to meet these challenges, CSIRO, as the

Getty Images

Another door opens: CSIRO strides towards a new era of innovation.

Pay for performance

Dr Geoff Garrett: "We have to pay people what they are worth. I'm a great believer in giving people performance-related remuneration. I don't believe we do enough non-pecuniary evaluation of people's performance. We also need to push the people who are definitely performing out of the box and we need to provide monetary rewards sooner.

"I don't think in the current climate we are using our existing systems adequately. There is also a groundswell of enthusiasm for more scientists sharing a slice of the action from the commercial benefits of their inventions. I'm a supporter of that but it's a debate that's ongoing in our environment and it's quite a complex one. It's important to recognise performance and reward individuals for great work. That's a key component of our plan."

national science agency, is focusing and delivering its research through three main areas known as: Flagship Programs, Emerging Science Initiative and Priority-driven Core Research.

Flying the flagship

The flagships initiative marks a new era for CSIRO and our research partners in science and industry. Flagships are very large nationwide partnerships in science that provide focus and critical mass around vital national challenges and opportunities.

There are six flagships:

■ **Preventative Health:** to help Australians live longer, healthier lives through early diagnosis and prevention;

■ **Agrifood Top 5:** to transform the international competitiveness and add $A3 billion in value to the Australian agrifood sector;

■ **Energy Transformed:** to develop low-emission energy technologies and systems that lead to the widespread use of hydrogen as an energy carrier across the economy;

■ **Healthy Country:** to achieve a tenfold increase in the social, economic and environmental benefits from water use by 2025;

■ **Light Metals:** to lead a global revolution in light metals, doubling export income while reducing environmental impact; and

■ **Wealth from Oceans:** to build on Australia's excellence in climate and ocean science to generate sustainable wealth from our marine resources.

These aims are deliberately large and bold, based on our view that limited goals seldom, if ever, deliver large outcomes, and that the scale of the challenge is intrinsic to the level of success.

The Federal Government has backed this program with an initial allocation of $20 million in the May 2003 budget. Flagships are a long-term initiative and by 2007 we plan to invest approximately $220 million – or some 40 per cent of our total resources – in them.

Research drivers

CSIRO is also delivering world-class science outcomes across our core capability in research areas such as agribusiness, community health, environmental and natural resources, manufacturing, information and communication technologies, radioastronomy and sustainable mineral and energy production. This is our single largest program and accounts for approximately half our total research expenditure.

Ten per cent of CSIRO's budget now goes to fund our Emerging Sciences Initiative, focusing the research base of our organisation on leading-edge technologies likely to be vital to Australia's future. These technologies are: complex systems science; socio-economic integration; novel biotechnologies; nanotechnology; and innovative information and communications technology.

CSIRO has set itself six priorities for the next three years. These are:

■ To focus our scientific effort on the most important issues facing the nation.

■ To deliver world-class science by attracting the best people and improving the management of research.

■ To partner as widely as possible within science, across industry and the community, so our research has the most impact.

■ To serve as a catalyst for industry innovation by deepening our relationship with both large and small companies.

■ To take full advantage of our capabilities and scale by building a one-CSIRO culture.

■ To build a secure financial foundation for growth through long-term high-value intellectual property streams, increased value-for-money for customers and lower overhead costs.

Building on its rich track record of achievement and globally-recognised skills, CSIRO today is on a journey from being an Australian research institution to a research enterprise with global reach. Through science and imagination, CSIRO's people are committed to helping create the type of future for Australia that Australians want. This Australia is a source of innovative, sustainable solutions, ideas and technologies to the world and a provider of knowledge exports worth billions of dollars. As a result of industrial and environmental innovation and rejuvenation, new jobs, new industries and fresh opportunities are created.

The most important changes are already under way. In the coming decade, CSIRO and its partners will help tackle some of the biggest and most challenging issues facing the nation. We will press the pace of commercialisation. Through vibrant partnerships with effective knowledge transfer, we will help drive business activity, exports and community benefits. Through unremitting excellence in both science and in the business of science, we will remain one of the world's pre-eminent scientific organisations. ■

Dr Geoff Garrett is the chief executive officer of CSIRO

The point of patents

Dr Geoff Garrett: "One of the problems is that for a bunch of clever scientists, this commercial stuff seems like kids' play – they wonder why they need outside help to do it. When you actually get out into the marketplace you know that things are very different. We are already doing significant re-evaluation of our patent portfolio – where are its strengths, where we are investing that we shouldn't be, and how we can exploit this in different ways.

"It is a much more effective, managed process. I believe unless you are prepared to back up your patent with legal and financial resources, what's the point of patenting?"

By Gayle Bryant

A foundation for success

→ South Australia's strategic investments in science, technology and creativity are providing a major impetus for the State's growth.

Over the past year, South Australia's base of research and innovation has been strengthened with the establishment of a new government department. The Science, Technology and Innovation (STI) Directorate within the South Australian Department of Further Education, Employment, Science and Technology was established in recognition of the fundamental role that these areas play in the State's economic and social life.

The STI Directorate was established to coordinate the development of policy and programs that would build the State's innovation and skill base and make a significant contribution to the overall wellbeing of the South Australian community.

One of the directorate's roles is to strengthen relationships with research institutions to help foster the growth of innovation in the State. One such relationship exists with the Centre for Environmental Risk Assessment and Remediation. Professor Ravi Naidu, director of the centre, is involved with research into contaminated sites that impact on environmental and human health, which he says are a major problem in Australia. Many are located in the high-value urban environment.

"There are more than 100,000 contaminated sites in Australia and more than 3 million sites in the Asian region," he says. "We know that more than 10,000 people in the Asian region have died of poisoning and potentially more than 100 million people are at risk of contaminant poisoning."

Getty Images

South Australia's STI Directorate is focusing on the building blocks of strong research and innovation.

Projects that are underway blend fundamental and applied research

Naidu says because of these alarming statistics, the centre was established to assess the risk that pollutants and contaminants pose, both to the environment and to human health.

"We are developing techniques or strategies that will minimise the risks," he says. "Cost-effective technologies are needed. Not everyone has the money to pay for some of these works. We also need to build capacity in Australian environmental risk assessment and remediation. The centre will produce Australians who are experienced in this area. At the moment, we are importing graduates from overseas."

All in a day's Wark

Another relationship exists with the Ian Wark Research Institute (the Wark) at the University of South Australia. The Wark is the Australian Research Council Special Research Centre (SRC) for Particle and Material Interfaces, and a National Centre for Nano and Bio Materials.

Dr Terry Wilks, centre manager at the Wark, says the projects that are underway blend fundamental and applied research.

"We have a global client base of companies and government agencies, and a world-class research effort that provides the basic science that supports technologies in existing and emerging industries," he says. "The Wark has received $1 million in Federal Government funding to work with overseas researchers to develop new products and materials using nano- and biotechnology."

Medical and dental implants is one area that the Wark is involved with. Researchers are collaborating with TGR BioSciences, Portland Orthopaedics and Flinders University on new implant materials for superior attachment to bone.

The research efforts mimic the human body's natural growth and repair mechanisms in what is called the biomimetic process.

The Wark is also leading a four-year international research and development project on advanced coatings, which involves access to a network of European academic laboratories with specific expertise in advanced coating phenomena.

"The $1 million advanced coatings project is funded through AusIndustry, via the Innovation Access Program under International Research and Development Technology Access," Wilks says. "Now in operation for 18 months, it involves four European universities and, importantly, five major European Union companies. This program has a major emphasis on clever coatings and involves research coupled with excellent technology transfer mechanisms, including international workshops." ■

By Tim Murphy

Driving biotechnology → business

BioMelbourne Network is creating an Australian biotechnology hub by establishing commercialisation links across the globe.

Melbourne – home to a visionary biotech hub.

Melbourne-based BioMelbourne Network is Australia's first independent state-focused biotechnology networking agency. Established by the Committee for Melbourne in 2001 to promote the specific interests of Victoria's biotechnology sector, it is expected to be the first point of contact for local and international companies looking to do business in Victoria.

BioMelbourne Network was established to confirm Victoria as a world-class centre of biotechnology by nurturing the environment for the growth of innovative biotechnology businesses. Melbourne is a leading hub in the Asia Pacific and Australia is the sixth largest biotechnology centre in the world.

One of BioMelbourne's initiatives is to forge a stronger relationship with China. In 2004, a delegation of biotechnology companies and organisations from Guangzhou in China will visit Melbourne, providing the opportunity to foster existing research and commercialisation links between biotechnology industries.

Members of the Guangzhou delegation will conduct meetings with Victorian biotechnology companies to discuss partnerships, alliances and trade opportunities. We also hope to lead a delegation of Victorian biotechnology companies to visit Guangzhou.

We believe it is vital for Victorian biotechnology to have a devoted agency to connect business and research. Victoria is home to leading research facilities including the Peter MacCallum Cancer Institute, Melbourne University, the National Stem Cell Centre, the Australian Synchrotron, the Howard Florey Institute and the Monash Institutes of Health. Nearly 40 per cent of all Australian biotech companies are based in Victoria.

The growing member base includes key industry organisations such as the Walter and Eliza Hall Institute of Medical Research, the Australian Genome Research Institute, GlaxoSmithKline Australia, Howard Florey, Monash Institute of Reproduction and Development, BioComm Services, IBM, Ernst & Young and Clayton Utz.

BioMelbourne also has solid credentials on the board. Nobel Prize-winning scientist, Professor Peter Doherty, is its inaugural science patron and will assist in forging links with the science community.

In May 2003, BioMelbourne Network announced its plans to build networks between science, government and business through influencing and project-managing industry development initiatives.

The agency will draw upon the collective interests of members to develop economic and scientific outcomes, promote Victoria's capabilities globally through branding Melbourne as a centre of biotech, and provide cost savings through ensuring coordination and sharing of knowledge and resources. BioMelbourne has been modelled on global trade agencies from biotech clusters.

One of the Network's key programs, and a world first, is the establishment of the "International Network of Networks". This will drive the formation of international collaborations and business matching through strong interaction between other industry agencies and their members.

On a local level, the BioMelbourne Network will provide a much-needed voice and support network for the sector. It will also act as a gateway for international businesses looking to access the best that biotechnology has to offer in the Asia-Pacific region. ■

Tim Murphy is the executive director of BioMelbourne Network.

By Tim Treadgold

Technology boost

→ New government policies that highlight the innovation taking place in Western Australia are attracting global attention.

With one-third of Australia's land and sea resources under its control, there has never been any question of Western Australia being a magnet for explorers, geologists and investors in mining and oil projects. But that hasn't always been the case with more academic matters – especially in the sciences – which is why a new set of government policies designed to encourage scientific innovation and technology is attracting international attention.

The central aim of the package of measures is to attract major research projects and a greater share of research funding, adding to the State's reputation as a leader in innovative mineral exploration, agricultural research and metal processing. The routes chosen include the creation of a research fellowship scheme, a research facilitation program and the establishment of a series of Centres of Excellence.

Fellows to follow

The first two fellows, professors Julian Gale and John Mackenzie, are pointers to what the West Australian Government wants to achieve. Gale is a world authority on nanotechnology, particularly nanochemistry. Mackenzie is a microbiologist who has played a leading role in research into viruses such as Japanese encephalitis and Ross River virus. Both men have been awarded $A1 million fellowships, spread over four years, to lead research programs at Perth's Curtin University of Technology that aim to improve public health and lift the State's international competitive position.

Gale, who is moving to Perth from London's Imperial College, has previously worked in collaboration with researchers at Curtin, where he will continue his nanochemistry work which studies materials sized between the molecular and the individual building blocks of bulk materials. To a layman, that means finding potentially smarter ways to process minerals, or to solve the problems associated with the hypersaline water of inland Australia.

Mackenzie's work will focus on improving the detection of emerging insect-borne diseases in northern Western Australia, and on identifying ways to predict and combat the spread of disease.

When announcing the fellowships, the West Australian Premier

To a layman, that means finding potentially smarter ways to process minerals, or to solve the problems associated with the hypersaline water of inland Australia

and Science Minister, Dr Geoff Gallop, said they are a key part of his strategy to boost technical innovation and research. Alongside the individual research appointments, there is a separate $21 million fund for the Major Research Facilities Program, and $3 million to create three new Centres of Excellence.

"This package is designed to cement our reputation as a world leader in innovation by extending our expertise into new areas," Dr Gallop says.

The hypersaline water of inland Australia is being investigated by scientists.

The funding arrangement complements a three-pronged strategy overseen by the Office of Science and Innovation led by chief scientist Dr Bruce Hobbs. The first of the three programs is called Innovate WA, which reaches out to schools, looks at ways to create jobs, and encourages new capabilities through mechanisms such as the Major Research Facilities Program.

The second leg is the creation of Centres of Excellence. The third is an international program that is actively encouraging the establishment of space research based on a new family of massive radio telescopes, for which Western Australia's safe and unpolluted farming regions and outback are considered to be ideal.

"In the past, WA has suffered from a lack of international-class research centres," Dr Gallop says. "The Major Research Facility Program aims to build a high-impact scientific research facility."

The categories earmarked by the Government for assistance under its program include oil, gas and minerals, health, agriculture and fisheries, information and communication technologies, environmental research with an emphasis on salinity, and the marine sector.

"The total budget for the centre will be a minimum of $60 million, with the State Government providing between 20 and 30 per cent of the funding and the remainder coming from business, research and public sectors," Dr Gallop says.

"WA is already a world leader in many areas of science and innovation, but we have the potential to lead the way in many other areas." ■

The important thing in science is not so much to obtain new facts as to discover new ways of thinking about them.

SIR WILLIAM BRAGG

By Brad Collis

Designer food

→ New technologies and tools are ensuring Australia stays at the forefront of developments in grain-based food products.

It isn't in the dictionary yet, but there is a new word entering common usage among people who are mapping the direction of Australian agriculture – in particular, the grains industry. The word is "pharmas".

Pharmas is the updated word for the traditional "farmer" and describes the way the role of these food producers has changed. It aims to encapsulate a revolution: new industries, new foods and even, in a sense, new people because of the fast-approaching capacity to match the genetic needs of the human body with the power and new manageability of plant chemistry.

A field of wheat will no longer be a "crop" but a vast assembly of tens of thousands of leafy biological factories producing the molecular ingredients for foods, pharmaceuticals, nutraceuticals and even natural chemical compounds for processing into bio-fuels, plastics and other products.

It reflects the new age of discovery that is emerging from the field of biotechnology and the accelerating worldwide accumulation of knowledge about genes and genetics, which is offering new ways to breed, grow and use crops that have been the mainstay of agriculture and food production for centuries.

A suite of new technologies and tools is offering pharmas the ability to lift the quality of food crops, to widen the range of end uses and markets, and to strengthen both economic and environmental performances.

Significantly, this is not a distant vision but is already here and Australian researchers have been at the forefront of its development. It already shapes the direction of research funded by Australia's principal investor in grain sciences, the Grains Research and Development Corporation (GRDC), and is underpinning ventures into new grains-based products.

In 2003, the Cooperative Research Centre (CRC) for Innovative Grain Food Products became the first CRC to be established as a commercial venture with a mix of corporate, government and university partners. Its objective is to create new grains-based industries.

The CRC was the brainchild of Dr Graham McMaster, the chief executive of food technology company BRI Australia, and his counterpart at the GRDC, Professor John Lovett. Both scientists realised that the time had come when science not only knew the molecular composition of grains – the proteins, minerals, sugars, fibres and numerous other nutritional compounds – but also the genes responsible for these components.

It meant that different varieties of a plant, such as wheat, could be bred with an emphasis on particular traits to produce, for example, compounds known to have therapeutic benefits for human conditions such as cardiovascular disease or diabetes. A bottle of pills could be replaced by a grain-based food product.

In the words of the CRC's chief executive, Dr Jan Mahoney, they are "designer foods to complement people's individual genetic make-up."

Food fantasies

In the past, this might have remained an idealistic fantasy, except for the emerging reality that Australia's bulk grain products, which are mostly exported, are going to become more difficult to sell against competition from emerging high-capacity countries such as India, Pakistan and the so-called "Black Sea" states, and the increasing amount of hard white wheat emerging from the United States.

In today's new age of farming, a field of wheat is now an assembly of leafy biological factories.

For decades, Australia has enjoyed a natural advantage because its growing conditions are ideal for the production of hard, white, high-protein wheats for which international buyers pay a premium. However, clearly no country has a monopoly on agronomy and plant breeding, and other countries with large areas of arable land are catching up.

"By moving to new grain-based food and health products, we can keep ahead of changing trading conditions by going down a new road and becoming the leader in new markets," says McMaster. He predicts a new $A40 billion grains-based food and pharmaceuticals industry within a decade.

Propelled by such vision and potential, it is not surprising that the ambitious new CRC quickly attracted wide support and was kick-started with $17 million from industry participants, $60 million of in-kind support from research organisations, and $24 million from the Federal Government.

This is a new story that is still unfolding but McMaster believes the Australian grains industry has reached the start of a new era because of the opportunities being opened up by advances in biochemistry and genetics – and he's far from alone in this vision.

Lovett says the grains industry is beginning to grasp the concept of biotransformation, which is defined as the plant being a factory not just for food but for products such as new pharmaceuticals, plastics and fuels such as biodiesel.

"Importantly, this combines 'science push' with 'market pull', which is why the momentum is building," he says.

Meanwhile, Mahoney says grains research has reached the point where it can begin to treat plants as a source of ingredients. "From now we will be looking at crops from the perspective of their functionality rather than their traditional purposes like to produce bread or beer," she says. "If you look at what we already know we can do with the biological components of grains and pulses in traditional and relatively immediate contexts, just imagine the potential associated with nanotechnology (processing and manufacturing at a molecular level, rearranging atoms instead of comparatively crude processes such as casting, grinding or milling).

> There are farmers growing different types of buckwheat to the precise specifications of particular Japanese restaurants

We haven't even begun to realise the scope of what lies ahead."

She adds, apart from treating the plant differently, there will also be new processing technologies to go with the new approaches to breeding, growing and valuing a crop for its biological components.

Lovett adds that the next generation of processing technologies will also, for the first time, enable much more of the value-adding of Australian crops to be done in Australia rather than at the other end of a bulk-commodity trade route.

"If the diversity of new products becomes what we expect, then much of this processing and value-adding is likely to be carried out on or near farms," he says. "The centres for grain-based manufacturing may well move away from the major ports and back

Looking to the past for future success

To cope with changing and perhaps deteriorating growing conditions, many of tomorrow's food crops may owe their productivity to the genes of distant biological ancestors.

Strip away 5000 years of domestication and breeding, and crops such as wheat, barley and rice are all basically grasses. So if there are wild grasses that flourish in harsh environments or are resistant to certain pests and diseases, there's a good chance they possess genes that can confer these strengths to modern crops.

This has been recognised for a while, but it is only now that biotechnology is giving scientists the tools with which to track down and harvest these ancient genetic resources.

The first successful example of this in cereals has been the discovery of wild grasses in the border region of Iraq, Iran and China that are resistant to barley yellow dwarf virus (BYDV) – arguably the most damaging cereal virus in the world. Yield losses from BYDV in Australia (mainly in barley, oats and the new winter wheats) have been as high as 40-50 per cent.

After 15 years of biological sleuthing, a group of CSIRO plant breeders, working with Chinese colleagues, have not only tracked down a BYDV-resistant gene but have successfully incorporated it into Australian wheat breeding lines.

In a world first, a new winter wheat – called Mackellar – will be released next year with BYDV resistance, and shortly afterwards a new spring wheat variety known as Glover.

The search for BYDV resistance in wild grasses began in the mid-1980s after it became clear there were no wheat lines anywhere with the resistance.

With a grant from the Australian Centre for International Agricultural Research, the CSIRO and Chinese researchers began screening grasses of the Triticeae family, a close relative of wheat.

The resistance to BYDV will have particular significance for winter wheat growers whose varieties have been more susceptible to the virus than spring wheat growers, and also for barley and oat growers. Early trials have shown a 26 per cent increase in the yields of BYDV-resistant varieties.

The "capture" of BYDV resistance for important commercial crops was achieved through a novel application of conventional biotechnology. It didn't involve any genetic modification, yet the researchers say it is still an important case study for the debate.

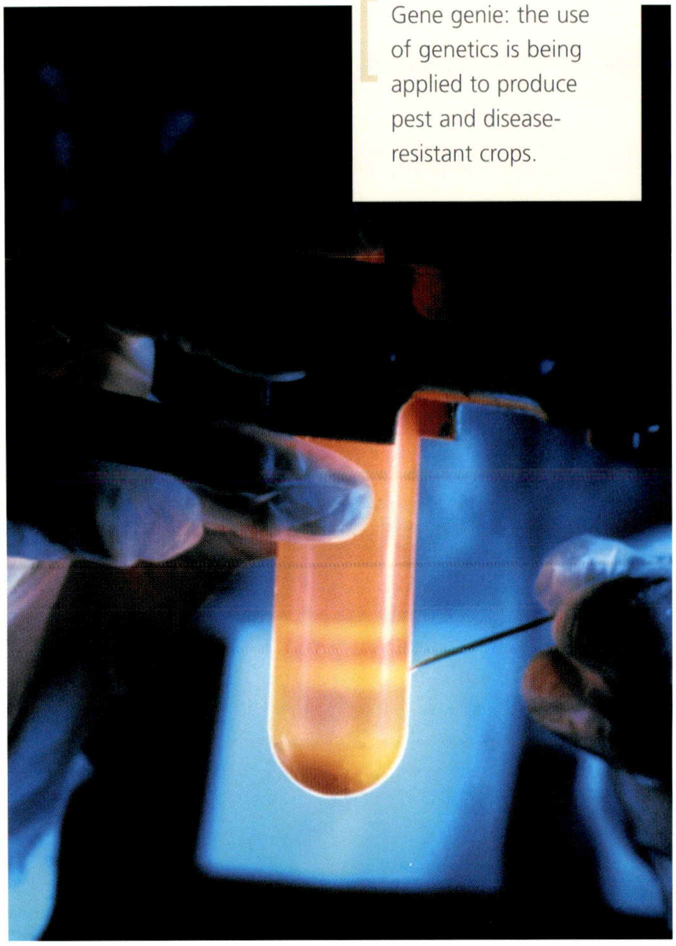

Gene genie: the use of genetics is being applied to produce pest and disease-resistant crops.

to regional Australia." He adds that, in many instances, the main processing is going to be happening in the plant, in the paddock.

"We're already seeing the start of this with growers who are now growing and packaging their own grain products for niche markets," he says. "There are farmers growing different types of buckwheat to the precise specifications of particular Japanese restaurants. The breeding and processing tools we're developing will make this type of specialisation more common."

Mahoney also believes that new genetic tools may open up opportunities to lift the potential of indigenous grasses that have been used by Australia's Aborigines for tens of thousands of years for food, but have never had the benefit of modern agronomy.

"Marry these with the advances that have been made in wheat since it was a wild grass and we're looking at brand new crops which are naturally suited to Australian conditions," she says.

Genetic grains

Another key aspect of the science being applied to food grains is the use of genetics to improve the basic quality of the crop to give plants an inbuilt resistance to pests and diseases and to minimise the use of potentially hazardous chemicals.

The first use of genetic modification has been in this area. It involved artificially altering a plant's genetic traits by introducing genes from other plants that are naturally resistant. It is a process that nature has been doing on its own for eons, but which plant breeders can now do in a single generation.

It is a proven, highly accurate way of "breeding" new traits into plants, but has been embroiled in controversy stemming from a range of commercial and philosophical arguments.

Growing potential: new plant-breeding technologies will produce crops with health benefits.

Gene modification crops offer the chance to grow crops that use less water, can tolerate salinity, don't need pesticides, are tolerant to new soil-friendly herbicides and can provide many of the health benefits being talked about by the CRC for Innovative Grain Food Products. However, for many researchers the controversy is a side issue because gene modification is only one of many new plant-breeding technologies.

CSIRO scientists, for example, have developed a "gene silencing" technology to produce healthier cooking oils and margarines. It is gene technology, but not gene modification. It works by using natural plant mechanisms to simply "switch off" the genes that convert oleic acid into cholesterol-raising fatty acids.

Nonetheless, it is still an example of how gene technology can provide consumer benefits through new food products that are not achievable with conventional plant breeding and that is what is at the heart of the innovative foods program.

"In the past we've tried to improve the industry's competitive position by improving on known products – by making the crop fit the requirements of known products," says Mahoney. "We've done well by improving noodle and pasta-making wheat varieties that are now in high demand around the world. But it's never long before competitors catch up. So what we're saying now is that instead of constantly trying to make our crops better fit the known products, Australia is setting out to create a whole range of new products."

It is ambitious, scientifically and commercially, but it is the very attitude that has made a comparatively small, isolated nation such as Australia a world leader time and again in new technologies. ■

Brad Collis is the editorial director of GRDC's Ground Cover *newspaper.*

International collaboration

Scientists from 10 countries are joining forces to develop the biotechnology platform for another green revolution – the next major genetic advance needed to lift the yield capacity of the world's most important food crop, rice.

The international collaboration – in which Australian scientists are playing a leading role – follows last year's mapping of the rice genome and has important implications for plant breeders working with other food crops, particularly cereals.

The CSIRO's Dr Liz Dennis, one of the Australian scientists involved in the project, explains that having the rice genome is a little like having a dictionary without an explanation of the words.

"If we're going to get anywhere we need to know what those words mean," she says. "The important factor for Australian growers is for Australian researchers to remain close to the project. We may only contribute 10 per cent of the information eventually uncovered, but involvement in the project will give us the chance to share in all of the information. The US has spent something like $US80 million ($A107.8 million) in plant genomic research in the past three years and the Japanese, Chinese and Taiwanese have also been spending considerable amounts. We've got the chance to share in that investment."

By Kathy Graham

Fields of the future

→ Increased pressure on the world's land resources has led to the adoption of radical crop protection strategies.

Bayer CropScience has a proven track record in fostering and managing innovation.

Companies and people wishing to shape the future have to be receptive to groundbreaking ideas. At times, this will mean far-reaching alterations and radical change.

Bayer CropScience has taken a considerable step towards achieving its goal of becoming a leading company in crop protection, biotechnology and seed breeding, as well as in non-agricultural pest control.

Its cutting-edge technological platform, sound experience and proven track record in the field of managing innovation will be instrumental in achieving its targets.

The major challenges for the crop protection industry today are world population growth, the decreasing availability of arable land and the worldwide shift from vegetable to animal protein in the human diet.

These changes mean that crop protection and crop improvement will continue to play an important role in the future. The research and development resources of Bayer CropScience place the company in an excellent position to address these issues. From the discovery of classical active ingredients to genomics, the organisation possesses one of the most complete crop-linked research platforms in the business sector.

Research target

According to Sam Howard, managing director of Bayer CropScience Australia, the company has a research and development spending target of about 9 to 10 per cent of sales. "We currently maintain one of the largest research and development budgets in the crop production industry," he says.

At international research centres around the world, scientists are using classical and cutting-edge approaches in their continuous pursuit of product innovations and enhancements to existing and new technologies.

If crop protection is to meet present-day expectations, new active ingredients must fulfil the following key requirements for the benefit of Bayer CropScience's customers, shareholders and stakeholders:

- Be highly selective in targeting pests, weeds or diseases;
- Be highly effective at low application rates;
- Be safe for humans and animals, environmentally compatible and biodegradable;
- The right value for customers.

Innovative inroads today are achieved by using test systems based on the identification of functional genes in plants, insects, fungi, bacteria and viruses. Bayer CropScience uses those test systems to check its extensive library of chemical substances and the continuous flow of new compounds for their biological efficacy.

The company's experts rely on process-oriented teamwork for the development of the products. This includes the field testing, the search for the best application form and the extensive evaluation of all safety aspects concerning human health, fauna, flora and the environment. In parallel, chemists and engineers are developing the manufacturing process from laboratory scale to industrial production.

Network of excellence

The individual research centres operate in multinational working groups, exploring new options and cooperating with other facilities in their projects. "The common goal is to find better crop protection products," Howard says. "Here, 'better' means enhanced benefit for the customers, safety in application and increased compatibility with the environment."

In the 21st century, a successful combination of cutting-edge technologies is pushing forward the development of innovative products with new modes of action. By pooling resources, Bayer CropScience is able to play a leading role in hit generation for new products using state-of-the-art platforms. A wide range of research activities focus on the discovery of new active ingredients. Gene analyses enable Bayer CropScience to search specifically for new targets in the metabolisms of various harmful organisms.

The company uses a technology platform encompassing everything from comprehensive substance libraries, UHTS processes (ultra-high throughput screening) and scientific computation to micro-screening and advanced logistics.

In its quest for new active substances, modern crop protection research uses automated processes to collect millions of pieces of test data every year. Automated processes – including the use of robots to test large numbers of chemical compounds for their biological effect – speed up the hit generation for new active ingredients. If this mass screening is to be applied effectively nowadays, work methods must include genetic science know-how.

According to Howard, on average only one in 50,000 tested chemical compounds fulfils the high standards for new commercial products. "Thirty years ago, the rate was one in 10,000," he says. "The reason for this is the permanent increase in the number of safety tests required before a product can be successfully placed on the market."

With its scientific computation systems, Bayer CropScience is well positioned to carry out the virtual screening of millions of existing data items. State-of-the-art computer systems calculate which molecules will match which specific structures and in exactly what way. The simulation of chemical structures enables characteristics to be researched and new compounds to be tested.

The BioScience business group of Bayer CropScience has added a new dimension to the research for classical crop protection substances by modifying the DNA of crop plants.

In this way it is possible to improve various properties of crop plants, not only in their resistance to pests and diseases but also in other properties or traits such as yield enhancement, reproductive biology (hybrid technologies), stress tolerance and carbohydrate modification.

There are many signs that biotechnology and crop protection in the future will be much more closely interwoven than in the past. Bayer CropScience is well equipped to address this development. ■

Bayer innovation

Bayer AG is a global enterprise with core competencies in not just crop science, but also healthcare, polymers and chemicals. Internationally, Bayer occupies leadership positions in 80 per cent of its business and it continues to strengthen its position through research and innovative products.

Research and development has always played a key role at Bayer. It is the basis for new products and processes, and thus contributes significantly to safeguarding the future of the company. Most of today's business activities are based on Bayer's own innovations. About 40 per cent of sales revenue comes from products that have emerged from Bayer's research laboratories over the past 10 years.

Bayer has a long history of milestones in research and innovation including:

- 1899 Aspirin: the world's best-known painkiller.
- 1910 Methyl rubber: the first synthetic rubber manufactured on a large scale.
- 1937 Polyurethanes: plastics made from isocyanates and polyols.
- 1953 Makrolon: the first high-performance thermoplastic based on polycarbonate (used in CDs, DVDs and CD-ROMs).
- 1975 Adalat: medicine for the treatment of coronary heart disease.
- 1987 Ciproxin: antibiotic for the treatment of bacterial infections.
- 1991 Gaucho: systemic insecticide with long-term effectiveness and very low toxicity to humans.
- 1996 Advantage: flea control product for cats and dogs (developed in Australia).
- 1998 ADVIA120: fully automated analyser for determining blood counts.
- 1999 Avelox: new respiratory antibiotic.
- 2001 Calypso: effective, environmentally friendly, broad-spectrum insecticide.
- 2003 Levitra: offer medicine to treat erectile dysfunction.

(All products mentioned in this article are registered trade marks of Bayer.)

By Gayle Bryant

An apple a day

→ Horticulture Australia aims to deliver commercial benefit to growers and has launched an initiative to promote healthy eating.

By increasing consumption of fruit to two serves a day and vegetables to five serves a day, Australians could prevent 20 per cent of all cancers, 16 per cent of all cardiovascular deaths and save the healthcare system $A180 million a year.

These statistics were quoted at the 2003 launch of a report titled "Better Health: It's Simple", written by the Australian Fruit and Vegetable Coalition (AFVC). The AFVC was formed to explore partnership approaches to increase the fruit and vegetable consumption of Australians.

One of the founding members of the AFVC is Horticulture Australia – the industry body for the fruit and vegetable industry. Other members of the AFVC – each keen to promote the benefits of eating more fruit and vegetables – include: Strategic Inter-Governmental Nutrition Alliance (SIGNAL); the Cancer Council Australia; National Heart Foundation; and the Department of Agriculture, Fisheries and Forestry.

With the help of industry representatives, Horticulture Australia has developed a strategic plan that forms the basis of its health initiative.

The health initiative recognises the need for the horticulture industry to have a consistent message, speak with one voice and maintain delivery over time.

The health initiative has enabled Horticulture Australia to take a leadership role in the AFVC, including being nominated as the Secretariat with the aim of increasing demand for and consumption of fresh fruit and vegetables as horticulture's contribution to improving the health of Australians.

You are what you eat

Sarah Pennell, senior marketing manager at Horticulture Australia, says having a national campaign to increase the amount of fruit and vegetables we eat is important. "Australia is one of only a few countries in the developed world that doesn't have a national campaign," she says. "The AFVC will be lobbying for government funds to launch one."

Another pressing issue that impacts on the horticulture industry is water. Horticulture comprises fruits, vegetables, nuts, nursery products, extractive crops, sports turf and cut flowers.

These crops need reliable supplies for irrigation and, in early 2003, Horticulture Australia implemented a water initiative to address ongoing access to water.

The initiative helps horticulture by articulating the economic and social benefits of horticulture water use.

It is also demonstrating the environmental credentials of horticulture and investing in projects that will further enhance the sustainability and reputation of the industry.

According to John Webster, managing director of Horticulture

Horticulture Australia: working in partnership with industry.

Australia, there is increasing scarcity of and competition for water. "There is also growing community scrutiny on the environment with regard to water users and increased pressure to allocate a greater share of river flows for environmental flows," he says. "Defining new water-sharing arrangements is a major part of the national water reform agenda."

He says horticulturists have already seen changes in their traditional level of access to water, with pressure for it to be lowered to accommodate environmental flows.

"Horticulture has been active in responding to these changes by adopting more efficient technology, trading water, increasing environmental performance and participating in natural resource management," Webster says. "These activities are vital for maintaining access to water."

The water initiative aims to grow profitable crops with less water by adopting and developing water-saving growing methods. It also wants to ensure policy-makers take horticulture and water access issues into account by maximising horticulture's voice in policy development. ■

The business of horticulture

- The gross value of horticulture production is $6 billion a year.
- In the five years to 2001-02, the gross value of production increased by more than 21 per cent.
- Horticulture employs 97,000 people, including 13,000 in the nursery industry.

By Sir Gustav Nossal

Winning gold with Australian science

Australia's agriculture industries are being supported with innovative strategies from Victoria's Department of Primary Industries.

These are exciting times for Australian agriculture. Globalisation, our expectations of the food we eat and our concern for the environment are driving demand in our food and agriculture sector as never before. Opportunities abound, particularly with our Asian neighbours where we are on the cusp of one of the biggest opportunities in our history – the chance to secure a significant share of huge, growing markets.

Our success in capturing these opportunities is directly linked to our ability to use science in conjunction with technologies that allow the creation of new products and new methods of production that increase productivity and facilitate change.

Far from being a "sunset" industry, the food and agriculture sector continues to outperform other key economic sectors, as it has done for the past 20 years. Simple ratios tell the story. If we take the ratio of exports to imports as a measure of competitiveness, agriculture is 10:1; food is 7:1; automotive is 0.4:1 and our information and communication technology rates just 0.1:1. Overall, food and agriculture contribute 18 per cent of our export performance for Australia (ABS, 2002). Clearly, in the food and agriculture sector we are punching well above our weight.

Growing new industries

Herein lies the true value of organisations such as Victoria's Department of Primary Industries (DPI). The DPI supports Victoria's fisheries, agriculture, minerals and petroleum industries with science, innovation and technology.

Its research and development underpins the future of these industries and that of more than 200,000 Victorians employed in the production, processing and marketing of food and agriculture products.

New technologies are opening doors we could only dream existed previously

Much of the work in the DPI's research and development arm, Primary Industries Research Victoria (PIRVic), is already paying off. From the development of a new chickpea variety resistant to blight to the development of a lentil industry providing half of Australia's annual $A140 million crop, the impact of the DPI's research effort is enormous. This research effort has also led to the development of a highly successful canola industry with an annual value of $130 million.

Improved grazing practices are generating, on average, some $20,000 per annum for participating dairy farmers. We now have a green tea industry and Australia is leading the world in strategies for combating fungal-borne diseases of vegetables.

Growing disease-resistant clover in test tubes.

The DPI is also helping the sector keep pace with changing community expectations of how we manage our environmental resources. Already we've experienced dramatic improvements, with innovative technologies and practices now achieving a 50 per cent enhancement in water use efficiency for some high-value crops.

The role of science in these gains is indisputable. Much of the food and agriculture sector's impressive record for achieving productivity improvements is largely a result of the application of new technologies and management practices.

However, even greater gains lie ahead. New technologies are opening doors we could only dream existed previously. The DPI is recognised for having the world's best pasture plant genomics group. This group has already mapped the bulk of the genes of our core pasture species – ryegrass and clover – and is using these maps to select plants with improved drought tolerance and nutrient utilisation. It has developed a white clover variety that is resistant to alfalfa mosaic virus and is developing cultivars resistant to two other major viral diseases of pasture legumes.

The DPI's research and development is at the forefront of applying new microsensing and precision farming technologies to keep our food and agriculture sector smart, flexible and enterprising. ■

Sir Gustav Nossal, former director of The Walter and Eliza Hall Institute, advises Victoria's Department of Primary Industries on science issues.

By Cathy Johnson

Agricultural innovation

→ The research initiatives of the Australian Centre for International Agricultural Research are making an impact across the globe.

They are just a few millimetres long, but they are one of the biggest barriers to health and prosperity in parts of the developing world. When tiny fruit flies lay their eggs in fruit and vegetables, the emerging maggots render the produce useless. In South-East Asia and the South Pacific, these crop losses are a significant barrier to economic development and can also contribute to malnutrition.

Developing new vaccines to take care of nest eggs.

But this is now set to change thanks to an innovative control system, developed using Australian research. It is a gelatinous bait that farmers use by simply dabbing a small amount on a tree. The gel is irresistible to flies, which eat it and die. Containing an ultra-low dose of insecticide, it is cheap and environmentally friendly. What's more, a key ingredient – protein – can be produced from industrial waste that would otherwise be pumped into waterways.

Tiny pests, huge impact

"It's a simple but great idea that's making a major contribution," says the research leader, Queensland entomologist Professor Dick Drew.

The project is one of many funded by the Australian Centre for International Agricultural Research (ACIAR), which aims to use Australia's strong capacity in agricultural research to benefit developing countries. In this instance, previous basic ecological research was pivotal to the work. Drew had already shown that flies in orchards were attracted to fruit because of certain bacteria growing on their surface. As part of the research, the bugs were grown in the laboratory on plates of a moist gel substance, to which protein had been added as food. "We were putting these dishes in our traps in the field and finding they were really good at drawing in fruit flies," recalls Drew.

Indeed, the dishes proved so irresistible that only the tiniest amount of insecticide had to be added to the mix to provide effective control. Farmers using the gel in their orchards use at least 1000 times less chemical than in a typical cover spray. In Vietnam, an agreement with the Fosters brewery has led to a factory being built to extract protein for the gel from a yeast waste solution. Previously, the waste, which causes environmental problems, had been discarded as there was no other use for it.

"We're looking at building these factories all over South-East Asia," says Drew, who hopes the gel might ultimately be used to reduce reliance on sprays back in his homeland.

Helping communities

Other ACIAR-funded research benefiting Australia focuses on the restoration of fisheries of the marine snail, the trochus. Trochus has suffered severely from over-fishing in recent decades but scientists working with communities in the Kimberley region in the country's northwest, as well as in the Pacific and eastern Indonesia, are hoping a revival of numbers may be in sight.

Early work aimed at enhancing fisheries by restocking reefs has shown great promise, according to research leader Dr Chan Lee.

A single trochus can produce more than a million eggs so the potential for a large increase in numbers is great.

Global achievements

Other ACIAR projects that aim to improve livelihoods in developing countries include:

- The extension of Australia's successful grassroots land conservation movement, Landcare, to the Philippines' southern island of Mindanao, substantially reversing both deforestation and land degradation
- The introduction of oilseed-based protein feed supplements to India to increase milk production by u to one litre a day in cattle and buffalo. A factory allowing mass production of the supplements was opened recently.
- Development of the world's first wheat variety which is resistant to Barley Yellow Dwarf Virus (a major disease of cereal crops) through breeding of a resistant gene from related wild grasses. The new resistant wheat varieties are being used in countries including Australia and China.

Building a better fly trap: a new gel helps alleviate the fruit fly problem.

In the Kimberleys, results of ACIAR's research have encouraged communities to form the Kimberley Aquaculture Aboriginal Corporation. This corporation recently funded a $A3.5 million multi-species hatchery to produce 36 million trochus juveniles for release on local reefs. The project work has also stimulated the construction of community-based hatcheries at One Arm Point in Broome, Western Australia and in the Pacific.

Efforts are also being extended to Samoa, which is committed to the establishment of a trochus fishery using hatchery juveniles and the release of adult trochus.

"A key aspect of ACIAR's approach – the active involvement of local people – has been critical to these successes," Lee says. "Building viable [trochus] populations in community-owned reefs is a long-term effort. It will probably take 10 years or more. But by partnering with communities in Australia and the Pacific, we've taken the first steps and that's always the hardest part."

Chicken vaccine aids developing countries

In many African and Asian nations, it's not marine snails but chickens that are a key contributor to many livelihoods. The birds are not just a nutritious food source but a type of feathered financial reserve, or, more literally, a "nest egg" which can be sold to generate money for medicines, clothes or school fees. But one of the main factors inhibiting successful poultry rearing is Newcastle disease – a viral illness that can wipe out entire flocks. While there is no cure, ACIAR-funded research led to the development of an effective vaccine that is ideal for developing nations.

"The vaccine can survive for some time without refrigeration and can be produced in small-scale laboratories using very basic facilities," says Dr Joanne Meers, leader of a project now applying the vaccine to villages in Myanmar. Previous projects have been successful in Cambodia, Malaysia, Vietnam, Bhutan, Mozambique, Senegal, Ghana and Tanzania.

Chickens and their eggs are now much more widely eaten in villages where the disease has been controlled and selling surplus birds is an important income generator for families. But the benefits go beyond health and economics. According to Meers, the work is also boosting the status of village women, who have control over the birds while the men seek work further afield. "In most of these places, women have very little power," she says. "But now it's the women who can make decisions about chickens that directly affect the family incomes."

Cathy Johnson is a Sydney-based science journalist who has reported for The Sydney Morning Herald, The Bulletin *and other publications.*

Be open-minded as to how the solution may come about

leaders SPEAK

Creative team-building

Back in my mining days, a difficult problem resulted in some outstanding innovation. Synthetic rutile, an industrial mineral we produced, suffered from radioactivity which we needed to remove to compete with overseas suppliers.

As a public company, Renison Goldfields had a good internal team of metallurgists and chemists but recognised the need for more expertise.

We funded collaborative research with CSIRO and the University of Western Australia. This approach was successful due to the brilliance of the individuals and the rigour of the approach.

In 18 months, we had evolved a world-class solution with other useful spin-off technology.

From this I learned how to guide an innovation process. First, identify very clearly the problem and the objective. Second, identify intelligent and highly motivated people and place them in a supportive, collaborative environment.

Finally, be open-minded as to how the solution may come about, this being an aspect of the inherent freedom. Creative people "sparking off" one another will inevitably come up with unusual approaches that need to be encouraged.

MARK BETHWAITE
MANAGING DIRECTOR AND CHIEF EXECUTIVE OFFICER
RENISON GOLDFIELDS

Venturing forth

Commercialisation, early-stage ventures and growth companies require investment to enable products and services to successfully reach end markets and generate returns. While some growth can be self-financed, many initiatives require the injection of patient equity investment to allow diffusion and deployment.

The Federal Government has instigated a number of initiatives designed to plug gaps in the commercial venture capital market through pre-seed funds, innovation investment funds, research and development grants and programs such as COMET and BITS Incubators.

However, in more difficult times venture capitalists tend to move "up the value chain" causing further funding gaps, and reducing the funds available to nascent and early-stage ventures.

Compounding this is the fact that Australia does not have a developed angel investment market and the Federal Government does not have the budget to sustain funding to the early-stage venture market.

The lack of an early-stage investment market happens for a number of reasons. One is that successful entrepreneurs grow up to become successful entrepreneur financiers. Most angel investors have made money out of their own ventures and are prepared to back others. Australia does not have a large pool of these people to draw from.

Angel and venture capital investment require the investor to measure returns over a portfolio, not just in relation to one investment. An angel and venture investor must be prepared to lose money and focus on obtaining a great return on the other investments in the portfolio.

There is also a need for active management, which is time consuming and requires the ability to deal with ambiguity, setbacks and to be strategically focused on creating sustainable value.

There is also a lack of investment taxation incentives and a lack of patient capital able to take 10- to 15-year time horizons before exit. To create new large corporations of the future we need a "funnel" of many small companies. In our funnel, a few will go on to great things and a few will fail, but the majority will create local employment opportunities.

Without a large funnel we will not get the next ResMed, Cochlear or Com Tech (now Dimension Data). Given the political pressure to have a "balanced" Federal Budget, we need to get other money from the investment market incentivised to invest for the future.

We need money that is prepared and able to take the high risk for the potential of high returns with a worst-case scenario of a tax write-off.

I am reminded of one of my favourite sayings from Robert F Kennedy: "To succeed greatly you must be prepared to fail greatly." Which investors and what initiatives are prepared to stand alongside our entrepreneurs to invest in our collective economic future?

LINDLEY EDWARDS
CHIEF EXECUTIVE OFFICER
THE VENTURE GROUP

By Abbie Thomas

Looking to the gods

→ A Greek deity is the inspiration for an initiative set to turn technology and research into commercial reality.

One of the legends of the Greek god Prometheus recounts how he stole fire from the god Zeus and gave it to mortals. Throughout the ages of Greek mythology, Prometheus has become known as the one who gave humanity the arts and sciences, as well as our means of survival.

A few years back while working in the United States raising venture capital, Rob Muir, director of business development at the Australian Nuclear Science and Technology Organisation (ANSTO), was inspired by this ancient legend to establish the Prometheus Alliance. The forum aimed to identify new technologies which would be of specific strategic interest to Fortune 500 companies.

"In marketing, you've got to get through the gatekeeper to the senior executive," Muir says. "We needed something that would catch their interest yet also embody some of the qualities of science and technology. Prometheus was the first guy to steal fire from Zeus and give it to mortals: this was the first instance of technology transfer without a licence."

The business of science

Knowing that science terminology is all Greek to many investors, and that business jargon is equally unfathomable to some scientists, Muir has established a new Business Lab, through ANSTO's Business Development arm, to turn the organisation's cutting-edge research and development into commercial reality.

"Although Australia is great at innovation, the fact is that there are very few Australian-based, high-growth, high-cap businesses established out of these ideas," says Muir. "Instead, there is an Australian pool of unexploited technologies in undercapitalised small companies struggling to make it across 'the Valley of Death' concept – moving innovation from the lab to the boardroom."

Muir's approach is to foster the researchers and their discoveries, and to introduce business skills and experience into the equation. Originally from Australia, he spent 20 years in the US raising venture capital through Fortune 500 companies for commercialising research. Part of this time was spent with some of the top research institutes – organisations whose function it is to interface between universities and industry.

> There is an Australian pool of unexploited technologies in undercapitalised small companies struggling to make it across 'the Valley of Death'

From this experience, Muir brings to ANSTO a market-driven, highly customer-focused way of operating. While the organisation has always carried out great, innovative research, he says a new environment has been created through the ANSTO Business Lab.

Muir says ANSTO's Business Lab, launched in March 2003, provides strategic business, market and financial resources, and expertise in technology development, commercialisation and venture management for clients, global partners and stakeholders.

With the idea of fostering an entrepreneurial culture within ANSTO, the ANSTO Business Lab provides a "cradle-to-grave" environment for ANSTO's intellectual properties and technologies.

Muir sees a number of parallels between the conventional science lab and the ANSTO business lab environment. Both start with a strong interest or goal and are driven by a hypothesis – the scientist designs experiments and collects data, as compared to the business person who develops a business model and carries out market research and validation. Each produce a result and both use tools, such as instruments, equipment and a methodology in the science lab, and analysis and diligence in the business lab. They also both offer incentives to their people: the scientist gains a reputation and career progression, while those in the business lab benefit from a boost in career and income.

Two components drive the ANSTO Business Lab: an internal peer review-based Patent Management Committee, and an external Technology Forum. The committee evaluates innovations and protects intellectual property, while the Technology Forum consists of a panel of outside experts who provide a reality check on the business model. These might include lawyers, marketers, intellectual property and industry experts or others with specific skills. A forum is held several times a year with researchers and key managers.

These experts give and get instant feedback, coaching and practical advice about new technologies under development and an appropriate business model for commercialisation.

Patent potentials

As well as fostering its own intellectual capital, ANSTO is also seeking industry partners to help develop and commercialise its technologies. One example is a new patented technology called "muCaps", which enables the encapsulation and release of a wide range of active molecules from ceramic particles. The particles have a defined structure, so they can be used for specific dose or release rates. The potential markets include drug delivery, active "paints" (tiny particles which are painted or sprayed on to release pesticides or anti-fouling agents) and even cosmetics, which could provide UV protection or carry abrasive materials.

Another opportunity is ANSTO Minerals, a range of solutions and processes for mine sites which has been developed to cut costs and improve environmental performance. These solutions and processes have particular applications in radioactivity control, hydrometallurgy, acid mine drainage and water conservation.

ANSTO has been producing radiopharmaceuticals for more than 30 years, and is also seeking strategic research and development

Far-reaching innovations: x-rays can be used to examine stresses in parts of jet engines.

partners to develop and deploy the next generation of personalised medical therapies and diagnostics. Muir believes that by harnessing ANSTO's expertise in this area, pharmaceutical and biotech companies can cut the amount of time and cost needed to develop and test products.

HIP to the heat

One technology developed through an ANSTO partnership that has recently found success in the marketplace is HIP (Hot Isostatic Pressing). An Australian medical device company called ASDM worked with ANSTO to develop the technology for use in producing artificial knees, and used a cobalt chrome molybdenum alloy.

When metals and other materials are shaped using casting, the material inside usually contains some porosity and a number of irregular flaws. The HIP technology applies heat and high pressure in all directions, removing flaws and producing a highly dense material which is strong, flexible and has a longer fatigue life.

Expecting to increase production significantly over the next few years, ASDM is targetting the multi-million-dollar international market. The company is now looking at producing other bio-implants using ceramics, and also at applying the technology to other high-quality products such as turbines in aircraft.

According to Muir, while ANSTO is still seen by some as a traditional nuclear technology provider, it has a much greater range of expertise – in environment, materials, nanotechnology and manufacturing.

"ANSTO is Australia's national laboratory for nuclear technology, but it can be so much more," he says. "We have a huge pool of intellectual capital and we want to engage and use our knowledge to help Australian industry become more globally competitive." ■

By Gayle Bryant

Fast-tracking ideas

→ Edison Mission Energy is overcoming the complex challenges of the energy industry with a progressive approach to commercialisation.

Getty Images

Edison: winning an uphill battle in the competitive energy market.

Energy companies face an uphill battle supplying economies with their energy needs from a dwindling supply of resources. But one energy company, Edison Mission Energy, has become a world leader in bringing complex and high-risk projects to fruition through its innovative approaches to business, both internally and externally.

The organisation is a global energy operator with projects around the world, including Australia, the UK, Indonesia, Italy, Thailand, Puerto Rico and the Philippines. To date it has developed and acquired a portfolio of assets with a net generating capacity of nearly 19,000 megawatts.

Edison's regional senior vice-president, Bob Driscoll, says a key competitive edge for the organisation is its employees. "We employ entrepreneurs," he says. "They are driven people who aren't afraid to assess and take risks. They oversee opportunities through to completion, from the original idea to the operation of the plant by Edison-trained personnel."

Get your hands dirty

An innovation in itself is Edison's management structure. While it employs 400 people outside power station operations, it has only four layers of management – an unusual structure for a large company. Driscoll says it is this four-layer management structure that provides employees with the autonomy to get things carried out.

"Having so few layers of management is a huge advantage," he says. "It allows us to fast-track projects through our approvals processes and due diligence to ensure we are able to move projects into commercialisation phases quickly. The structure means everyone gets their hands dirty. There is no suffocating bureaucracy."

Around the world, countries are demanding more energy supplies. Driscoll says that whenever possible, Edison looks to use renewable sources. It also runs a number of programs that help address land degradation, one of Australia's most urgent environmental problems.

In Europe, Edison has developed the Doga Energi Cogeneration project – Turkey's first independent power project under its Build Own Operate program. The new technology will displace coal home heating in 14,000 homes in Turkey's Esenkent housing complex with clean and efficient heating energy.

"Increasingly, energy companies are being encouraged to be more efficient, not just with emerging technologies but also by greater use of existing technologies such as wind power," Driscoll says. "New technologies are taken up as the technology improves and organisations such as Edison are always keen to work with government and other organisations on increasing its use of renewable energy." ■

Regeneration programs

Edison operates a $A1 million landcare program in the Gippsland region of Victoria as part of its Greenhouse Challenge action plan. Its Landcare Program was developed in partnership with the Victorian and Federal governments and provides funding to the landcare movement in addition to State and Federal funds.

Projects sponsored by Edison Mission Energy across the Gippsland region include:

■ **Wellington Greenprint:** a large-scale revegetation project, including the establishment of an indigenous seed bank at Maffra, to the east of the Latrobe Valley, to supply local seed for direct seeding revegetation activities throughout the region. The project aims to rehabilitate land degraded by erosion and salinity, assisted by the introduction of whole-farm planning concepts.

■ **Powlett Project:** a project in South Gippsland, which addresses land, water and coastal degradation in the Powlett River catchment between Korumburra in the upper catchment and the Bass Strait coast at Kilcunda.

■ **Phillip Island Wildlife Corridor:** linking the plantings by Phillip Island Landcare Group with remnant native bush on private land and native reserves across Phillip Island.

■ **Churchill:** in co-operation with Monash University, revegetating land near the university's Gippsland campus.

By Abbie Thomas

Growing in the wind

→ Australia's commitment to reducing greenhouse gas emissions has opened up opportunities for energy producer Hydro Tasmania.

Hydro Tasmania's Woolnorth Wind Farm in Tasmania.

Once the target of angry campaigns by the green movement, Hydro Tasmania is embracing the mood of the times and spreading its wings. It has its eye on being Tasmania's world renowned renewable energy business with three lines of business: Asset Backed Trader, Renewables Developer and Global Consultant.

When the Federal Government introduced its Mandatory Renewable Energy Target (MRET) in 2001, it created a major opportunity for producers such as Hydro Tasmania, in the business of fostering renewable energy. A new law requires the Australian energy industry as a whole to buy an additional 9500 gigawatt hours of electricity from renewable sources by 2010 (enough for 4 million people).

Already generating 60 per cent of Australia's renewable energy, Hydro Tasmania is seizing opportunities to expand its business and develop new technologies such as wind and hydrogen power.

Hydro Tasmania has $A3 billion in assets, with a total generating capacity of 2262 megawatts and more than 50 large dams, making it the largest dam owner in Australia. These assets provide the basis for Hydro Tasmania's Asset Backed Trader role in the National Electricity Market.

Basslink

Following approval in 2002, construction is underway for Basslink, which will result in a 350-kilometre undersea cable link connecting the electricity supply between Tasmania and mainland Australia.

Basslink will enable Tasmania to compete in the National Electricity Market. John Titchen, the company's manager of renewable strategy, says through Basslink, Hydro Tasmania can supply electricity to Victoria during periods of high usage, such as on cold mornings or hot days.

"This is when power is most in demand," he says. "Basslink can also help 'drought-proof' the State, which is accustomed to frequent power outages that cause major disruptions to homes and businesses.

"Hydro development was once the company's main development focus, but most suitable hydro sites have now been developed. A new focus for Hydro Tasmania is the development of wind farms, both at home and on mainland Australia."

The first large-scale wind project is located at Woolnorth, Tasmania. Although already operating, when completed it will generate nearly 140 megawatts of electricity. Two other wind projects are also planned for Tasmania: one at Heemskirk (160 megawatts) and the other at Musselroe (120 megawatts). The Woolnorth project has already benefited the local community, with the Danish company Vestas establishing a factory in 2002 and employing more than 70 people to make wind turbines.

Hydro Tasmania is also constructing its first large interstate wind farm, a 66-megawatt facility at Cathedral Rocks near Port Lincoln in South Australia.

Wind farms

Titchen says Hydro Tasmania was well prepared by its years of experience in generating electricity from water to take on wind power and all its vagaries.

"We've deployed our technical expertise in hydro in assessing uncertain natural resources, and drawn these into the wind arena," he says. "We evaluate the returns over the economic life of the project, which for a wind farm is around 20 years. The returns on this basis would be comparable with a coal-fired power station, partly because producers of renewable electricity get an additional revenue from MRET certificates."

Hydro Tasmania continues to seek out new opportunities. Titchen believes hydrogen power holds great promise for the future growth of the company.

"Proven hydrogen technology could further enhance the viability of wind power by providing an efficient way to store the electricity generated by wind turbines," he says. "It's not well developed yet, but there are good reasons to be involved. There is a view that hydrogen will be the fuel of the future and we see a very good link between hydrogen and our business in renewables." ■

Abbie Thomas is a science writer based in Sydney.

By David Salt

Going nuts over macadamia power

A Queensland company takes a closer look at how to convert waste into energy – with spectacular results

Here's fuel for thought. What do you do with a mountain of rock-hard macadamia shells left over after the kernels have been removed? According to an Australian energy company, the answer is to convert the shells into electricity and use the energy in the processing of the macadamias while contributing to a better environment.

Could it be that simple? Yes and no. Macadamia shells are an excellent source of fuel but if you want to convert them into a commercial source of electricity, there are a few hurdles you have to overcome first.

The company which took on this challenge is Ergon Energy, a leading energy retailer based in Queensland. About three years ago it began searching for sources of renewable energy in line with the Federal Government's drive for electricity retailers to source an extra 2 per cent of their electricity from renewable sources by 2010.

Ergon Energy began discussing possibilities with its customers, including the possibility of converting waste into energy. After more than 30 years of working with nearly all the State's sugar mills to take "green electricity" produced from waste sugar cane, Ergon Energy looked at waste streams from meat processing, timber and peanuts.

The industry that offered particular promise was the processing of Australia's own macadamia nuts. The shell is a good fuel and the Australian industry is experiencing incredible growth and is the global leader in production.

Demanding nuts

Macadamia nut production is booming with growing demand from the snack, bakery and confectionary sectors both here and overseas. There are now more than 4.5 million macadamia trees used for commercial production in Australia and numbers are climbing dramatically. However, as the industry grows, so does its waste stream.

Suncoast Gold Macadamias is a macadamia processor based in Gympie (two hours' drive north of Brisbane). It is the world's third-largest macadamia processor and generates about 5000 tonnes of waste shells every year, with an expectation that this could double in the next few years. This is one massive pile of shells.

The company was attempting to turn the shells into a resource by combusting them to heat a boiler to generate steam. The steam was used to heat and dry the macadamia nuts to assist in the process of removing the shells. While this was a fantastic use for the waste shells, it only consumed a portion of the pile. As production rose, so did the quantity of excess shells.

Ergon Energy's customer solutions general manager, Ross Blundell, says macadamia shells are a fantastic fuel source, better than most people realise.

"Fuels are often rated by their moisture content and burning quality," he says. "For example, the calorific value of black coal is given a rating of 26 megajoules per kilogram (MJ/kg); brown coal 24 MJ/kg and timber 14 MJ/kg. Macadamia shells are right up there with a rating of 21 MJ/kg, while the kernel is 35 MJ/kg."

Suncoast Gold Macadamias had a growing waste stream with high fuel potential and Ergon Energy was looking for a source of renewable energy. It seemed like a perfect partnership. However, it would be wrong to suggest that macadamia shells are the perfect fuel.

"The energy content of the shells is quite variable," says Blundell. "This is largely due to fragments of the nut kernel still being attached to the shell, although how much is left varies quite a bit. The kernel is oil rich and energy dense. When it burns it provides a lot of heat so we needed a boiler that was self-regulating and could cope with a fuel with a variable energy content. Ergon Energy's engineers designed a mini power plant with a 6 megawatt high pressure steam boiler and a 1.5 megawatt turbine."

Gerard Macauley, an engineer with Ergon Energy says the boiler needed to be "smart". "It is equipped with a computer control that can sense when the energy is higher or lower based on the heat and the pressure generated inside," he says. "It then makes appropriate adjustments. We're proud of what we've developed. It's the first of its kind of this size anywhere in the world."

The boiler produces high-pressure steam at temperatures of between 380 to 400 degrees. The steam is put through a conventional steam turbine, a small 1.5 megawatt unit, which generates the electricity. The steam is then cooled and returned into the feed system for the boiler.

The plant includes a 400 tonne capacity storage silo – the world's biggest macadamia shell silo – that cost $A3 million. During any hour of operation the plant can convert 1680 kilograms of waste shell into 1.5 megawatts of electricity. This is enough renewable energy or 'green electricity' to power more than 1200 homes for a year. It will initially process 5000 tonnes of shell per year with the expectation that it might double this by 2005, which is likely to double the electricity output. About 20 per cent of the electricity will be used to power Suncoast Gold Macadamia's factory, with the rest being exported into the national electricity grid.

This is enough renewable energy or 'green electricity' to power more than 1200 homes for a year

"Under the arrangement Suncoast Gold Macadamias gives us the waste stream which is the shell and we give them back energy for free," Ergon Energy's retail group general manager Kate Skilleter says. "They also get the steam which they use in their processing

Ergon Energy finds a cracking solution to sourcing energy from renewable sources.

plant. Above a certain amount we pay them a royalty. They get their power needs for free and we can sell some of the environmentally friendly energy left over into the national electricity market."

Skilleter adds that by 2005, the facility is expected to double its power output with more than 150 growers throughout Queensland and northern New South Wales supplying Suncoast Gold Macadamias annually with about 10,000 tonnes of macadamias.

"That is equivalent in volume to filling about five Olympic-size swimming pools," she says. "Over the 20 years that this project is expected to run it is anticipated the facility will reduce greenhouse gases by about 9500 tonnes annually – that's like taking more than 2000 cars off the road."

Companies in Japan, Europe, South Africa and the United States are already showing interest in what has been achieved so far. Meanwhile, Ergon Energy is looking for new opportunities across a range of other industries including sugar, peanut, timber, wheat and grain processing where waste streams could generate heat, electricity and revenue.

David Salt is a science writer based in Canberra and the former editor of Newton.

In a nutshell

Who: Ergon Energy with Suncoast Gold Macadamias.
Where: Gympie, Queensland.
What: Converts 5000-10,000 tonnes of macadamia shells each year into "green" electricity (via combustion and steam generation) to power 1200 to 2400 homes.
Greenhouse savings: Reduce greenhouse gases by about 9500 tonnes a year, which is the equivalent of taking more than 2000 cars off the roads.

By Gayle Bryant

Under the sea

→ The Great Barrier Reef is set to enchant marine-lovers for generations thanks to a team of scientists dedicated to keeping it alive.

One of the natural wonders of the world is Australia's Great Barrier Reef. No one can doubt this magical area off the Queensland coast is an awe-inspiring feat of nature. However, the facade of the blue sea, exotic sea-life and beautiful coral reef hides an area that is increasingly threatened by a range of human activities such as global warming, prolonged and increasing fishing pressure and runoff from silts and fertilisers from the land.

The Great Barrier Reef: beautiful one day, at risk the next.

To preserve this natural gem, a team of scientists – with diverse expertise ranging from climatology and oceanography to ecology and biotechnology – work on innovative, world-class scientific and technological research aimed at keeping the reef alive.

Headquartered at Cape Ferguson, 50km from Townsville in north Queensland, the Australian Institute of Marine Science (AIMS) was established in 1972 by the Australian Government to generate and transfer the knowledge needed for the sustainable use and protection of the marine environment. With experts in everything from box jellyfish and plankton to tides and seafood toxins, AIMS is home to leading marine science research.

Heating up the reef

AIMS supports research into the links between physics, chemistry and biology to predict the impact of climate and climate change on marine ecosystems. In 1998, a global episode of coral bleaching (loss of colour in coral) highlighted that many reef corals live near their upper thermal limits. Where corals bleached, sea temperatures were hotter than normal and among the highest on record. Such patterns are consistent with predictions of climate change due to the burning of fossil fuels (the enhanced greenhouse effect), which create an urgent need to understand whether global warming threatens the viability of coral reefs.

Dr Janice Lough is a climatologist working with scientists on climate change and the impact on coral reefs. One of her projects is extracting proxy climate and environmental records from banding patterns in massive corals.

"To gain a better perspective and to understand climate change from more than 150 years ago, you need proxy records which can come from biological and geological sources affected by climate and leave a record of that influence," she explains.

Lough uses coral bands to see how coral has grown over time. "Corals contain annual growth bands similar to tree rings," she says. "Living corals can grow to seven metres high and may contain records covering several centuries – before the time of human impacts. You take a slice from a coral core and x-ray it to reveal annual density bands. These reveal a lot about how the coral was growing. If they are growing unusually slowly or there are breaks in growth then this might be a sign of stress." Massive corals are retrospective monitors of coral reefs.

Lough uses the results of her research to help explain the effects of processes such as coral bleaching on coral growth rates. "There have been two major coral bleaching events on the Great Barrier

AIMS' diverse research capabilities

■ AIMS researchers have built international reputations in three broad areas of science – marine biodiversity and conservation, coastal processes and marine biotechnology.

■ The two research vessels, RV Cape Ferguson and the RV Lady Basten, log an average of 250 days at sea each year, plying the tropical waters off Queensland, the Northern Territory and Western Australia.

■ The institute employs more than 160 staff, and satellite bases in Fremantle and Darwin supplement research at the main facility.

■ Within the three science groups there are 18 research projects running which address areas of relevance and importance to the nation.

Solutions to combat the effects of global warming are being researched by scientists at AIMS.

Reef in the past six years," she says. "Average water temperatures of the reef have warmed by about 0.4 degrees Celsius since the end of the 19th century – consistent with global warming of the tropical oceans due to the enhanced greenhouse effect."

Lough adds that the current rate of warming is outside the natural range of climate over at least the past millennium and tropical ocean waters are projected to warm by between 1-3 degrees Celsius by the end of this century.

"This is of serious concern for the health of the world's coral reefs especially when combined with regional-scale threats to coral reef health," she says.

The ecological bottom line

Another scientist whose focus is on coral – namely coral communities – is reef ecologist and principal scientist Dr Terry Done. He is involved with the long-term variability in reefs including the natural changes.

"I am now working on coral bleaching and am trying to document and understand what is going on from an ecological point of view," he says.

Done's research confirms Lough's findings that lower rates of global warming are better for coral than higher rates. "However, lower rates of global warming do not look like they are sufficient to stop the reefs deteriorating unless there is quite a bit of adaptation to higher temperatures," he explains.

In particular, Done's group is trying to learn how to be specific about global climate, especially its effect on coral bleaching. "We know that bleaching kills coral but we want to determine which places are more – and less – vulnerable," he says.

"If we can predict where the best places are for survival, then this information can be taken into account by reef managers."

Done says the research he has been involved with provides a more realistic understanding of the natural variability of the system. "There is no such thing as a standard reef; there is a lot of diversity within coral communities," he says. "I have also been involved with understanding the resilience in ecological systems and how they can bounce back from disturbances. They have a natural capacity to bounce back and we need to learn what circumstances lead to this resilience so we can then better manage the area in a way to bring those circumstances about."

He adds that one objective they are working towards is understanding what the ecological bottom line might be for conservation.

"This is a very important notion to capture," he says. "We don't just do descriptive science and hope the managers pick it up. We give them principles based on ecology which they then try to apply to their management to make sure the reef is maintained."

Taking maths to the ocean

Another project that aims to better understand the impact of human activities on coral reefs is being conducted by leading scientist and oceanographer Dr Eric Wolanski.

Wolanski believes science can help save coral reefs by providing land and marine resource managers with accurate and adequate data on key threats, specific indicators of reef resilience and science-based models with which to predict the impact of various decisions about land use and reef fisheries on reef health.

"We have developed a mathematical model for the Great Barrier Reef for which extensive physical and biological data are available," he says. "We wanted to model the region believed to be the most susceptible to anthropogenic impacts from land runoff."

He explains the model uses coral cover as a proxy measure of reef health. "Without human influences, coastal runoff degrades the reef in a zone whose width and degree of impact vary with latitude, with maximum damage in the Cairns region," he says. "The model predicts that with human activities on land, the zone of damage has already grown much larger than the natural state and will increase in size and intensity in the future unless the effects of human influences are curtailed."

Crucial connections

The model developed comprises 261 reefs in a 400-kilometre stretch that extends from Lizard Island in the north to the beautiful Whitsunday Islands in the south.

Wolanski says science has a crucial role to play in demonstrating the connections between land and reef ecosystems and the profound effects those connections can have.

"The model provides the ability to quantify the effects of various scenarios for the control of land use activities," he says. "It offers decision-makers and the public a science-based tool to decide what activities should be controlled on land and at sea in order to produce a desired state of heath for coral reefs."

Thousands of tourists visit the Great Barrier Reef each year. The research carried out by the scientists at AIMS goes a long way in ensuring that there will still be a reef to visit. ■

leaders SPEAK

Encouraging inclusiveness

The Australian Bureau of Statistics enjoys an excellent reputation for advanced thinking among its global colleagues. Sometimes it has been said that we are travelling ahead of the pack with our management thinking. Twenty-five years ago we began a process which was, at the time, considered unusual.

We wanted to run our organisation in a more business-like fashion. With this in mind we extended an invitation to our chief information officer and to our head of methodology to join our executive team.

At the time many of our sister organisations in other countries still viewed these people as "odd bods" who filled a vital if somewhat arcane role and often in a somewhat isolated fashion.

We abolished the thinking of separateness. The key to efficiency and innovation in delivering statistical information to the community lay in creating a unity between these functions and our business units.

We brought our specialists into our core business of producing statistics and a cross-pollination of ideas began to emerge.

Our top managers had always had an alignment with the business but not so with our vital specialists. How many organisations still make this mistake?

In the rapidly changing environment in which we work today, my suggestion is to review inclusiveness, seek cross-pollination of ideas and bring the "odd bods" in. It works!

DENNIS TREWIN
CHIEF STATISTICIAN
AUSTRALIAN BUREAU OF STATISTICS

All men by nature desire knowledge.

ARISTOTLE

ABIE

The education of innovation

There is much debate on whether entrepreneurship and the drive to achieve are innate qualities – or if these intangible traits can be learned. *ABIE* explores if innovators are born – or made.

Sponsored by Australia Post Macquarie Bank

By Mark Kestigian

Nurture not nature?

→ The ability to innovate is not just bestowed on a few lucky people. With the right approach, it is a process that can be learned.

Ask 100 people for a definition of innovation and you're likely to get 101 different answers. Ask these same 100 people on whether innovation is something that comes naturally or through nurturing and you'll wind up in the middle of a philosophical scrum that would make the Wallabies front row run for cover.

Amid the innovation information overload, however, several sizeable chunks of useful data have emerged, which not only reveal how and why it works, but offer ways in which organisations can pump up innovative principles into their daily business routines.

It may first prove useful to establish just what innovation is and what makes up its DNA.

One way of looking at innovation is to differentiate it from the use of this term as it pertains to someone involved in the arts. When someone is said to be truly "artistic" – see innovative or creative – it generally means they have come up with a great idea. In business, however, the floors and dumpsters of corporate labs are littered with good ideas.

According to Teresa Amabile of Harvard University's Business School, the trick is not only to be original, but "useful and actionable. It must somehow influence the way business gets done – by improving a product ... or by opening up a new way to approach a process," she said in a 1998 *Harvard Business Review* article titled "How to Kill Creativity".

> Innovation can be learned and follows rational, regular, systematic patterns

Many surveys paint bleak pictures on just how few employees get the innovation bug. Dr William Klemm, a professor of veterinary medicine at Texas A&M University, cites a survey of 1300 research and development scientists and engineers in the mid-1970s, which showed that 50 per cent of the engineers had no patents in the previous five years and 40 per cent of the junior scientists had not published anything in the previous three years. Another report by noted science historian Derek de Solla Price has shown that "scientific research papers come from a small elite whose number is calculated to be about the square root of the total population of scientists".

Why is getting good ideas from the lab bench and into the marketplace so difficult? For one thing, few people can agree on just where innovation comes from. Many experts claim that the innovation process is magical or mystical and therefore end up saying there is very little any leader – no matter how great or motivated – can do to foster it in the workplace.

Other authorities say there's no magic at all. In fact, they say there's no "Eureka"-style brainstorm that leads to technological breakthroughs. Rather, innovation is something that can be learned and follows rational, regular, systematic patterns – particularly in successful companies.

A Canadian-based management consultancy says that the biggest pitfall in fostering innovative workplace behaviour is too much success.

"Success encourages us to stick with present methods and to do the same things more effectively rather than invent new ways," argue members of the Streamlined Management Group. "Continued success can keep us from changing direction in spite of signs that indicate change is needed."

To overcome this hurdle, the Streamlined Group suggests companies should openly encourage innovators. Rather than just talk about it, give an employee a dinner for two or a new computer system for coming up with the best suggestion for new products or services or new ways in which to deliver them. Another suggestion is to make an innovation project part of each manager's performance review. (See examples of encouraging innovation in the workplace on page 181.)

There are many reasons for pursuing such bold, knowledge-based innovations but they are perhaps best summarised by findings from the Australian Bureau of Statistics (ABS). "While businesses that undertake technological innovation only represent about a quarter of all manufacturing businesses, they contribute about two-thirds of the total employment and three-quarters of the total turnover of all manufacturing businesses," reported a 1996-97 ABS study titled "Innovation in Manufacturing".

"On average, a business which is undertaking technological innovation has over five times as many employees as a business which doesn't undertake any technological innovation and almost 10 times as much turnover."

Fostering creativity

Aside from appointing innovation champions, how else does an organisation foster creativity in the workplace?

Dr Jim Fox, founder and CEO of Vision Systems Ltd, one of Australia's leading technology-based manufacturing companies, says there are many factors necessary to creating a sustainable business – particularly in the hurly-burly world of international technology manufacturing – but high on the list is people. "It's all about people – leadership, risk-taking, celebration of success, support for the failures. People build businesses and people trade – not companies, not governments," says Fox.

At an internal company level, Fox is equally as adamant on ensuring staff are well looked after. He says while financial incentives such as competitive salary packages and shareholder investment schemes are critical variables, these factors alone won't keep good people forever. "There must be job diversity – new challenges coming up regularly that keep

A questioning mind should be valued by modern management.

staff motivated and willing to learn new disciplines." He cites Vision's research and development engine room – Invetech – as a prime testament. Invetech has completed about 5000 different projects over the past 20 years, ranging from bicycle crash helmets to complex cancer detection systems and the beds for the Sydney Olympic Village.

Another person who has spent many years looking at how best to get good ideas out of laboratories and into the marketplace is Dr Greg Smith. A Melburnian, Smith spent nearly 20 years in the US and Europe working for AlliedSignal (now Honeywell), Alcoa and Exxon Chemical Company before returning in 1999. He now manages the SciVentures Pre-Seed Fund, part of the Commonwealth's Backing Australia's Ability program, aimed at providing funding for promising start-up companies emanating mainly from universities or public sector groups like the CSIRO.

"Innovation is a dynamic process or attitude highly sought after by many companies – the end products of which generally end up being minor or incremental changes," Smith explains.

"While these are fine for maintaining their current customer base and market share, the sort of change that venture capitalists, governments and company directors are seeking, are ones that create major breakthroughs or what are known as disruptive technologies."

Smith cites as a prime example the advent of the internet overtaking faxes and snail mail as people's preferred way of business communication.

"These disruptive technologies represent the Holy Grail, because they give rise to a range of new products and services," he adds.

While most employers talk the talk when it comes to extolling the virtues of disruptive technologies, few walk the walk, preferring to stay well and truly in the incremental change comfort zone.

Tom Peters, who skyrocketed to fame aboard the best-selling *In Search of Excellence*, has told seminar attendees about a flight attendant for American Airlines who suggested reducing the number of olives in on-flight martinis from two to one as a major cost-saving exercise.

"The idea produced a surprisingly ample saving of $US42,000 ($A57,200) a year, but obviously the airline couldn't cut costs the next year by further reducing the number of olives, so it was merely a good idea with an exceedingly short-term benefit," Peters declared.

The management change guru went on to say that lack of passion and/or belief in one's capability as major impediments to achieving genuine innovation. "When you say, 'My job stinks' or 'My boss stinks' or 'My company stinks,' and you stick around, guess what? You stink. You don't want that on your tombstone," said Peters.

Innovation and incompetence

Another observer, Dr Carol Steiner, the principal of Jitter Philosophical Services in Melbourne, puts it more succinctly: "If you want to be truly innovative – be incompetent."

Before anyone starts reaching for the "What the…?" button, step back and consider her definition of incompetence. It's based on research conducted by leading philosophers and change agents who argue that most people become cosily ensconced in their particular field of endeavour, preferring to maintain the beliefs and values held within that profession – be it dentist, doctor, baker or candlestick maker.

"To be incompetent then does not mean to be unskilled or careless. Rather, it means you're less committed to your profession's party line, choosing instead to question and doubt it and to work outside the square. To be incompetent means to see oneself as a unique and free individual rather than to accept a discipline-defined role and identity," Steiner argues.

> Before scientists can abandon conformity and embrace incompetence, they need courage – the courage to be individuals rather than pack followers

"However, before scientists can abandon conformity and embrace incompetence, they need courage – the courage to be individuals rather than pack followers. The obvious reason that non-conformity is rare among scientists is that science demands conformity with its paradigms," she says.

"Scientists who do not conform lose their professional credibility and status and can be dismissed as bad scientists or, even worse, as cranks and crackpots. Modern management theory now suggests that organisational peace and such conformist behaviour may be undesirable. It is now understood that conflict can be productive, creativity breeds innovation and bureaucratised, hierarchical organisations are less flexible, less likely to be innovative and less likely to empower staff with the courage to be incompetent."

In the end, Steiner says we should follow the lead of famous mathematician Warren Weaver, who said: "What every creative organisation needs is one damn fool who doesn't know what you can't do. That damn fool is someone who cannot be limited by competence." ■

Mark Kestigian is a journalist and corporate communications consultant.

Innovation can be taught

Encouraging innovation

Progressive companies that want to foster an innovative culture have structured programs in place to facilitate this process. For one local company, innovation is as easy as i-2-B.

Getty Images

i-2-B: Three easy steps to surfing the innovation wave

Taking ideas to business

That's the name of a program initiated by Carter Holt Harvey (CHH) – one of Australasia's largest forest products companies – to encourage employees to generate ideas that may lead to new products and services.

As part of the company's New Ventures division, which is the venture capitalist and caretaker of CHH's investment portfolio, the i2B (Ideas to Business) program generated about 500 ideas in the year 2000 – leaping to more than 1500 ideas from CHH's 11000 employees one year later.

According to CHH, 55 of these ideas were turned into business cases, of which 10 ultimately underwent a detailed due diligence process prior to investment from the company.

Climate was the first i2B business to become established. It's based around a powder-coating process for medium-density fibreboard and plywood and is believed to be the first of its kind in Australasia. The paint-like Climate surface has the colour and design flexibility of paint with the durability and strength of melamine.

An artful partnership

While oil and water may not mix, Woodside Energy Ltd – Australia's largest independent oil and gas exploration and production company – proves oil and the arts blend quite nicely.

Indeed, since teaming up with the Perth-based AWESOME International Arts Festival in 2001, Woodside has brought art to life – and not simply as part of a corporate collection.

A photographic montage was created by the children of Woodside employees with the help of an artist-in-residence. The kids were asked to create a visual representation of their ideas about the company. The result was the creation of an image that has been adopted by Woodside as representative of its activities.

This led to a second project – developing site-specific artworks in the company's finance department. This time, staff members worked with the AWESOME artists-in-residence transforming five sections of the department into creative working environments.

Staff can appreciate and interact with the artworks – for example, artist Stuart Clipston's large-scale magnetic rubber shapes can be moved around on the wall like fridge magnets. A staff member commented: "Having the work continuously present in our workspaces is a permanent reminder of the positive side of the project – reinforced by the fact that we are 'part owners' of the works."

The unique collaboration gained further recognition by being named a finalist in last year's national Australian Business Arts Foundation Awards. The judges said: "The Woodside-AWESOME partnership is an unusual, innovative and creative partnership, leading to cultural change within the business."

Innovation in education

The world's richest businessman regularly travels the globe preaching to others about how to turn their organisations into world beating companies.

One place global moguls like Bill Gates have never visited, however, is a small academic institution in Sydney's West – Noumea Public School – which was classified as one of New South Wales' disadvantaged schools.

Big business solutions

In tackling this challenge, the school's principal, Jenny Lewis, used a bit of lateral thinking and figured that some of the principles employed by leading business organisations could be applied to Noumea's day-to-day operation.

"As I looked at different approaches to school reform, I found a lot of answers were coming from large companies," explains Lewis, who was identified as one of the top 100 smart thinkers in Australia by *The Bulletin* magazine in 2003.

"I've always been interested in new research. As I looked at different approaches to school reform, I found a lot of answers were coming from the business community. I was interested in companies like IBM and Microsoft, at how they turned their organisations around. At the time I was going into a leadership position [Lewis was appointed principal in 1994] there were a lot of 'good-to-great' books being written about leaders who had significantly changed their organisations.

"I did a lot of research in that area and used a lot of the best practice principles I was reading about to guide the school into changing in many radical ways."

Lewis' bold plan revamp included the development of an innovative knowledge management system which contained comprehensive student performance data.

"We can interpret up-to-date – sometimes daily – data on a student. It's like what a doctor may do for patients. We needed information that would enable us to make intelligent decisions about a child's learning for that day or week."

The system was recognised in 2003 by the actKM, a national learning community dedicated to building understanding about public sector knowledge, with a platinum award for development of a sound organisational culture.

"Facilitating change is not hard. But it involves keeping up with the latest research on learning organisations and being able to share that with the community and staff."

Getty Images

Facilitating change is easy

The expert at anything was once a beginner.

ABBY HAYES

By Dr Colin Taylor

Running the science race

→ Excellence among inquiring young minds is celebrated at the annual International Science Olympiads.

Science and technology have an important and increasing role in every aspect of our lives, whether it is with the environment, economy or leisure activities. Encouraging the pursuit of scientific knowledge is crucial if we are to meet the challenges that lay ahead. One way to do this is to ensure students of today are given the opportunity to compete in science initiatives.

Competition brings out the best in people and competing on a scientific level is no different. One event that aims to challenge today's students in the field of sciences is the International Science Olympiads – otherwise known as the Olympic Games of the Mind.

The International Science Olympiads began more than 30 years ago as a formal test between government education authorities behind the Iron Curtain.

Since these serious beginnings, the Science Olympiads have developed into vibrant competitions involving around 80 countries fielding teams to at least one Olympiad.

Along with an evolving academic program with a constantly rising standard of competition, the Olympiads celebrate the power and potential of youth and the ability of science to build bridges between nations.

The best of the best

Unlike the sports-based Olympic Games, the International Science Olympiads are held every year and are open to students who have not yet enrolled in university. Typically they are Year 11 and 12 students.

Olympiads are held in three areas of science: biology, chemistry and physics. They are independent of each other and held under the auspices of UNESCO. The host nation for each Olympiad changes each year.

Each Olympiad lasts for one week and consists of a series of theoretical and practical examinations which test knowledge and skill in all aspects of the relevant discipline. However, as only the best from each country can compete, there is a need to identify and prepare the science students. Usually this is carried out through local programs. Australia's program is known as the Rio Tinto Australian Science Olympiads (RTASO).

The RTASO is an independent non-profit organisation that is supported by industry, government, schools, teachers and students throughout Australia. It has a long-term partnership with Rio Tinto and is supported by the Department of Science, Education and Training. Support is also received from many businesses, institutions and professional societies.

The aim is to allow every participant to learn in a fun, challenging and rewarding manner.

RTASO provides students throughout the nation with opportunities to enhance their knowledge, understanding and skill in biology, chemistry and physics. It conducts programs in these areas and organises Australia's presence at the International Olympiads.

The overall aim is to provide high-quality programs for students and teachers and to promote science in a positive and effective manner. A key emphasis of the RTASO activities is not what an individual achieves but whether or not they have done their best in developing and realising their potential.

A breeding ground for achievers

The alumni of the Australian Olympiad program boasts a Young Australian of the Year, several Rhodes Scholars and countless successful postgraduates from some of the most famous and prestigious universities in the world. The students that take part are seriously motivated people – highly driven, highly able, ambitious, and competitive. They have their eyes firmly fixed on an international gold medal.

Each year the students have the chance to travel to Canberra to compete at Australia's Science Boot-Camp – an intensive learning experience where they study two years' worth of school and university lessons in a little over two weeks.

The potential rewards of attending are huge because the best of the best will earn a place in Australia's Olympic teams at the International Science Olympiads.

Stacey Peters, a student who won a bronze medal in biology at the International Biology Olympiad in Belarus completed her final year of secondary school in Melbourne in 2003.

Her comments are typical of many Olympians. "I have accelerated my knowledge, developed invaluable study skills, can work under intense pressure, and am able to sit ridiculously long examinations," she says.

"This is only the practical side of my experience. I have also travelled to a country I would not otherwise have thought of going to, made numerous life-long memories and been given the amazing opportunity to meet students from all over the world with similar aspirations."

Teachers of science Olympians agree that, when one student stands among the best in the world, that student's network is carried along with them. Achievement becomes the achievements of classmates, teachers, schools and regions.

Michael Bishop of Sydney Grammar School has taught several Olympians. He says the benefits of Olympiad participation extend beyond the individual into the classroom.

The students that take part are seriously motivated people – highly driven, highly able, ambitious, and competitive

Getty Images

Heading straight for success: the Olympiads motivate students to be the best.

When one student stands among the best in the world that student's network is carried along with them

Bishop says, "In my experience, the classmates of Olympiad participants appreciate the success of their peers and the existence of what they see as a source of academic authority.

"This changes the dynamic of the classroom in interesting and positive ways. Teachers and classmates alike are stimulated by the presence of more sophisticated pupils and the interesting questions they raise in class." ■

Dr Colin Taylor is executive director of Rio Tinto Australian Science Olympiads.

Olympic qualification

The Australian Olympiad Program was first established in 1987 with the Physics Program. This was followed by Chemistry in 1988 and Biology in 1992. From the humble beginnings and hard work of many people, the Rio Tinto Australian Science Olympiads (RTASO) involves many thousands of students and hundreds of teachers and schools each year in the Olympiad Programs. It works very closely with the Australian Mathematical Olympiad Committee, which conducts a similar program for mathematics.

To qualify for the Australian Olympiad team, students need to pass several levels. The first level is at the school level where a National Qualifying Examination is conducted. If successful, students can then work up to a national level – Scholar Training Program – and then international level – the Australian Olympiad Team.

Every secondary school in Australia can nominate and support students for entry into the National Qualifying Examinations. From each National Qualifying Examination, around 20 candidates are invited to become scholars.

The Scholar Training Programs are accelerated programs of learning that consist of a series of assignments and tutorials by correspondence and a 17-day residential school in Canberra. They aim to enhance knowledge and skill as well as provide students with skills in time-management, communication and effective study.

By Tracey Evans

It's all in the delivery

→ In a world full of struggling postal agencies, Australia Post has successfully evolved to meet the changing demands of its customers.

Cutting corners: the national mail carrier has found high-tech ways to lift efficiency.

Australia Post achieved its best delivery performance ever with a 96.5 per cent early or on-time delivery result in the year to June 2003. This was achieved while maintaining one of the cheapest standard postage rates in the world at 50 cents.

Its after-tax profit continues to grow at $A330.8 million (last year it was $297 million). Its return on assets (14 per cent) would be envied by virtually any commercial organisation in the world.

In the face of extraordinary change in methods of communication, Australia Post has achieved these results by bringing about its own revolution in mail delivery. The demand for postal services continues, thanks largely to direct mail companies anxious to get their messages out to households and businesses around the country.

A $600 million investment in new technology has led to Australia Post now having one of the best bulk-mail delivery and sorting systems in the world. Jim Marshall, general manager of Australia

Post's Mail and Networks Division, says the vision for the organisation had been mapped out for a long time because it was such a dramatic change and had a substantial gestation period.

"Machines had to be designed, land found and purchased, and buildings built," he says. "An intense campaign to educate staff and customers was also necessary to ensure a smooth transition."

The project to transform the mail network began in 1995 with an investigation of what was happening in other countries.

"We went to the United States and Europe and cherry-picked what we saw because we felt if we simply implemented what others were using it wouldn't necessarily work for us," says Marshall.

Back home, the brief was to design a system that would be efficient as well as financially viable, using the best technology available. In one of the most significant breakthroughs of what became known as FuturePost, the project team came up with a bold plan to assign a unique number (known as a delivery point identifier or DPID) to every delivery address in Australia as a way of improving the accuracy of mail sorting and delivery.

The result is one of the biggest and most important databases in the country. The postal address file (PAF), as it is known, contains more than 9.6 million addresses.

Benefits of barcodes

In a major departure from the old methods of mail sorting, machines read the DPID and sort the mail into piles for each postie. In the past, letters were sorted into postcodes then manually sorted into smaller groups. With as many as 5000 addresses in each postcode, the reduction in time and resources has been huge.

The new mail system, which is used by organisations that send bulk mail, is based on barcodes. Businesses, including direct mail companies, use software to match their own mailing lists against the postal address file and assign the DPID for each address in their records. Then, as each address is printed on an envelope, the DPID prints as a barcode and it is this barcode that is read by Australia Post's sorting machines.

While the barcodes and DPIDs are fundamental to the efficiency of the new system, the long planning and implementation stages have yielded other benefits.

"What we were doing was fundamentally changing the way the mail industry operated," says Marshall. "It's not just about barcoding, it went right down deep into addressing rules and accuracy, and addressing profiles and databases. Out of that has grown a much more accurate set of disciplines about how the industry operates and provides for potential development and growth in the industry."

It was a serious rethink for Australia Post about how it ran its business, how it organised the mail network, dealt with customers, the value customers might add to the process as well as pricing issues.

The machinery that runs the new system is impressive, expensive and massive. Australia Post built two large facilities, one in the Melbourne suburb of Dandenong, the other in Sydney's Strathfield, to house the biggest pieces of equipment – flat mail optical character readers. Stretching to 80 metres, the machines are almost 10 metres longer than a Boeing 747. Similar, but smaller, machines were also purchased for the other State capitals.

Using more than 15,000 parts and complex technology that combines the use of light and sound, the machines measure and read addresses of larger, non-standard articles that are passed through them.

> What we were doing was fundamentally changing the way the mail industry operated

The flat mail optical character readers may be the most impressive machines, but they are just part of a long list of new equipment designed and adapted by Australia Post to meet its needs.

"We also introduced a video encoding capability that helps sort mail for non-business customers," says Gary Lee, Australia Post's group manager, Letters.

The video coding units take over the identification process when the optical character readers cannot interpret an address (on non-barcoded mail) or a barcode (on mail incorrectly barcoded).

"The units take an image of the front of the envelope", says Lee. "People sitting in front of video screens check the image to try to make sense of the address."

We came through that whole experience with minimal industrial disruption and a stronger relationship with our workforce than we had before we started

The new equipment has seen a shift in emphasis in Australia Post to a more technical workforce. "It's a very different network today because so much of it is information-controlled. It's about passing information through the network and having that information available to people," says Marshall. "Whereas in the old days people in the smaller centres were able to manage by line of sight, now it's the trading of information and images. It's a much more complex set of processes but a much more productive set of processes.

"The operators and the managers of the new centres are now far more technically savvy people than they were only a few years ago – they operate in a very different world," Marshall says.

Posting a profit

Some of Australia Post's highlights for the 2002-03 financial year include:

- After-tax profit of $330.8 million.
- Return on assets of 14 per cent.
- Handled 5.26 billion mail articles and serviced 9.4 million delivery points.
- Completed the construction of new parcel processing centres in Melbourne and Sydney.
- Expanded business banking services through giroPost to 494 Post outlets (from 320 last year).
- Commissioned three of the eight new machines that will automate the processing of large letters.
- Customer Service Award at the 2003 World Mail Awards in Rome.
- Received rating as the number one company for employee management in the 2002 Good Reputation Index.

Delivering the mail has become easier for Australia Post and its customers.

Communication is key

Communicating these massive changes to customers and staff during the implementation was an important part of the project. Australia Post began discussing the project with both staff and customers three years before the system began in 1999.

"The project affected the work of more than 20,000 people," says Marshall. "Every employee in our mail processing centres and every postie in the capital cities and most provincial centres had their work changed, in many cases quite fundamentally, by what we did."

More than 8000 mail processing and technical staff underwent extensive education and training to learn the new technology.

There were no forced retrenchments. Some 2500 employees chose to take voluntary redundancy packages, says Marshall. "We came through that whole experience with minimal industrial disruption and a stronger relationship with our workforce than we had before we started," he says.

For business and the direct mail industry, the new system was good news. "The new sorting equipment enabled us to allow more innovative designs on envelopes," says Lee. "In the past, our machines required a certain amount of white space around the address. But with the new machines we only have to read the barcode, so more colour and graphics are now possible on envelopes."

As a further enticement, bulk mail customers were offered a 10 per cent discount if they used the barcoding system. This has paid off with more than 80 per cent of business bulk mail now using barcodes, the fastest uptake of any country where barcoding has been introduced.

Lee attributes the success to both the discount and the widespread consultation. "It took a lot of effort," he says. "We worked with mail houses, because most business bulk mail goes through them; we worked with industry and software vendors to make sure that appropriate address-matching software would be available in time; and we worked with businesses to fine-tune the system to meet their needs." ■

I have not failed. I've just found 10,000 ways that don't work.

THOMAS ALVA EDISON

By Gayle Bryant

Picture perfect

→ A digital imaging research organisation in Sydney is developing cutting edge technology which is being used around the world.

The issue of security is one facing businesses all over the world and a little-known Australian research and development organisation is helping them to beat the problem.

The organisation, based in Sydney's North Ryde, is Canon Information Systems Research Australia (CISRA) – the Japanese giant Canon Inc's Australian research arm. Established in 1990, many of its technologies are finding their way into the products consumers and businesses use every day.

In 2002, CISRA initiated collaboration with Canon Australia to develop market-driven solutions that can be developed into products for the global Canon group. It is now realising the first product from that alliance, focusing on video surveillance systems.

CISRA has developed software – the Canon Network Video Recorder VK-64 – for video surveillance using Canon's web-cams. Canon's surveillance cameras enable the capture of clear images in low-light conditions. The software allows users to record live video in various modes from up to 64 networked cameras.

The launch of Canon's new surveillance camera and network video recording software marks the company's full-fledged entry into the video surveillance device market, responding to increasing consumer demand for anti-crime and general surveillance equipment.

According to Phil Robertson, a CISRA director and general manager for its solutions division, this is an excellent example of developing a product in Australia based on leading-edge customer requirements and then taking it to the global market through Canon's global distribution channels.

Foolproof films: new technologies turn video clips into home movies.

"CISRA is already starting work on the next version and being asked if it will undertake specific customisations for global customers," he says. "This will bring more business back into Australian research and development."

Canon Inc (51 per cent) and Canon Australia (49 per cent) jointly own CISRA. The company was formed in 1990 following Canon's purchase of a digital imaging company called Integrated Arts which specialised in animation and rendering.

CISRA's core competency is in digital imaging technology. It is responsible for the software that comes with every Canon digital camera and bubble-jet printer and more than 20 million copies of its software have been distributed worldwide.

The launch of Canon's new surveillance camera and network video recording software marks the company's full-fledged entry into the video surveillance device market

The company's history of innovation is impressive. In 2001, it won the Australian Design of the Year award with its uiCard (user interface card). This novel technology enables easy internet access with a low-cost, hand-held remote control that can communicate with a TV set box or computer. Strategic discussions are underway with organisations within the interactive television industry for a commercial release of the technology.

People make the difference

The products and services that come out of CISRA are a result of the innovative people that work there. There are 250 staff based at its offices, an increase that has led to the organisation expanding over three floors.

CISRA runs a number of programs to ensure the innovative culture is nurtured within the organisation. Robertson says one way is to mix the new graduates with experienced staff.

"The graduates bring in their fresh ideas and enthusiasm and gain invaluable experience by working with our senior staff," he says. "Over time, we move the graduates to different projects so they can build their capabilities."

One project underway is building prototypes for new software applications. "This also builds up experience in understanding the importance of patents, both in filing patents to protect new techniques and in searching to make sure no one else has patented them previously," Robertson says. "After the prototypes are built, they are shown to colleagues and potential customers for feedback. Planning new software applications to add value to existing Canon products helps to develop a good grounding in all Canon products and also in thinking about how they can be extended with new features or services."

To ensure that the products developed by CISRA are not only innovative but also user-friendly to the end consumer, there is a separate user interface design team that works closely with the engineering staff on all projects that involve a user interface.

"The designers drive the look and feel of the user interface, run usability tests and ensure the user interaction is intuitive and consistent with the expectations of the user of the product," Robertson explains. "This is regardless of whether it is a consumer product such as a photo album application or a business product such as an office business application."

Patenting innovation

An area where CISRA is very proactive is with its patents. At the end of November 2003, it had been granted 127 US patents, 271 Australian patents and others in Europe and Japan. A total of 702 provisional patents have been filed in Australia.

"We file between 50 to 100 patents each year," Robertson says. "We need to protect what we are developing from companies that may copy our technologies. Patent protection is very important in our industry and an area where we spend a lot of time teaching graduates."

An internal program running within CISRA and known as 'Level 8' encourages staff to come up with ideas for new products and then to write proposals for why they should be taken up by the company. The ideas with the most potential are selected and moved into a pilot project stage. After further development, CISRA management decides whether to turn them into fully financed projects.

"Some of these ideas are incremental to projects we are already working on and others are for new products," Robertson says. "Several Level 8 ideas are now being developed for inclusion into Canon products."

CISRA is enjoying the fruits of its innovation with an increase in turnover. "In 2002, turnover was $A36.5 million and we expect it to increase to $40 million for the 2003 year," Robertson says.

Most of CISRA's revenues come from research and development contract fees although it also receives royalties from licensing its software applications. The company takes a small profit from all the projects it does for Canon and that money is put back into developing its own products.

The excellence of CISRA technology is evident by the awards it receives. Late in 2003, CISRA received a Canon President's Award for its Ultra Fast Rendering technology, which forms a core component of imageRUNNER iR C3200 digital copiers, the current generation of office business machines. Two awards received earlier in 2003 were awarded by Canon for CISRA's intellectual property and patents work. One was an award for a particular patent associated with its rendering technology and the other was for a process it has developed for improving and streamlining the patent writing and filing process.

CISRA has a strong affiliation with Australian universities, in particular with the University of New South Wales where it sponsors a number of prizes and also contracts research. In late 2003, CISRA brought out Canon's chief technology officer from Japan for meetings with several groups as part of a push to forge stronger links with academia and other research organisations in Australia.

Robertson says CISRA undertakes research and development in Australia at a scale that would not be possible without being part of a large international company with global marketing channels.

"While Australia is a good lead market with forward-thinking people, it is not big enough to support the amount of R&D required for developing advanced products and solutions," Robertson says. "We develop technologies and solutions here and these are integrated with Canon products and taken into the worldwide market by our parent company."

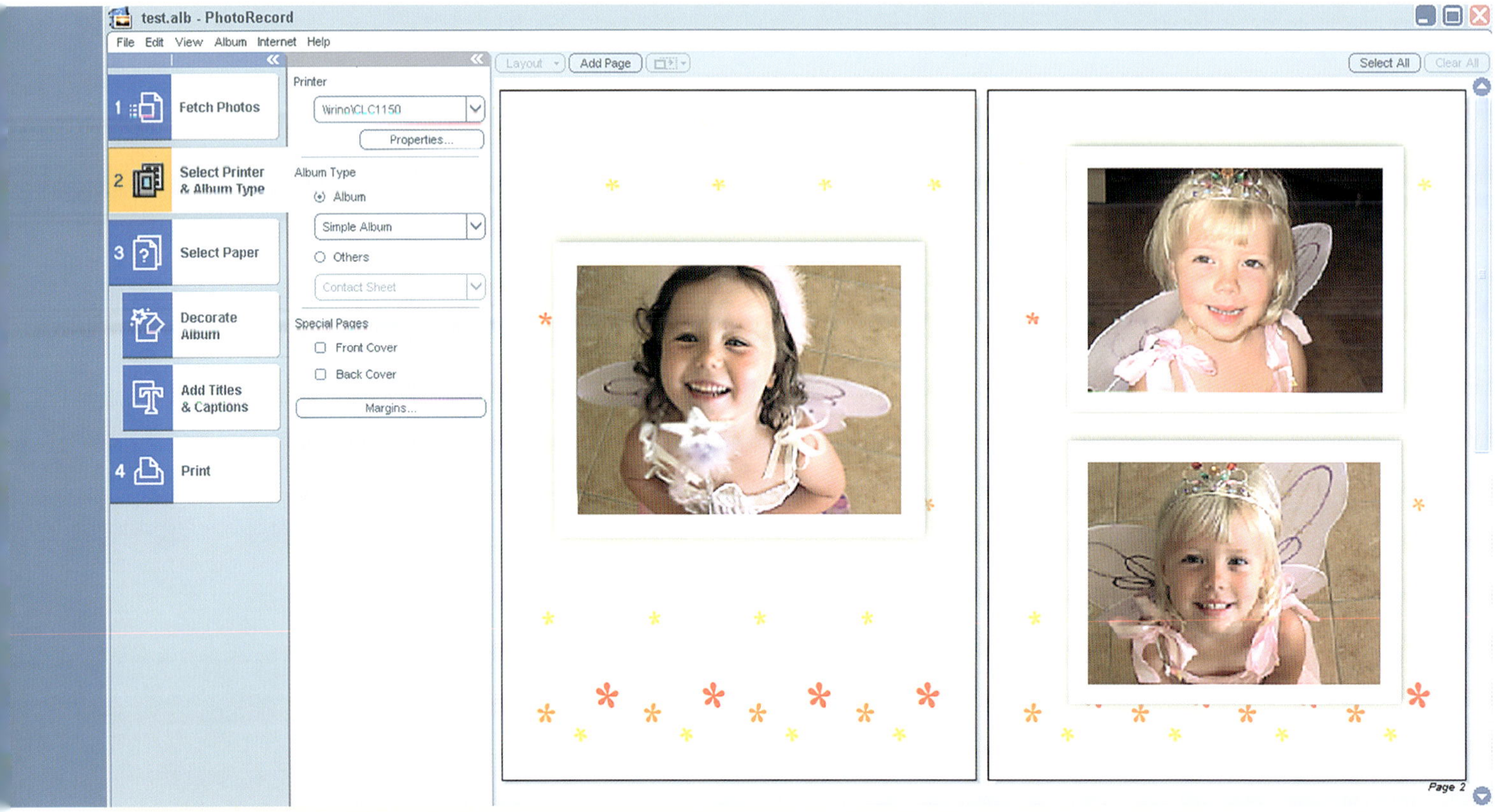

Photo flexibility

Many of the innovations developed by Canon Information Systems Research Australia (CISRA) have been commercialised and used in products that are sold all over the world.

According to Jim Metcalfe, a CISRA director and general manager for the Applied Technology Division, one of the most exciting technologies to come out of the group is Ultra Fast Rendering (UFR) which is now used in Canon's imageRUNNER iR C3200 series of digital copiers.

"The multifunction device was launched early this year and has been extremely successful," he says. "UFR is an embedded technology that provides high-quality and high-speed colour printing. This is world-beating rendering technology which enables much faster performance than conventional methods without the need for expensive hardware."

UFR incorporates an application-specific integrated circuit known as the SURF chip. The chip enables faster rendering of two-dimensional images which is a crucial function of a photocopier or printer. The technology enables the copier to print at the same peak speeds, whether the material being printed is black and white or in colour."

Canon is also applying UFR technology to other types of Canon printers including wide-format and some desktop lasers.

Photo flexibility

CISRA technology can also be found in Canon cameras. "Whether still or video all Canon cameras come standard with a Canon Solution Disk that features the software applications: ZoomBrowser, PhotoRecord, and PhotoStitch," explains Stuart Poignand, marketing manager with Canon Australia. "ZoomBrowser enables the user to store and manage digital photos, including downloading them from the camera. It has a simple and intuitive user-interface that incorporates a unique zooming browser."

PhotoRecord makes printing images and designing photo albums trouble-free. The program guides users through creating each photo album page, allowing them to select a layout pattern, insert captions anywhere on the page and choose the image size.

"You can also add backgrounds and borders and have multiple pictures of different sizes on the same page," Poignand says. "PhotoRecord makes printing effortless with auto-layout printing that starts automatically when you select 'print' from the ZoomBrowser menu bar."

Movie making

Canon digital video cameras sold in Australia come with the VideoPresenter software, developed by CISRA. VideoPresenter turns video clips from a digital video camera into professional looking home movies by providing a template that automatically creates a movie. Version 2.0 was launched in mid-2003 with an expanded set of templates. Poignand says with all CISRA innovations the aim is to make Canon customers' lives easier. "While Canon makes great hardware, the software developments coming out of CISRA are adding capabilities to Canon products that add so much more to the experience." ■

Hire good people and then leave them alone

Fostering creativity

At 3M, creating innovative products and services has always been a way of life. William L McKnight, former president of 3M, believed that one should "hire good people and then leave them alone". This approach has worked well for 3M for the past 100 years. 3M's culture has fostered creativity and given employees the freedom to take risks and try new ideas.

This environment has resulted in 3M's diverse technology platforms which form the basis of a broad range of products. However, in the increasingly competitive business arena, time to market and probability of commercial success become all the more important.

Therefore, the innovation process hinges on a delicate balance between organisational creativity and operational discipline.

DR FRED BRADNER
TECHNICAL DIRECTOR
3M AUSTRALIA

Without doubt, the greatest opportunities and challenges exist in addressing the route to export markets

leaders SPEAK

Clear vision

Australians have an international reputation for creativity across a wide spectrum, from entertainment and the arts to science. This has delivered many internationally significant inventions and discoveries. However, we are not so well known for capturing our innovations in the development of growing, branded, internationally-based businesses.

Yet businesses based primarily on the export of high-value products and services will deliver the greatest returns in our standard of living and in job creation. It is the ultimate test to sell products or services in remote markets off a domestic base of only 2 per cent of the world's economy. Without doubt, the greatest opportunities and challenges exist in addressing the route to export markets.

Our company, Vision Systems, has more than 85 per cent of its sales of high-value products and services in export markets, running at a rate of $A120 million a year. This has come from a zero start 15 years ago and growth has been driven by investment in research and development and route-to-market resources. Last year, our R&D expenditure exceeded our sales in Australia, given the international nature of our customer base.

For our country to capitalise fully on our public science spend and capability, the existence of companies that have a route to export markets and a vibrant industry-based R&D resource is an essential precursor. In this, there are a small number of stand-out companies that are models of the way forward. We need 200 more like them.

JIM FOX
MANAGING DIRECTOR
VISION SYSTEMS

ABIE

By Jane Breusch

Leveraging Australia's creativity

Global science company DuPont's operations in Australia are at the leading edge of science and continue its 200-year history of innovation.

According to Hutch Ranck, DuPont Australia's managing director, the scale of the majority of Australian enterprises gives Australia an edge in research and development.

"Because of the scale of Australian manufacturing, many researchers and technicians have to master several different disciplines," he says. "In large global organisations, technicians focus on one particular discipline for their whole career. As a result, Australians tend to be more creative because they have a multi-tasking, multi-discipline approach."

Ranck says it enables researchers to think "out of the box". "When people work in smaller organisations, they have broader experience and they see more possibilities. I think it gives Australia an edge."

A history of innovation

DuPont was founded in 1802, primarily as an explosives company. In 1902, its focus shifted to chemicals, materials and energy. Today the company delivers science-based solutions in areas as diverse as food and nutrition, healthcare, apparel, safety and security, electronics, construction and transportation. Its registered trade marks include Teflon, Kevlar, Tyvek, Corian and Nomex, to name a few.

The company's most recent focus is on sustainable growth – creating commercially successful products but with a smaller environmental footprint than existing technology.

A smaller footprint often means less waste. "We look for technology improvements that allow us to make more with less," says Ranck.

Often this is done by improving yield and plant uptime. "DuPont believes that if it works to achieve sustainable growth, it will be a better environmental steward," says Ranck. "But also our overall business will be less cyclical, grow faster, incorporate more knowledge content into our offerings and achieve greater productivity."

This knowledge content has manifested itself in the creation of DuPont Safety Resources and DuPont Consulting Services. Besides sharing proven safety management practices, DuPont runs seminars in decision and risk analysis and credit management for growth.

Building blocks of science

To achieve sustainable growth, DuPont has recognised the need for integrated science or an interface of different branches of science. "For us, integrated science means adding biology capabilities to our traditional strengths," Ranck says. "It's bringing on that additional capability and then looking for opportunities where more than one science comes together. "That's where we will find our future opportunities. We will look for disruptive, breakthrough-type technologies."

Disruptive technology has been the subject of a vast amount of study. It's known that if you are the number four entrant in a specific product space, for example, you will almost certainly be confined to a 7 per cent market share. So disruptive technology is one that leaps over the existing platform to overtake the competition. It puts all competitors at risk and it allows a company to make more money.

This is where DuPont's partnerships with other scientific research organisations come in. Ranck believes firmly in not only leveraging Australia's creativity but also combining and building

Powerful partnerships

The technology alliance between DuPont and CSIRO has generated more than 50 worldwide patents in just 10 years, providing huge possibilities for commercialisation of products driven by Australian research.

"Clean, green car paints, tyres with better rolling resistance and flocculants for water purification are just some of the applications of the technology that have come out of this highly successful research partnering," explains Leo Hyde, DuPont's R&D manager Australia and New Zealand. One of the technologies is RAFT – a revolutionary means of controlling the way polymers form and making them more stable. This gives researchers the ability to tailor-make polymers for many different applications, including automotive finishes and paints. The DuPont/CSIRO partnership is now the world leader in this field. "It controls the architecture of polymers very, very precisely," says Hyde.

Other targeted, existing and potential applications of the technology include:

- Pigment dispersants for high-performance inks. The technology means dispersants can be made more economically and improved versions can be developed.
- Higher performance rechargeable batteries.
- Compatibilising agents for plastics that presently cannot be mixed, to allow more plastics to be recycled.

Hyde says the partnership with CSIRO will undoubtedly lead to new commercial opportunities.

DuPont Australia's managing director Hutch Ranck says: "Our aim is to have DuPont synonymous with innovation and sustainability."

on developments in Australia and taking them worldwide.

DuPont's largest and most visible partnership is with the Massachusetts Institute of Technology in the area of bio-based materials.

DuPont believes that if it works to achieve sustainable growth, it will be a better environmental steward

In Australia, the company has a long-standing partnership with CSIRO as well as the University of Sydney and the University of Melbourne. "All knowledge does not reside in North America, contrary to what some Americans might like to believe," says Leo Hyde, R&D manager, Australia and New Zealand.

"We work with some of the best people in the world. CSIRO happens to be very strong in polymer research and the universities are strong in other areas."

Looking into the future

As it moves into its third century of operation DuPont, whose slogan is "The miracles of science", has several focus areas in which it will seek new developments.

They include biotechnology, new materials made from renewable resources, new foods with higher nutritional value, new technology such as electro-luminescent polymers and a continued focus on mature products such as automotive paints and ink. ■

Smaller-scale operations in Australia enable DuPont's researchers to think outside the box.

Getty Images. "The miracles of science" is a registered trade mark of the DuPont Company.

By Tracey Evans

Stuck on ideas

→ For more than 100 years, 3M has been developing products that have made our lives easier and safer.

They were the most damaging and destructive bushfires New South Wales had seen. For two weeks of January 1994, hundreds of fires were alight along the length of the coastline. More than 600,000 hectares and 200 homes were destroyed with a damage bill of more than $A100 million.

At the peak of the emergency, some 10,000 firefighters from around the country were battling the blazes in searing temperatures.

Meanwhile, in a 3M factory in the Sydney suburb of St Marys, chemist Ted Schaefer and his team were working around the clock to produce vast quantities of an important new weapon that would give the weary fire crews a fighting chance. After years of work in the laboratory, Schaefer had just perfected a chemical that would turn water into firefighting foam, dramatically increasing the effectiveness of fire tankers.

3M was working on plans to launch the new product, to be known as Fire-Brake™, but the bushfire emergency saw the company ramp up its production process to produce tonnes of the chemical months ahead of schedule. Fire-Brake was so successful that demand in Australia, New Zealand and Canada skyrocketed.

3M Australia was established in 1951. It was the first subsidiary established by 3M Company, which now operates subsidiaries in more than 60 countries. It is a diversified company that invents and markets a broad range of products that fall into 40 operating divisions.

The importance of Australian research and development to the larger 3M group is primarily in the fields of pharmaceuticals and healthcare. "We really are the healthcare R&D hub for Asia," says Fred Bradner, 3M Australia's technical director. "Although there are labs around Asia, most of the healthcare R&D expertise is here in Australia."

Tamper-proof passports

Another of 3M's fields of expertise, namely its Security Products Division, has been a key player in a worldwide breakthrough in passport security. The Department of Foreign Affairs and Trade asked 3M to come up with a way to prevent criminals tampering with passports. The now-retired 3M Australian scientist Graeme Mann came up with a technique to scan all information on a passport, including the colour photograph, into a clear adhesive layer that sticks the plastic cover onto the first page.

"If the adhesive layer is peeled off, the information is destroyed, and if an attempt is made to change the photograph, a torch light will reveal the tampering," says Bradner. The innovation is now being marketed around the world.

New ideas have been a hallmark of 3M's operations since it was established as the Minnesota Mining & Manufacturing Company in 1902. It took two years and some setbacks before the company's first product – sandpaper – was launched. Since then, the company's commitment to research and development has led to more than 50,000 new products being developed. In the past five years alone, 3M has committed $US5.3 billion ($A7.2 billion) to research and development. Sales for the 2002 financial year exceeded $US16 billion worldwide.

The Post-it® Note was a solution no-one realised was needed until it came along

Possibly one of the best known of all the 3M products, the Post-it® Note, was a solution no-one realised was needed until it came along. It almost failed to make it out of the laboratory, except for the tenacity of new product researcher Art Fry. Fry had been impressed with a "re-positionable" adhesive first discovered by 3M scientist Dr Spence Silver in 1968. Silver knew the adhesive was going to be useful for something, but wasn't sure what. He badgered various 3M product people for five years before Fry came to hear of it. Fry, annoyed that scraps of paper he used as page markers kept falling out of his church choir hymnbook, thought the special adhesive might be used to make a better bookmark. From there it was a short step to the Post-it® Note.

However, having found a problem for the solution, Fry then had

Product roll call

3M has fostered a culture of innovation since it was first established in 1902, producing more than 50,00 products over the past 100 years. Some of the most well-known and valuable inventions are:

1904	First sandpaper is made
1925	Scotch Masking Tape
1930	Scotch Cellophane Tape
1947	Scotch Magnetic Tape, the first commercially acceptable magnetic audio recording tape
1960	3M Steristrip wound closures are introduced, the first medical wound closure tape
1972	3M Data Cartridge technology revolutionises computer data storage
1980	Post-it Notes introduced
1985	First successful optical disks for information storage, video and audio reproduction
1991	3M Scotchshield Window film, offering shatter-, heat- and cold-resistant window protection
2001	A halon replacement alternative to hydrofluorocarbons and perfluorocarbons for extinguishing fires, and 3M Inflata-Pak air-cushion packaging.

Getty Images

Although 3M is synonymous with Post-it Notes, it has developed a diverse range of products including ones that combat bushfires and prevent passport fraud.

a job to convince the engineering and production people it was a feasible product. They were concerned that measurement and coating difficulties in producing the Post-it Note would be insurmountable and create too much waste. However, he wasn't thwarted, telling them: "If it were easy, then anyone could do it. If it really is as tough as you say, then 3M is the company that can do it." Twelve years after Silver developed his unusual adhesive, the Post-it® Note was launched.

While 3M might be best known for Post-it® Notes, sandpapers and tapes, it is a diverse operation with interests in seven major business sectors: healthcare; industrial, consumer and office; display and graphics; electro and communications; safety; security and protection services; and transportation. New products are launched every year in each of these areas by the 32 laboratories and 29 manufacturing operations around the world.

3M's early success was built on the concept of "patient money". Introduced by early investor (and later president) Lucius Ordway, patient money describes the long-term investment in an idea, technology or product that shows promise, even when others argue otherwise. Even more radical in a short-term-focused business arena is 3M's "15 per cent rule". Employees are encouraged to devote up to 15 per cent of their working hours to their own independent projects. One example of the benefit to 3M of this policy is the enduring product Scotch® Masking Tape.

Dick Drew, a young engineering school dropout, was ordered to stop work on an experiment to develop a tape for auto spray painters that would not lift paint when it was removed. Drew ignored the instruction and continued work on it until he found success. His certainty about achieving a result, despite management's lack of faith, led to the introduction of the 15 per cent rule in which technical employees could disagree with their bosses and continue to experiment.

Over its century of operations, 3M has been rewarded many times for its patient approach to innovation. ■

By Tim Treadgold

The complete package

→ Packaging isn't just a matter of luck. Ask Amcor what is involved with giving consumers an easier way to untwist a bottle top.

Few consumers pause to think about the bottle they're holding when having a drink on a hot day, or the screw cap on the wine they're about to pour. It's a pity because much innovation, planning and design has gone into each of those products.

Just how much thought is involved in making sure that the package holding the goods fulfils its function is only appreciated when someone realises that good design not only enhances appearance, it can also improve functionality, add to shelf life and help sell a product.

One company that understands the importance of packaging is Amcor, an Australian business which has gone global to claim a position among the world's top three packaging companies.

Just about everything handled by a consumer, from corrugated boxes and cartons to aluminium and steel cans, plastic and glass bottles, closures and flexible plastic packaging, falls under the umbrella of Amcor.

"Amcor has established market-leading positions across a number of its product segments," Amcor's managing director, Russell Jones, says. "We are the number one producer of PET (polyethylene terephthelete) containers and tobacco cartons globally, we are the market leader in European flexibles and we are the number one packaging company in Australasia."

Good design not only enhances appearance, it can also improve functionality, add to shelf life and help sell a product

Reaching the top is one part of the Amcor story, which was achieved by selling its original interests in paper manufacturing,

focusing on the packaging industry, and acquiring the German-based Schmalbach-Lubeca PET and Closures business in 2000. It was this deal which catapulted Amcor up the global packaging ladder and created a business which today employs 29,500 people around the world, posts annual sales of $A11 billion and operates in 42 countries.

However, having reached the top, the trick now for Amcor is staying there. To do that the business, which specialises in the humble art of packaging, operates one of Australia's biggest research and development programs. In 2002, for example, Amcor invested $106 million on product innovation and development.

Jones describes the heavy investment in research and development as "the Amcor process of product leadership and innovation" which involves "working with our customers in all stages of the process, including understanding their needs, packaging their products and helping with the commercialisation of a new idea".

Innovation identification

Within Amcor, the process has been dubbed PLI, short for Product Leadership and Innovation. This process is not something restricted to the upper echelons of management or a room full of boffins and marketing executives locked in deep thought. The company's PLI charter makes it clear that every employee – all 29,500 – can play a role in "identifying packaging issues and needs".

The PLI charter reads, in part: "As consumers we come into contact with packaging every day, and most of it works well. That is, it looks great, stores well, it opens easily, it closes again, and it keeps fresh. But sometimes there is packaging that does not work as well as it could. For example, beer twist crowns are painful to remove, cereal bags split during opening and certain packaging materials are not easy to recycle. Every time we encounter packaging not working at its best, that represents an opportunity for Amcor to apply PLI and develop an innovative solution."

Getty Images

Innovative design is the key to perfect packaging at Amcor.

In effect, Amcor's PLI process seeks to turn every one of its employees into a product tester, and a player in the first stage of the assessment procedure which it calls "needs identification". From this stage "solution development" is implemented, followed by a "review process".

Examples include the development of the ergo-grip PET bottle, push-pull caps for sports drink containers, wine bottle screw caps, pressed corrugated fibreboard baking trays, slimline aerosol cans, TearStick packets for instant coffee, Ezi-pour cement mix sacks and one-piece trays for mango farmers.

Amcor's annual report, a document which most companies use as a financial performance communications tool, is something different. Rather than being presented from page one by a chairman's statement and the managing director's critique of the past 12 months, shareholders find themselves reading about the PLI process.

From the start of the 56-page document, the 110,000 shareholders (the owners of the business) are instructed in how PLI works, and why it is becoming part of the company culture.

The document states: "Customers expect Amcor to constantly innovate and develop new products to meet their changing market needs." Pages of examples follow, with the constant theme being that Amcor will succeed only while it "anticipates the packaging needs of the world".

Some readers may find the approach corny, but this is the heart and soul of a business which has grown rapidly because it is thinking and innovating in its chosen field.

Consider these examples of innovation:

- A resealable flexible bag that is convenient, simple and avoids the need to transfer food from its original packaging to a secondary container once the product is opened. The solution is the Amcor TapeTop Resealable Flexible Bag which involves a patented manufacturing process that cuts and positions a tape on the bag with an adhesive not previously used for food applications.
- A foil piercing sports drink closure, which was developed after observing that many drinks have a foil cover to prolong shelf life but which needs to be removed before drinking. The solution is a built-in piercing mechanism which works when the spout is pressed down. When retracted, the contents are ready to drink.
- The TrayBon baking system, a patented baking tray made of corrugated fibreboard which fulfils three functions: the cake is baked in the tray, decorated and transported to customers, saving time, money and improving hygiene.

Perhaps the most important part of the PLI process is in the final review stage. It is one thing to identify a need, another to develop a solution, but the real value comes from reviewing how it works in the field. This is when Amcor extends its work with a customer to see if the bright idea, the innovation and the design are actually taken up by the customer. If it isn't, then it is back to the drawing board – which is what PLI is all about: constant questioning, constant observing and constant innovation. ■

By Nic Svenson

Lucrative linkages

→ Accessing new science and technology means this Australian chemical giant maintains a competitive edge.

Getty Images

Keep it fresh: products less than three years old account for roughly a third of sales in Orica's decorative paint division.

Maintaining market share means keeping up with cutting-edge science, according to Australian-owned chemical giant Orica. And it believes the best way to do that is to get involved.

"Getting access to new science means being able to ensure that, over the long term, we continue to have new products," says Roy Rose, Orica's Technology and Environment manager.

Orica forms linkages with the best people it can find

Rose is responsible for coordinating the company's overall approach to science and technology and he says it's very much product-driven. "In our decorative paints division, for example, 30 to 35 per cent of sales were of products less than three years old," he says. "We set some key performance indicators to measure sales of products less than three years old. If we're really trying to make sure our product market is being refreshed, I think it should be 30 to 40 per cent – and questions do get asked if numbers are less than we would like."

Orica's 2002-03 annual report states the company invested $A30 million into research and development, but Rose believes the company is dramatically understating its R&D spend. "What we're counting at the moment seems to be the cost of running our R&D labs," he says. "We're failing to report the amount of R&D in other areas, such as engineering, for example."

Orica products

- Chemicals for purifying drinking water and for swimming pools
- Paint, sealant and varnish
- Fertiliser
- Adhesives in MDF, particleboard and plywood
- Resin for fibreglass and laminates
- Commercial explosives and initiating systems for mining, quarrying and construction
- Industrial chemicals.

He says it is difficult to get this data because work occurs in so many places and the practitioners do not recognise that they are undertaking research and development work. "They say they're just solving the problem in the plant," Rose says.

Orica has plants around Australia and overseas making products for the mining sector, industrial and agricultural chemicals, and for hardware stores under brand names such as Dulux, Selleys, Berger and Cabot's.

Orica has consistently viewed technology as providing communities with a product or service as well as a social contribution, because the two go hand in hand.

Diverse idea sources

There is no one centralised research laboratory for product development: rather, each business has its own. "Also we have a strategic technology fund which Orica businesses bid for on a competitive basis," Rose explains. "These funds go to projects lasting five to 10 years, with an emphasis on new science and technology. These projects are carried out in partnership with CSIRO, a university, or a research organisation."

He says Orica often forms linkages with the best people and uses the science to turn out new products.

Patrick Largier, general manager of Orica ChlorAlkali, agrees: "We're strong on the D side and we work with partners who are strong on the R side," he says.

The development and commercialisation of MIEX (Magnetic Ion Exchange) resin is one example of such a partnership which began in the late 1980s and involved CSIRO.

'Organics', or dissolved organic carbon to use the technical term, means tiny particles of living, or once living, things such as algae and ash. It is important to get rid of these particles because they can clog up filters, become a breeding ground for bacteria, and react with disinfectants such as chlorine to produce harmful chemicals known in the trade as disinfectant by-products.

"Conventional methods struggle to take out this small stuff," Largier says.

Going global

Largier says the first large-scale plant opened in Australia two years ago and processes 110 million litres a day. Plants are under construction in the US and there is great interest in Europe and Japan.

"For 60 or 70 years we've been supplying chemicals for water treatment," Largier says. "MIEX was spawned by working with clients to solve their problems."

When Orica has a problem, they call in their experts. "Usually we have a target ahead of time, for example a new paint latex, and we need the science worked up," Roy Rose says. "A lot of the work on the colloid science behind latex binders in paint is done at the Key Centre for Polymer Colloids at the University of Sydney, which we support with an Australian Research Council Linkage grant."

> It's an emerging technology and we expect it to really take off

Orica's mining division carries out the predominant share of inhouse research and development. At laboratories in Kurri Kurri, New South Wales, and Denver, Colorado, scientists refine their mathematical models of blast control.

Rose explains the modern blasting business is less about the chemistry of explosives and more about detonation physics – that is, what really happens to the rocks when you push the button.

"It's what they call rock on ground," he said. "Do you want giant boulders or fist size lumps? And do you want them over here or over there?"

The industry is also moving away from old-fashioned detonation techniques made famous by Warner Bros cartoons: such as lighting a fuse or pushing a plunger. Nowadays electronic detonators, such as Orica's i-kon system, developed by Orica's technical team in Germany, are the popular choice of detonator because, as with comedy, timing is everything.

The i-kon blasting system is an electronic initiation system able to deliver high levels of accuracy and reliability in blast initiation. Firing times for each blast-hole can be programmed to a detonator in millisecond increments up to 15 seconds. Full two-way communication between detonators and control equipment means the shot firer knows each detonator is programmed and will fire.

Rose says higher precision means greater control and that equals safer blasting. "It's an emerging technology, and we expect it to really take off," he says.

While Rose believes most of the innovations to come will be in improving the formulation of existing products, Orica is not likely to stop supporting fundamental scientific research.

"Unless there's new science and technology coming, we won't have new products," he says. ■

How MIEX (Magnetic Ion Exchange) works

Imagine a channel through which water flows and is cleaned along the way: MIEX beads are in the first section, spending 10 minutes to half an hour being thoroughly mixed through the water before it reaches a settling area.

The beads are slightly magnetic so they stick together when there's no stirring keeping them apart, making them easy to recover. Most MIEX is fed back into the top of the channel; only 10 per cent or so is taken out and refreshed by rinsing it in brine. MIEX is a registered trade mark of Orica.

By Gayle Bryant

Too much information

→ An information management solutions company is developing smarter ways to store and manage today's data.

The truly paperless office has been an elusive concept for years and companies have turned to data storage and management experts for answers. Leading player EMC, which has evolved from a data storage company to offering comprehensive information management solutions, has used innovative marketing to capture a substantial share from its competitors.

It has also been documented in the business text *Radical Marketing From Harvard to Harley, Lessons from Ten that Broke the Rules and Made it Big*. Published in 1999 by HarperBusiness, the book discusses what the authors call "radical marketing" and shows how particular companies, which tend to be smaller in size, adopted approaches to product development and developing customer bases that were different to more traditional approaches.

EMC continues to use innovation to grow its market share. In April 2003, it launched Centera – the world's first content-addressed storage system, which offers a new way of storing and retrieving data.

An earlier survey EMC conducted found that 75 per cent of all the information in the world is "fixed content", that is data that is written once and never changed. This includes invoices, purchase orders, financial statements, archived emails and x-rays. While this information needs to be stored, it cannot be changed.

EMC invented the phrase "content-addressed storage" when it launched Centera. It refers to providing a digital fingerprint for a stored piece of information. This fingerprint ensures that only that piece of information is saved and no duplicates are made.

The take-up of Centera has been huge globally with many early adopters coming from large imaging organisations, such as banks and telecommunications companies. Clive Gold, EMC Australia's marketing director, says Centera provides online access to help with content retrieval for fixed content items.

Getty Images

[EMC has transformed the process of data storage and management.

Self-healing systems

"Centera has many innovative features such as the ability to self-heal, which means the solution continuously monitors to detect and repair soft errors," he says. "It also automatically reconfigures itself and replicates information if hardware failures occur."

EMC is now moving into a new market opportunity in storage, which is that of information life-cycle management. The idea behind this is that information should be managed throughout its entire life cycle – that is, from the cradle to the grave.

This concept is being helped with EMC's purchase of the company Legato Systems. The purchase allows EMC to fold in Legato's email archiving and information protection solutions to boost its own storage software lines.

"Through acquiring Legato we can assist companies more actively in managing their information," Gold says. "There are so many disparate systems in the marketplace and we are moving to being able to offer companies one tool which can encapsulate all the tools being used."

Gold adds that EMC is now competing in many markets. "We have the traditional hardware that goes into the storage market," he says. "We have the software that we run in conjunction with the hardware and we also offer open software that runs in any environment. Finally, we have a services business to help companies with their business issues. We view EMC's role as assisting people to maximise the use of their information while helping them to reduce the costs."

EMC is also moving into what it calls the mid-tier market. "With many of the products we bring to market, we have already carried out the research and development with higher-end customers," Gold says. "We are able to bring this technology and expertise into the mid-tier and this is a sort of 'trickle down' innovation. They are saying EMC is allowing them to now do what would have cost them five to 10 times as much if they had carried it out only two years ago." ■

By Tim Treadgold

Bugs, bacteria and BioHeap

→ The use of new strains of bacteria to revolutionise the mining process has opened up a wealth of opportunities for Titan Resources.

Revolutions, contrary to the popular view, can often take time. But when they succeed, change can be dramatic. That is the nature of a revolution in mineral processing being plotted by West Australian miner Titan Resources as it works on new strains of bacteria to extract base metals such as copper, nickel and zinc from difficult and low-grade ores.

The theory behind bacterial extraction of metals is not new. Goldminers have been using "bugs" to liberate gold from difficult ores for more than a decade. In essence, the bacteria "eats" or oxidises the sulphide component of the mineralised material to make it soluble in a mild sulphuric acid solution.

The process, which is generally confined to large, climate-controlled tanks, avoids what can be an expensive and pollution-generating step of roasting sulphur-rich ores.

Titan's bacterial breakthrough is in the use of bugs in the open, on low-grade ores which have simply been pushed into heaps, and in the development of new strains of bacteria which work on non-gold material, such as nickel and copper. Titan's revolution is called BioHeap.

Trials on nickel ore have proved the technical merit of the process to the point that we have been able to expand the work to mineralised material provided by other companies

"Our early work has been significant," says Titan chief executive Bill Ryan. "Trials on nickel ore have proved the technical merit of the process to the point that we have been able to expand the work to mineralised material provided by other companies. We have also been able to enter into an arrangement with one of China's biggest metal companies to assess the suitability of low-grade ore deposits in China for the use of BioHeap."

Evaluation of BioHeap by the Jinchuan Group in China is just one of a number of investigations Titan has underway into the bacterial process.

Samples from three Chinese nickel deposits have been sent to Australia for testing in Titan's Perth laboratories. Once that work is complete, it is likely that a large-scale test will take place in China, possibly on ore at the Hami nickel field where management is keen to boost nickel production.

"Work on nickel ores has produced good results but there are also indications that one of the classic ores of copper, that is chalcopyrite, is yielding particularly exciting results," Ryan says. "The copper industry, which is much bigger than nickel, may prove to be the best target market for BioHeap."

Technical work on the bacterial process has included trials with ore provided by WMC Resources from its Western Australia-based Leinster nickel mine, and on copper ore sent to Australia from mines in Mongolia and South America. "We have made excellent progress with the recovery of copper from chalcopyrite," says Ryan. "Interest in the process has grown to the point where several new parties have submitted copper ores for preliminary investigation."

Titan Resources' BioHeap has harnessed new strains of bacteria to revolutionise the extraction of valuable base metals from mined ores.

Bugs in the test bed

Titan is also working with ore samples from its own nickel projects. After successfully operating a BioHeap trial at its Radio Hill mine in the north of Western Australia, the company is investigating the use of BioHeap on new nickel assets it is planning to develop in the south of the State.

The aim is to mine nickel at the Armstrong, Zabel and McEwen projects, initially sending ore for conventional treatment at WMC's nickel smelter.

As these projects get underway and generate cash to sustain the company, trials will be conducted on lower-grade ores using a portable BioHeap test bed.

"We see the real rewards of BioHeap not just in creating a biological alternative to smelting," explains Ryan. "Rather, we feel it is in the ability of the process to go direct from the ore stage to a post-smelting product, by-passing several stages in the processing of minerals with the potential to convert low-grade and difficult ore bodies into commercially viable projects." ■

Since more than 90 per cent of pre-commercialisation projects fail, you have to learn to fail, cut your losses, encourage your people and move on

leaders SPEAK

Art and science do mix

I've worked for 25 years for DuPont both in Australia and Asia. The company's size, diversity and its foundation in science and research and development have made it a remarkable learning place. Now that so many more companies are actively engaging in research and development and commercialisation, and innovation has become a much more talked-about corporate topic, it is good to be able to draw on more than two decades of experience and share some thinking and some lessons learned. Some of them were learned thard way.

You can analyse corporate behaviour and innovation from any number of vantage points, as well as its result, whice hh is successful commercialisation. However, for convenience and clarity, I would break it into three units: first, the ingredients for innovation; second, what can kill innovation; and third, what innovative organisations consistently do.

Ingredients

Some ingredients seem quite obvious, but there are others that don't. First, you must have a core competency consistent with the area of innovation. If you don't have it then you need to acquire a company that does. We have learned that 10 per cent of the journey is in the invention and 90 per cent is in the commercialisation. You need an intimate knowledge of the technology and market space to be successful in the 90 per cent.

Second, you must have a supportive culture. This is not always easy, as it requires leadership tolerance and patience. I describe it as a "listening culture". I'd go even further: you need a mindset that almost celebrates failure.

This is a delicate balancing act, but not an exaggeration. A supportive culture can handle failure of a project and then move on. Since more than 90 per cent of pre-commercialisation projects fail, you have to learn to fail, cut your losses, encourage your people and move on.

As a leader you need to develop skills in risk and failure management. Managing failure may be one of the least understood skill sets. This skill set involves much questioning, not only of researchers but also of the market. Elegant technology that doesn't fit real customer needs is the biggest trap of all. Recognising this as a failure (even though the technology works) takes strong leadership and a market focus. Companies with a technology or product focus commercialise products that customers don't value. Over time, these projects never yield a positive return.

Therefore, leadership is a key ingredient and it must be consistent. If any of the underlying processes of innovation are divorced from leadership, projects will fail at a higher rate. The business model I like is one that is multi-functional and multi-layered. A team made up of "hands-on" people plus the strategic perspective of leadership is a powerful combination.

Leadership must also recognise that creative people are different and they need to be handled differently. In today's slick, corporate management culture, this has become increasingly difficult. Creative people don't necessarily play by the same rules. They may be disorganised, ortheir labs a mess and safety precautions ignored. But they must be given the room and the support, along with some discipline. Leadership must understand their mindset, because these are very special and valuable people.It is no surprise, therefore, that the most successful innovation companies have patient, market-focused, disciplined leadership.

What kills innovation?

If you run a business only by the numbers you risk killing off innovation. Obsession with shareholder value or net present value can lead you to abandon projects too early. This does not mean you should disregard numbers. Companies which have successfully developed new technology know that a strict dollars-and-short-term mindset will not work. Keep questioning the strategy, testing the market and controlling costs, but understand that innovation in the early years is cash negative.

Hierarchical management that creates too many layers – thus separation from the market interface – will not aid and may destroy innovation. These leaders may not understand what drives value and cannot recognise the merits of potential projects.

Political leadership, or as I call it "management by appearances", will not work either. In these companies no one likes to deliver bad news. This culture affects the company's ability to adjust and change. Bad news delivered in real time actually saves money.

How to innovate with success

Companies that innovate successfully leverage small teams with passionate leaders. People work better in teams of about five to seven people than in groups of, say, 15 or more. Decisions are made quickly and information flows freely.

These successful companies have a willingness to fund new ideas. Along with this, they become expert in managing failure as well as success. They are prepared to kill a project that is not meeting technology milestones or strategic targets and move on. (But at the same time they allow some projects to continue to incubate.)

Perhaps, above all, they are not rigid in defining their business. A flexibility of mind is necessary – many commercially viable businesses have come about as a related departure from an initial project. There are cases where open or lateral thinking has been applied and the result is a business turnover three or four times greater than originally intended.

Leadership must constantly ask the question: "What business am I in?" There may be an opportunity to push the boundaries, take your core competencies further down the value chain and turn a product or core competency into a service business.

Some of the above are simple, commonsense ideas but are not practised enough. However, none of these concepts is mutually exclusive. It is a case of where business practice – in this case, research and development and innovation management – requires a balance, a certain delicacy and understanding which, you might say, in its highest form is the art of intelligent innovation management.

J. HUTCHINSON RANCK
MANAGING DIRECTOR
DUPONT AUSTRALIA

By Tim Treadgold

Technology improves mine productivity

A major resources company has revolutionised mining with a new

Safe, precise, efficient, innovative and technically advanced. These are words normally associated with a modern medical procedure or the development of a new computer system. They are not traditional descriptions of mining – but they soon might be, as companies such as WMC Resources lead a revolution in the way minerals are extracted from deep underground.

The latest breakthrough in modern mining came on August 23, 2003 at WMC's Olympic Dam copper and uranium mine. In a world-first, a single operator took control of two independent rock loaders (called LHD vehicles – load, haul and dump) working in the same area to achieve a breakthrough in the speed and effectiveness with which they hauled material away from a mine face.

Safely housed at the surface, far from the dangers of the freshly blasted rock, the operator was able to manipulate his two machines to quickly achieve a productivity level which matched that of traditional mining techniques without all of their inherent dangers. In time, the productivity level of twin-loader automated mining will more than double those of manually operated equipment in some applications.

In many ways, the work by WMC and its team of associates that includes the CSIRO, the Australian Mineral Industry Research Association, Caterpillar Elphinstone, Dynamic Automation Systems, Remote Control Technologies and the Centre for Mining Technology and Equipment, is the ultimate practical application of technology and science.

Drivers take a back seat

Most people, unaware of how tough conditions are in underground mines, are equally unaware that great changes are being made and that cutting-edge technology is being used underground. The trial of twin, driverless loaders involved the use of highly sophisticated traffic management controls and mine planning, and represents the most recent high point in seven years of assessment.

The WMC system at Olympic Dam is made up of computers on the LHD vehicles, surface computers, underground radio cells, independent remote control systems, scanning lasers, articulation sensors and cameras. It does not require underground guidance infrastructure (such as markers on walls) except for communication networks, and can operate in extreme dust and poor surface conditions (known underground as road conditions).

The actual process performed by the remote equipment is, in essence, the same as that conducted in underground mines for centuries. This involves drilling a rock face, placing explosives, clearing the shattered rock (bogging), loading it into trucks, hauling it to a central collection point (tramming), dumping and then delivering it to a processing centre on the surface.

WMC says that in some mines, which use a technique known as long-hole stoping, a bottleneck in mine production is the lack of speed of remote bogging. The use of automated remote tramming and dumping has increased this cycle speed by up to 100 per cent with the ability to use two loaders in the same area possibly leading to a 180 per cent improvement over current tele-remote systems.

Copper anodes are lifted from the casting wheel at Olympic Dam's copper smelter.

Smart as it may be, the use of twin, remote-operated loaders is really just one example of the radical changes underway in mining as it adapts and develops the best in new technology. With the use of a crystal ball it is possible to see mining taking on more the appearance of a surgical procedure, not unlike keyhole or arthroscopic surgery, in which a small hole on the earth's surface allows the extraction of the minerals the world needs. The CSIRO has referred to this as "keyhole mining".

It is possible to see mining taking on more the appearance of a surgical procedure, not unlike keyhole surgery

For WMC, the road to its latest world-first in automated underground mining started with experiments with single unmanned loaders. While they worked well, and became progressively more efficient, there was sometimes an issue when a hold-up occurred, such as the need to remove oversized rocks.

WMC's work has shown that operators of the loaders can achieve up to 80 per cent of the productivity of an experienced manual operator within two weeks of training. After that, and as technology plays an even greater role in the mining process, efficiency levels rise for a number of simple reasons, such as a machine being able to operate seamlessly for hours (or days) without a break, an economic gain which pays no regard to the vast improvement in mine safety because unmanned machines can go where humans cannot.

Development of remote/autonomous loaders and other technical improvements in the mining process has been high on the WMC agenda for more than 20 years. Early work used line-of-site wireless remote control vehicles and graduated in the late 1990s to tele-remote vehicles.

In 2002, a world breakthrough was achieved at Olympic Dam with the use of the first surface-controlled laser automated hauling and dumping loader, followed in August 2003 by the first multi-machine application with a surface operator working two machines in the same area of the mine.

WMC's work has shown that operators of the loaders can achieve up to 80 per cent of the productivity of an experienced manual operator within two weeks of training

WMC has found that wireless remote control technology has a huge impact on costs. Before the systems were developed, it was necessary to prepare an ore extraction level so that a manual operator was not exposed to falling rocks. It is now possible to dispense with extensive extraction procedures and to use the access way prepared by the drilling team to extract the ore, saving millions in capital development.

From being a self-described "fast follower", WMC has grown to be a world leader in underground mine automation, confident that what it is achieving will be felt throughout the mining industry for many years to come.

Efficient systems

The main advantages of the twin-loader system include:

- increased worker safety and improved workplace conditions;
- higher productivity than a manual bogging system (20 per cent to 40 per cent more productive in trials over 400 metres);
- an ability to operate in restricted areas with the twin units which do not require large areas of clearance and are not confined to pre-determined paths;
- reduced maintenance because of fewer wall collisions;
- the ability to operate unmanned in harsh underground conditions; and
- the underground layout not having to cater for people.

By David Salt

Inspired steel

→ Technical know-how and an innovative approach to customers, suppliers and staff is a winning combination at BlueScope Steel.

Changing the face of architecture.

Steel power poles are increasingly being seen as a viable alternative to the traditional wooden poles in Australia, providing a lightweight, durable alternative to a finite resource.

The steel power pole system SURELINE is the brainchild of innovators at BlueScope Steel, Australia's leading producer of flat steel products serving customers all over the world in the building and construction, manufacturing, automotive and packaging industries.

"To stay number one in the highly competitive steel industry is no mean feat," says Kathryn Fagg, president of Market and Logistics Solutions at BlueScope Steel. "It's not enough to simply produce quality steel. You have to offer innovative solutions with steel products. Innovation in this area is one part technical know-how and one part effective customer interaction."

Fagg says BlueScope's track record of success demonstrates it has the technical know-how. It has remained at the forefront of technical innovation for decades.

"Back in 1966 we made a major breakthrough with COLORBOND steel, the world's most advanced pre-painted steel product," she says. "It revolutionised the building industry and changed the face of architecture. That breakthrough was followed in 1976 by ZINCALUME steel, a zinc/aluminium alloy coating that extended product life and led to the development of more reliable steel products for the construction and manufacturing industries."

While technical know-how is important, Fagg says what possibly distinguishes BlueScope Steel's innovative approach is its ability to apply this know-how and work with customers to meet their specific needs. "We are customer-focused and this has allowed us to stay at the front of the pack," she says.

The SURELINE system is a further example of BlueScope Steel's approach to providing steel solutions. The company collaborated with the power distribution industry to develop a steel power pole that provided operational security for the life of the asset. Only a systems approach could meet the complexities of this key industry need. Hence, BlueScope Steel designed innovative fittings, logistics and sourced in-service inspection methods to be used in conjunction with the pole.

Innovation is fundamental to everything we do

"Some companies equate innovation purely with research and development," says Fagg. "BlueScope Steel invests heavily in this area but we see innovation in a much broader light. Innovation is fundamental to everything we do. It's built into our products, our systems and processes, the way we work with our customers and the way we regard our own people."

Building brands

It is easy to see how innovation is built into the company's products. BlueScope Steel is a longstanding leader in metallic coating and painting technologies, and has developed some of the world's most advanced coated building materials.

However, the company has also put considerable effort into creating and promoting named brands for steel products and it has done it very effectively. Everyone in the building game instantly connects COLORBOND steel, one of BlueScope Steel's longest-standing brands, with tough, long-life, durable, coloured, metallic-coated flat steel. Meanwhile, ZINCALUME is the trusted brand for zinc/aluminium coated steel.

"Our success comes from our people," says Fagg. "They are our strength. The company acknowledges that if it creates a safe and satisfying environment and we treat each other with trust and respect, our experience and teamwork will continue to deliver steel-inspired solutions. Our operations are among the safest, most cost-effective found anywhere."

She adds that the company's success also depends on its customers and suppliers choosing it.

"Our strength lies in working closely with them to create value and trust, together with superior products, service and ideas," she says. "If you regard your customers as your partners, it is much easier to tailor solutions to meet their needs." ■

By Jane Breusch

Winning idea wins the deal

→ A rough sketch on a napkin led to a $2 million order for Australia's leading plasterboard manufacturer.

Getty Images

[Eureka! Boral got it.

Melbourne's Eureka Tower, the tallest apartment project in the Southern Hemisphere, provided the catalyst for thinking outside the square at Boral Plasterboard. The project led to the development of a new plasterboard system that brought in a winning deal for the company.

The landmark 88-storey Eureka Tower presented builder Grocon with some specific challenges, including the need for lightweight construction, high acoustic specifications, complex material handling arrangements and managing multiple trades on a tall building.

The plans for the inter-tenancy wall system, like many other high-rise apartment developments, were based on an autoclaved, aerated, concrete product that is widely favoured by architects. An inter-tenancy wall system provides the divisions between neighbouring apartments.

However, an opportunity for Boral to become involved in the project arose following a lunchtime meeting between John Bajkai and Dave Stephens from Boral and two Grocon representatives. Boral was involved with Grocon on other developments and had established a reputation for providing sound technical advice and support.

Serious scribbles

The discussion over lunch focused on alternatives to the concrete product and Stephens began scribbling some ideas down on a napkin. Bajkai, an architect and now Boral commercial sales manager, says architects have long favoured the aerated concrete product because it is perceived as a masonry product, although builders are less enthusiastic about it.

"With plasterboard, when you tap on the walls you don't get the solid feel that you get with masonry products and this is where the concrete product has the psychological advantage," he says.

Boral was determined to change this mindset. The drawings on the napkin came back to Boral and from that the EurekaWALL system was born.

EurekaWALL has a number of advantages, including being cost-effective and more easily installed.

The new product is based around Boral's successful single membrane PartiWALL system that is used within low-rise multi-residential buildings.

Boral acoustic development engineer Mark Debevc says Boral got to a point where it was confident about presenting something to Grocon because with Eureka Tower being an already specified project, there was not a big window of opportunity to change the specifications.

"We used EurekaWALL to build several prototype walls to enable the contractor to assess the product," he says. "It's easy to draw things on paper but if it's difficult to build and there's contractor resistance, you might as well forget it. However, we got the contractor's blessing and then everything else just fell into place."

Boral was awarded the contract – worth more than $A2 million – to supply the inter-tenancy wall system and plasterboard requirements for the entire Eureka Tower project, a major coup for the Boral Plasterboard business and one that it hopes to use as a springboard for future projects.

The company believes EurekaWALL will create a fundamental shift in construction techniques for internal partitioning and it is well-positioned to be the market leader for wall systems in the high-rise segment.

The company is already pushing the product to the specifiers and sales staff are promoting it to builders throughout Australia. A number of potential projects are in the pipeline in New South Wales and Western Australia.

"The fact that Grocon is using it has led people to think there must be something in it," Debevc says. "It will not be long before EurekaWALL is used in many other high-rise projects." ■

By Kevin Childs

Spot the difference

→ From its humble beginnings as a one-store dry-cleaning shop, Spotless has expanded into a plethora of diversified services.

In 1998, the Department of Defence's brief was clear: businesses that were awarded contracts must use subcontractors. Spotless – a business that had grown from dry-cleaning to offer a host of specialist services with a multi-billion-dollar turnover – accepted the challenge.

Out of this has sprung an army of small to medium businesses across the nation – trained, overseen and guided to a new level of efficiency. In effect, it represents nothing less than the coming of age of the humble subcontractor. Just how Spotless did this and used it as a template for other dynamic activities, such as catering for the athletes at the Sydney Olympics, demonstrates inspiring innovation.

Working with Defence meant that Spotless began talking to small-to-medium businesses and, if they were suitable, training them, enabling them to grow, honing their financial management and other skills, and signing them up for a plethora of work.

The result to date is 2000 thriving subcontractor businesses that help Spotless in managing buildings and fixed plant and equipment, as well as work with their staff to provide catering, cleaning, grounds maintenance and accommodation services at Defence bases across Australia and New Zealand.

There are many other impressive figures that highlight Spotless' success. For example, productivity is up 20 per cent since this model was implemented. Quality has rocketed, with satisfaction rates increasing from between 65 to 70 per cent to 85 to 90 per cent. One plumbing contractor, since starting work for Spotless, has built its staff from four to 12 in three years and now services other firms besides Spotless.

The story is the same elsewhere, says Lewis Kaerger, general manager development, Property & Facilities division. "With up to several thousand Defence personnel on some bases, efficient operation is critical," he says. "The investment we put into training preferred subcontractors has enabled Spotless to support local communities while still servicing the needs of customers in remote regional areas."

Dishing up

Spotless used an innovative supply chain solution to serve up more than 1.5 million meals over three weeks during the 2000 Sydney Olympic Games. The nutritional value of every dish for athletes from about 200 nations, with a range of culinary and religious needs, was available through a unique electronic menu management system so that the required recipe would come up at the click of a button. Chefs were then able to click on to a range of approved suppliers for ingredients.

The use of online buying developed by Spotless as its vital link with suppliers at the Olympics has become a key part of other burgeoning businesses. For example, Australian Leisure and Hospitality Group Limited who owns more than 100 independent pubs and hotels in Australia uses the system to order food for its hotel chain.

The Spotless Electronic Buying System now leads the way in online procurement and has helped make Spotless a world leader through the application of this business model.

From Australia's Department of Defence and the Olympics, it may seem quite a step to the humble coat hanger. Through a logical leap, however, the model serves to produce yet more clever innovation.

> The company has consistently applied complex logistic solutions to cottage businesses such as catering, cleaning, laundries and maintenance so that it can effectively service larger-scale global customers

As Michael Givoni, national general manager, group marketing, explains: "Rather than provide coathangers directly to department stores, Spotless supplies them to garment makers around the world. This is a significant advance on the old system of clothes arriving into department stores in boxes. Now they arrive uncreased on hangers and are colour-coded according to size."

This hanger system has eliminated much manual work on the shop floor. "Spotless took something that was generic, like a coat hanger, and made it into a sophisticated global business," says Givoni. "The company has consistently applied complex logistic solutions to cottage businesses such as catering, cleaning, laundries and maintenance so that it can effectively service larger-scale global customers."

These innovative businesses have grown from one dry-cleaning

Spotless snapshot

Spotless provides a number of services including:

- Service solutions to business, industry and government;
- Market expertise in sectors such as healthcare, education, defence and resources, apparel retail, as well as social infrastructure projects such as public and private partnerships;
- Support services including property and facilities management, hospitality, catering and other food services, open-space and infrastructure cleansing, building and asset maintenance, cleaning services, print management, work-wear rental, linen supply and laundry services, and pharmaceutical distribution; and
- Retailer services including the sourcing, supply and manufacture of garment hanger systems, labels and packaging for garment manufacturers and retail industries globally.

Humble beginnings: the first store opened by Spotless in 1946.

shop that was established in 1946. Founder Ian McMullin initially started acquiring more dry-cleaning businesses but then, in a move that was decades ahead of its time, he started franchising his Spotless business.

The genius of McMullin was in recognising that the way to grow was to move from servicing the general public to having businesses as clients. A decade after that first store had been set up, Spotless, led by current chairman Brian Blythe, made a bold takeover of Ensign Services, a business three times the size of Spotless.

Ensign had grown from a tin-shed, T-model Ford business in the tough docks area of Port Melbourne, where it washed and ironed overalls for local firms, to generate an annual turnover of almost $100 million. With this acquisition, Spotless took Ensign into commercial, large-scale laundering.

This was the start of a dynamic, highly innovative business that now has 20 laundries cleaning thousands of uniforms for a national supermarket chain, as well as supplying them with barcodes for each staff member, across Australia and New Zealand.

Along the way, Spotless moved into commercial catering and property and facilities management. It has a global staff of 27,000 and an annual turnover of $2.4 billion.

The pattern established by McMullin, now aged nearly 80 and still a board member, has continued, with Spotless taking small regional businesses and expanding their capacity for big national or international customers.

"Innovation is driven through our culture at Spotless," Givoni says. "It is how we developed our business model to overcome the challenges of effectively servicing large-scale customers." ■

Kevin Childs is a Melbourne-based freelance journalist and author.

The road to innovation

Three years ago we embarked on a journey of transformation for PricewaterhouseCoopers based around the theme of innovation.

It began with our recognition that we needed to communicate our core values to the young people within the firm. Our goal was to create a more unified, high-performance culture where the best of our corporate thinking was continually refreshed and regenerated with the latest ideas.

Our methods and thinking were developed in Australia but are now being considered for adoption around the world.

This innovation drive also had a marketing impetus. It was a way of differentiating ourselves from our competition. Ultimately, of course, it was all aimed at finding ways of bringing new value to our clients.

We began with a core group of senior partners who worked together for eight months in a series of meetings and workshops. These sessions were intense and vigorous, ranging in style from the semi-formal to the informal.

We were after culture change. For, while we were already doing many things well, we needed to bring more unity to a diverse and multi-disciplinary working environment. We wanted to bring a common footprint to the many services we offer.

This is vital in any business but perhaps more so in one like ours, which is genuinely based on human capital. Our people are our business. Considering the youthfulness of the firm, we devised a three-tiered structure of activity and responsibility. These became our platforms for activity and for a shared philosophy of all our people but particularly our partners as leaders of the firm.

We want all our people to "develop", "achieve" and "inspire". These actions define how we behave internally and how we focus our energy for the benefit of our clients.

After those eight months, we widened the circle and discussed our thinking with a wider group of 50 champion partners. This was done through a great variety of environments and collaborative meetings and workshops. We created four leadership teams of about 15 people.

leaders SPEAK

This innovation drive also had a marketing impetus

Each of these had responsibility for a particular platform in line with the initial philosophy: develop, achieve, inspire. And then we added a fourth: "values". And for a year these groups met and worked on reinventing our PwC culture. This method ensured that the changes did come from within and that a very real engagement took place.

After this very fruitful year, we took the initiative to all partners nationally. I was convinced that, if we got the processes right, the output would naturally follow and impact favourably on the bottom line.

I like to explain the process as being one of "refloating the currency of leadership". It worked.

Now, as a result of our own experiences, we're beginning to work with clients on cultural change and how to instill values of innovation, leadership and change.

A new service has been created and the early results look promising. Based on our experience, we have now created a "People, Culture and Performance" practice to help our clients through the critical thinking and implementation issues associated with winning in today's fast-moving marketplace.

We don't profess to know all the answers but feel certain the talent and methodologies we are gathering in this new team will substantially support our clients into the future.

TONY HARRINGTON
CHIEF EXECUTIVE
PRICEWATERHOUSECOOPERS AUSTRALIA

Innovation is the specific instrument of entrepreneurship... the act that endows resources with a new capacity to create wealth.

PETER F DRUCKER

By Karen Boalch

Financial firsts

→ Constantly adding new products to its mix has kept Macquarie at the forefront of financial services in Australia and around the world.

Macquarie Bank is Australia's only independent investment bank and in less than 30 years has become one of the country's top 25 listed companies. This has occurred through a process of constant evolution and in expanding its services to those that are well beyond the scope of traditional investment banking markets.

Thriving in tough economic conditions requires focus, ambition, hard work and innovation. These are qualities for which Macquarie Bank is renown. Established in 1969 as Hill Samuel Australia, the company's aim was to provide advisory and investment banking services of an international standard to the Australian market.

This was an ambitious goal at the time. Corporate finance was extremely under-developed in Australia, to the point where anyone who seriously wanted to raise significant capital caught a plane to New York or London. At the time, Hill Samuel Australia had a capital base of barely \$A2 million and a staff of 12. It had little or no presence in stockbroking, foreign exchange, corporate advisory or securities trading – the bread and butter of traditional investment banking.

Today, Macquarie Bank has a market capitalisation of around \$7.1 billion and employs over 5000 people in 18 countries. Through innovation, it has pioneered new asset classes and leads the world in several key areas of business.

> Macquarie has established a unique presence in the provision and financing of world-class infrastructure

It is the number one project finance adviser in the Asia Pacific and Americas and number two globally. It is also one of Australia's top mergers and acquisitions advisers, a leading debt issuer and world-leading arranger of cross border leasing. Indeed, it takes great pride in the fact that it is an Australian investment bank competing successfully on the world stage.

One of the main areas where the bank has established a unique presence is in the provision and financing of world-class infrastructure. Macquarie launched the first listed infrastructure vehicle in Australia in 1996 and today the Macquarie Infrastructure Group is among the world's largest owners and developers of toll roads. It is also one of Australia's largest companies with assets worth more than \$10 billion in eight countries.

Sector specific specialists

Following on from this success, Macquarie has developed a range of other specialist funds that manage and own airports, communications infrastructure and sector-specific property assets. These funds have more than \$23 billion in assets under management and are a key driver of Macquarie's growth. Its listed funds have achieved strong growth and consistently outperform the benchmark S&P/ASX 300 accumulation index.

Macquarie believes its experience and expertise in these areas gives it a competitive advantage in acquiring and managing assets. For this reason, it has established several international infrastructure funds. These include the:

- Macquarie Essential Assets Partnership, which invests in utility assets in Canada;
- Macquarie Global Infrastructure Fund, which has invested in the Detroit Windsor Tunnel between the United States and Canada and the Soojungsan tunnel in South Korea;
- South Africa Infrastructure Fund, which targets equity investments in sub-Saharan Africa; and
- Korean Road Infrastructure Fund, which invests in South Korean toll roads and tunnels.

> Innovation means the Macquarie business is constantly evolving

Some of Macquarie's most successful specialist funds invest in property. The bank manages more than \$9 billion in property assets, with sector-specific funds covering the office, industrial, retail and leisure property markets. The company has successfully exported this expertise and been pivotal in introducing listed property trusts as an asset class to key Asian markets.

Macquarie's international strategy is to expand selectively, seeking to offer targeted products and services in niche markets. This enables it to respond to the special requirements of individual markets around the world. As the above examples demonstrate, this strategy has enabled Macquarie to establish leading positions in a diverse range of international markets.

Supporting the community

Macquarie's innovation does not stop with its product offerings. As one of the pioneers of corporate philanthropy in Australia, Macquarie has provided support to community programs for more than 30 years. It does so in the belief that a company is a member of the society in which it operates. It follows that one of its important duties is to work for the betterment of that community.

Companies today are assessed on more than their bottom lines. They are assessed on their culture, their ethics, the way they treat their staff and the extent to which they engage in the wider community. It makes little sense to talk about successful companies in a society where areas of the community are neglected. Whether we act as individuals, in government or in business, we all have a role to play in building our social capital, now and in the future.

Daring to be different and embracing innovation keeps Macquarie at the head of the investment banking industry.

Innovation means the Macquarie business is constantly evolving. Today, Macquarie is involved in businesses that weren't even heard of 10 years ago. No doubt, a decade from now, it will be developing products undreamt of today. This innovative approach has enabled Macquarie to remain at the forefront of a fiercely competitive industry and has delivered more than a decade of consistent profit growth to shareholders. ■

Karen Boalch is a communications executive at Macquarie Bank.

Macquarie has always understood the importance of this relationship with the community. This founding philosophy was formalised in 1984, with the establishment of the Macquarie Bank Foundation. Each year, the Foundation is allocated funds based on a formula determined by Macquarie Bank's annual profit. This figure has grown consistently, with the Foundation currently contributing more than $2.5 million a year in community grants to around 200 organisations.

The Foundation focuses its resources on the core areas of education, health care and research, welfare, the environment and the arts. Within these areas, it looks for opportunities that are innovative and genuinely responsive to the community's needs. The Foundation also seeks to involve Macquarie staff in community activities, contributing time and expertise, as well as financial support.

In recent years, the Foundation has focused on building strategic relationships with community organisations and developing stronger links between the corporate community and the not-for-profit sector. One of its senior private equity specialists, for example, runs Social Ventures Australia, an organisation which aims to get important social programs off the ground by applying venture capital principles to the community sector.

As a financial institution, Macquarie is also keen to support programs that develop the financial skills of those who participate in the non-profit sector, who don't always have access to the same training available to corporate employees. These are just some of the innovations Macquarie has brought to corporate philanthropy in Australia. Developments such as these have earned Macquarie a reputation for best practice in this area.

By Gayle Bryant

An innovative challenge

→ A focus on goals, values and people is the strategy behind this dynamic professional services firm.

PwC: stretching the boundaries of innovation.

Professional services firm PricewaterhouseCoopers (PwC) places a high emphasis on the importance of innovation.

"In order to gain a competitive edge, organisations must implement new ideas and business approaches that lead to productivity gains and overall improved performance," says the firm's chief executive, Tony Harrington.

People represent the only competitive difference between organisations – the rest is just infrastructure

With this in mind, PwC decided several years ago to re-engineer the firm with an explicit focus on innovation.

"Our methods and thinking were initially developed on a local level but were so successful they are now being contemplated for wider adoption around the world," Harrington says.

"At PwC we have a lot of young people. Our aim was to create a more unified, high-performance culture that inspired younger members of the firm. We also saw it as another way of differentiating ourselves from the competition.

"Our culture is defined by three verbs, develop, achieve and inspire, that bring to life the firm's values of teamwork, excellence and leadership.

"These values reflect how we behave both internally and externally to the considerable benefit of our clients and our people," says Harrington.

People, culture and performance

As a result of PwC's own success with its cultural change work, the firm is now beginning to assist clients on cultural change and innovation through the creation of a new "People, Culture and Performance" practice within the firm. The leader of the practice, Chris Blake, says research makes it clear that people represent the only competitive difference between organisations – the rest is just infrastructure.

"Over the past two years we found that there were no providers in the market offering assistance across all of the critical elements required to create cultural change in organisations. Most providers focus on just one key component of change", he says. "We felt that to create change in a way that would be sustainable requires the alignment of all people processes – not just attention to one or two."

The People, Culture and Performance practice focuses on the four critical areas of leadership development, change management, reward and performance, and people strategy to support sustainable change on a holistic basis.

Our ability to identify and attract innovators is a crucial part of the recruitment process at PwC

According to Blake: "Discussion of these areas isn't just rhetoric, we are acting on them. For example, the change management focus considers ways of transforming an organisation in a managed way by beginning at the leadership level and then aligning the organisation through all people processes rather than just setting directions and hoping they get noticed.

How does I-Challenge work?

I-Challenge runs for five months between August and December each year. In 2003, about 350 graduates from PwC offices all over Australia participated. The graduates are divided into teams which are then told to research three possible business issues that are affecting organisations. Clients participate as well.

Over the five months, the graduates proceed from identifying issues that require solutions to understanding the use of team synergies and learning how to proactively use people networks, research tools and techniques.

The teams are introduced to experts in different fields. These range from industry experts who discuss the major issues affecting their industries and experts on creative thinking who introduce practical thinking techniques to presentation experts who talk about communicating new ideas effectively.

The teams are supported by a "buddy", someone who participated in I-Challenge the previous year. They also have access to senior management within PwC who will coach them and sponsor their ideas.

"The Ideas and Solutions Group ends up with 50 new ideas by early December," Umberto Righetti says. "We then decide upon around five to 10 that we can finance immediately. Others that are not time critical will go into our ideas bank and are implemented when the time is appropriate."

He adds that I-Challenge is highly regarded by all in PwC. "It teaches the graduates about business issues and how to deal with them," he says. "It is a combination of education and innovative thinking. While it is an Australian initiative, it has now been adopted by our Singapore office."

A great idea that was born out of I-Challenge was to introduce the initiative to PwC clients who were looking at ways to introduce innovation into their own organisations. Financial services companies were first to embrace the idea and have been so impressed that they are putting more and more teams through it every year.

"With reward and performance, we look at ways of rewarding people in a wider context as opposed to assuming money is the key. We look at what a business is trying to achieve and how rewarding people can help meet strategy. In today's environment, companies need to understand what drives people and these drivers are different to those of 15 years ago."

Harrington strongly agrees that the main differentiator of any firm is its people. "Our ability to identify and attract innovators is a crucial part of the recruitment process at PwC," he says.

"Our people are our greatest resource. We look for candidates who possess the unique blend of knowledge and innovation that has become synonymous with our brand. The strategies coming out of the People, Culture and Performance practice fit neatly with what the firm is doing as a whole."

The practice is building up capabilities that are now being applied to external clients. "We are putting together teams that previously wouldn't have existed before," Blake explains. "Our teams bring together people with an eclectic mix of skills such as change management, leadership and reward packaging. For example, we have been working with Daniel Petre – the former Microsoft Australia managing director and founding chairman of Ecorp, who has enormous experience in the area of strategic advice on people and leadership issues."

Integrated and accountable solutions

In many organisations, senior management, boards and the HR function are looking for more "hard-edge" accountability and management around the strategic value of their people. Organisations are looking at ways to better connect their HR departments with their wider business strategy. PwC is able to combine a deep understanding of business and finance issues with the specialist skills within the People, Culture and Performance practice to develop and deliver integrated and accountable people solutions.

The People, Culture and Performance practice is all about helping businesses achieve tangible results through their people. Forward-thinking organisations sustain their success because they invest in their people over the long term. Improving performance and unlocking the potential in people requires strategies designed to develop and achieve goals, and to inspire people.

The practice works with organisations to develop people strategies, assess people risks, enhance and change organisational culture, develop talent management and retention strategies, and build HR tools for measurement and reporting.

One area where the People, Culture and Performance practice has assisted many organisations is in mergers and acquisitions. Realising optimal value from a merger or acquisition is traditionally difficult. In fact, recent research has shown that a staggering 80 per cent of deals fail to deliver the intended shareholder value.

By looking at not only the financial issues around a transaction or restructure, but also considering all the people issues, PwC have helped organisations through the transformation process to ensure that maximum value is extracted from the transaction.

The People, Culture and Performance practice has access to a range of tools, programs and methodologies that are at the leading edge globally, which means that Australian companies are able to benchmark their people practices against the best in the world.

The innovation challenge

One program to come out of the new thinking is I-Challenge, which stands for Innovation Challenge.

"We introduced I-Challenge several years ago for a number of reasons," says Umberto Righetti, the business development leader of PwC's Ideas and Solutions Group. "The market is constantly changing and we recognised the need to develop our people to be adaptive to our clients' businesses. I-Challenge supports our ongoing commitment to fostering ideas and innovation."

> PwC is a clear market leader in its field but it is innovation that will increase the distance between us and the others

Righetti says the main aim of I-Challenge is to identify and develop a capability or solution that adds value to clients' businesses and to PwC.

"Ultimately we are building a culture within our own organisation that supports innovation and new ways of thinking to show our clients that we are at the forefront when it comes to the issues that they face every day," he says.

"It demonstrates to them that we have invested in thinking up new ways of dealing with business issues and this also shows them how innovative we are, which is essential if you are going to survive in today's environment."

Meanwhile, Harrington emphasises the importance of combining knowledge with innovation. "PwC is a clear market leader in its field but it is innovation that will increase the distance between us and the others," he says. "Our clients expect us to bring new ideas and solutions for their complex business problems and we need to do it every day to consistently prove our abilities as innovators." ■

Global recognition

Two ideas from last year's I-Challenge were presented to global chief executive officer Sam DiPiazza during a visit he made to Australia in 2003.

One team presented a very specific solution set, developed for reviewing employee reimbursement.

"This is an area in which GST is often applied incorrectly and potentially at great cost," says Umberto Righetti, business development leader of the Ideas and Solutions division. "Post I-Challenge, the team worked with the wider GST group to launch it on to the market."

Another team presented a big-picture idea, which it called the PwC Spiderweb – an ambitious real-time software solution designed for maintaining control and minimising the business and security risks of the increased mobility of people.

At PwC, I-Challenge has been a tried and tested initiative to embed innovation into the firm's corporate culture. Now, together with the work being done by the People, Culture and Performance practice, PwC is assisting many other large Australian organisations looking to embed innovation into their corporate culture.

By Jane Breusch

Learning to make sense

→ The Commonwealth Bank of Australia has launched a website to help young Australians learn about financial literacy.

Getty Images

The CBA is helping kids learn to save.

The Commonwealth Bank of Australia has developed a unique financial literacy education website to provide practical information about money management for young Australians aged between 14 and 17.

"The website is innovative because it was the first interactive financial literacy site developed by a major Australian bank that could be used by young Australians and teachers as a valuable learning resource," says Bryan Fitzgerald, general manager, media and community.

The DollarsandSense.com.au website is the first in Australia to be mapped to the secondary school mathematics and business/ commerce curricula across all Australian States and Territories – a mammoth task, given the myriad individual requirements.

"As each State and Territory education department sets its own curricula, we consulted with each throughout the development stage to ensure it was a relevant tool that could be used independently by young people and teachers in the classroom," Fitzgerald says.

The website includes budgeting tools, financial skills tests and tips, and interactive forums with business leaders. Forum guests have included the Taxation Commissioner Michael Carmody, Commonwealth Bank chief executive David Murray and Telstra Businesswoman of the Year Di Yerbury.

The website has been designed for students to use independently while also providing a resource for teachers to use in the classroom. Some of the innovative features include a curriculum library, research questions to set for students and lesson plans.

The Commonwealth Bank has been involved in school banking and financial literacy since 1931, providing passbooks and money boxes to young students to enable them to do their banking at school.

However, with the advent of electronic banking, increasing numbers of children have been using EFTPOS and online banking, and this was one of the reasons why the Bank decided to develop the website.

Electronic age

"Electronic banking has made banking and access to money accessible for people of all ages," Fitzgerald says. "We saw that there was a gap in the teaching of money skills and that kids needed to be educated, not only about their current financial position but also about future finance matters such as credit cards and housing loans."

He says that if people don't learn finance skills at a young age, they could be disadvantaged when they leave school. "We saw there was a gap, we spoke to the experts, undertook research and development of the site followed." ■

Teaching our society

The Commonwealth Bank has a strong belief that financial literacy education is the key to a more secure and financially stable Australia. Inspired by this belief, the bank launched the Commonwealth Bank Foundation early in 2003 to help educate and thereby empower Australians of all ages with financial literacy skills.

The bank's board contributed the initial funding of $A70 million to the Foundation and income earned from investing the capital will be spent on education and, in particular, financial literacy.

"Regardless of the age group, there are a lot of people who don't understand financial literacy, finance or budgeting," says Bryan Fitzgerald, the bank's general manager, media and community. "We see it as very important that education takes place across the whole range of our customer base."

It is practical, knowledge-giving innovations like these that have a direct impact on society. This is a fine demonstration that a good idea – one which is a straightforward extension of a core business – can provide great benefits capable of reaching a wide proportion of the community.

By Valerie Khoo

Evaluating risk

→ An Australian insurance giant is promoting an innovative approach to the role of risk management and insurance.

The recent volatility in insurance markets, combined with regulator requirements for proactive corporate governance, means that companies are now taking a serious look at the business of risk management.

"Most companies didn't really understand top-down risk management," says executive director of Aon Risk Services Bill Peck. "While some companies had positioning statements indicating that they managed risk, they were not proactive in their risk analysis or management. This situation is no longer acceptable to the investment markets. These days, you have to align your actions with your words."

The result is an innovative approach to the risk management process and a fresh look at the role of insurance. "Analysing and embracing the right kinds and levels of risk will not only decrease volatility but also have a significant impact on share price, return on investment and the bottom line," Peck says.

With insurance traditionally purchased as a product, this new approach is helping companies to recognise the value of insurance as a financial tool.

"We wanted an approach that followed good risk management and risk financing principles," Peck says. "Simply using our market clout to get a cheap insurance product may have been simplistically appealing but it wasn't really helping our clients identify and finance their risks with any level of confidence."

A foundational concept is the "cost of insurable risk". "The best driver of how to use insurance should be based on the cost of insurable risk, not simply the cost of the premium," Peck says. (See "Smart Insurance", right).

Dumbing down

"Competitive sustainability", Peck says, should be etched on the desk of every chief executive. "One of the main problems I see in corporate Australia is that risk management today is being elevated up the executive agenda, but being dumbed down as a 'process' and a 'system'," he says.

"It's essential that executives understand the difference between various risk management functions and what is required at different organisational levels." Peck says executive risk management is about identifying killer risks, incorporating their management into corporate strategy and building an optimal risk/reward profile that is understood by investors and the market.

"While risk management systems and sophisticated models to predict the future are undoubtedly valuable tools, executives must appreciate that many killer risks are simply not 'in the square'. Killer risks are like tomorrow's jobs – many of them have not yet been invented."

Peck believes it is important for executives to spend time thinking outside the constraints of day-to-day management, to speculate in an environment where there are no stupid questions and where contemplation of the future is not constrained by what has gone on in the past.

"My first rule for executive risk management is that the future is not what it used to be," he says. ■

Valerie Khoo is managing editor of Australia's Nobel Laureates.

> [It's important to take the right kind of risk.

Smart insurance

Executive director of Aon Risk Services, Bill Peck, says the cost of insurable risk is the cost of insurance plus the cost of retained losses – that is, those costs incurred within the deductible or above the insurance limit. He says this needs to be added to the external and internal costs of managing and controlling that risk.

"CFOs and risk managers should be working together to target the lowest sustainable cost of insurable risk," he says. "It means identifying where the best point is between retaining risk as a deductible and insuring it. It means the cost of risk management and risk controls need to generate a return by reducing the future cost of insurable risk – either by reducing the cost of retained losses or the cost of insurance."

Once the mind has been stretched
by a new idea, it will never again
return to its original size.

OLIVER WENDELL HOLMES

By Gayle Bryant

Building life into work

→ Adopting a partnership approach has led to the construction of one of Australia's most innovative buildings.

A quiet but profound revolution is taking place. One which may alter the way in which office buildings are understood and created. Indeed the new concepts espoused by a unique team of collaborators have redefined the meaning of people at work and how a building can leverage human sustainability, happiness and productivity. The building – the National Australia Bank's new flagship building in Melbourne's Docklands – is a stunning and imaginative work of 21st century architecture. However, it is the collaborative process of a group of project directors for the National, Lend Lease, General Property Trust and the architect

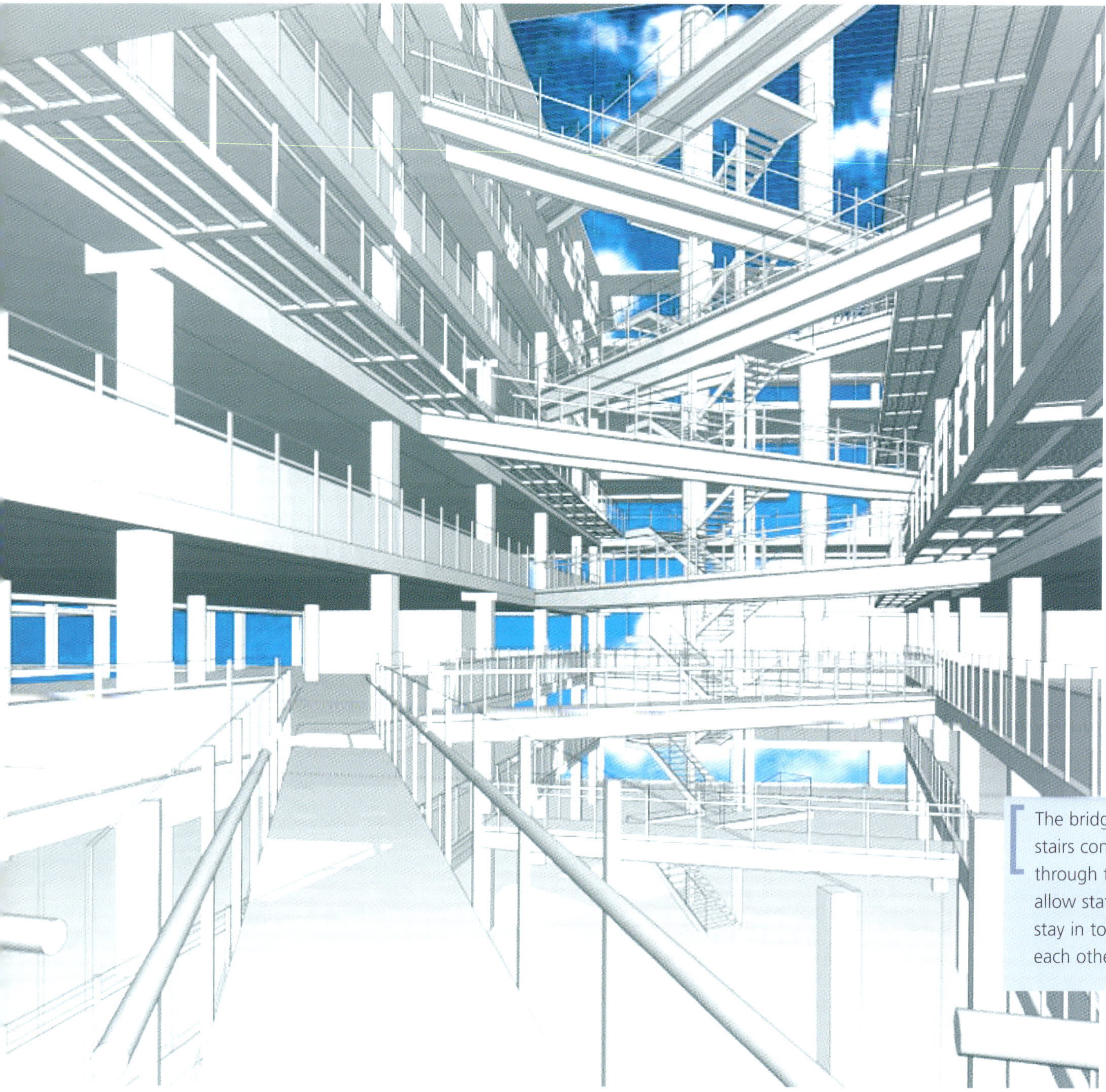

The bridges and stairs connecting through the atrium allow staff to easily stay in touch with each other.

Bligh Voller Nield that has broken new ground in the way that a multi-disciplinary team acts on an office development.

National@Docklands (N@D) is not just a new building for the National, it is the embodiment and a new catalyst for a profound cultural shift that chief executive Frank Cicutto is seeking at Australia's largest and most profitable bank. Creating an energised workplace, a community without boundaries between different banking businesses is essential if the National's strategy of "integrated financial services" is to work.

"We embarked on a cultural change several years ago. Docklands is part of this journey," Cicutto says.

This is a story of innovation, achievement and of architecture but above all it is a story of collaboration

In the annals of innovation, N@D stands apart in many respects but above all, in contrast to the "accepted" methods of design, building and finance, it has reshaped many precepts common to the traditions of the managers' professions.

This is a story of innovation, achievement and of architecture but above all it is a story of collaboration, one which sets new standards both in Australia and worldwide for its fundamental concern for people, their opinions and ultimate fulfilment within a purpose-built physical space.

N@D's design is more than open plan. Created from the workstation out, it is about environmental sustainability, not just because that is the way all buildings must go but because it is better for people.

David Mooney, the National's general manager office of the EGM, says such a bold statement in infrastructure, in both design and function, will add to the growing momentum for change at the bank. And it is more than just symbolic.

Putting people first

"In the traditional building, closed rooms or separate floors keep people apart. Even traditional open plan offices don't always encourage community," he says. "But great things can start from one idea or from a chance conversation. Docklands will encourage people to meet, to bump into one another and this will increase the chance for that flow of ideas."

This new workplace with its bold colours, connected atriums and naturally ventilated workspaces is the product of a revolutionary approach to property development.

"The entire process of making this workplace has been based around partnerships," explains Rosemary Kirkby, joint project director for N@D. "Usually, the tenant would not be involved in the design of the building but here we have had input into every design decision. With hindsight, it is incredible that it hasn't been done before; it is so logical. N@D is about a partnership process and all partners bring to the table their own specialist skills and a willingness and capacity to collaborate with like-minded people."

The $A240 million facility is owned by listed property fund General Property Trust (GPT), developed by Lend Lease Development and project managed by Bovis Lend Lease. Lend Lease Corporate Solutions worked closely with the National as its property adviser and then helped manage the broader project within the National. The architect is Bligh Voller Nield.

The size and scale of the project meant a partnership approach was the only way to balance a diverse range of stakeholder needs and to avoid the time and cost issues that arise when disagreements occur between parties.

Paul Cull, Lend Lease group executive and account manager on the project for the National, says the Docklands project has adopted an open book philosophy which provides the amount of flexibility required to deliver a project like this. "Trust is a precursor to making this flexibility work," he says.

"We also work holistically across all project elements driving from business needs and with exceptionally high levels of user engagement, all within tight commercial parameters. The management of the process demands creativity, flexibility and cohesive teamwork."

Kirkby agrees. "We also know that the flexibility we have will not be at the expense of a loss in accountability for time and cost outcomes," she says, adding that all partners had to contribute a passionate commitment to developing the intellectual capital on the project. Team members were selected on the basis of their capacity to contribute to the aspiration as well as their ability to control time, cost and quality risks.

"Every design decision was heavily reworked to ensure alignment with the principles in a value-for-money environment," Kirkby says. "Because the motivation behind the project is to provide work sustainability, it is a people project first, a business project second, and a property project third."

All partners had to contribute a passionate commitment to developing the intellectual capital on the project

Key to the success of the process was the weekly design review meetings of all partners that allowed trust to develop and innovation to flourish. Each partner had an overlapping sense of ownership in all aspects, reflecting the truly integrated nature of such projects.

N@D will accommodate staff from the original National Australia Bank headquarters at 271 Collins Street, Melbourne, as well as many from the National headquarters at 500 Bourke Street, also in Melbourne, and several other locations. In all, about 3500 staff will relocate during 2004. The building is a dramatic new one for a radically new type of bank.

Face of Melbourne

Surrounded by water and right at the front door of one of the world's most liveable cities, Victoria Harbour is a 30-hectare precinct and centrepiece of the Melbourne Docklands. It has already been called the new face of Melbourne. The location was also chosen to reflect the needs of the National today and into the future.

International experts such as Professor Franklin Becker agree that the levels of innovation and sophistication evidenced on the project are rarely seen. As director of International Workplace Studies, Cornell University, he says what has impressed him about the Docklands project is that it appeared to understand that the building has this potential to act as a catalyst for significant organisational change.

A model office: N@D from the north-east.

"Not a lot of companies do that," he says. "I'm impressed with the fact that there's been so much effort to involve people in the kinds of decisions that have been made about the work environment. One reason that Docklands is different from other workplace projects around the world is that it's so integrated.

"It's not just a building project, it's not just about new and different technology, it's simultaneously looking at how you can use technology, design and location across the board for branding, for involving and energising employees, changing the way in which people interact and the way that information flows."

Mark Rada, Lend Lease development manager, commercial, Victoria Harbour, says N@D was the most remarkable collaboration he has been involved in over nearly three decades in the business. "The contracts had all those clauses you normally have about dispute resolution, conflicts, who pays – but not once did we have to refer to them. It was a breath of fresh air," he says.

He adds that the key to the process was continuity in the design review group and a very realistic attitude to compromise, together with a clear understanding by each party of what they wanted.

In the end, the National achieves its aspirations for its people in the new workplace, Lend Lease seeds the Victoria Harbour development with one of the finest examples of sustainable development in the world and GPT gets an environmentally and socially responsible investment that provides a strong return to its unitholders.

Design departure

The National came into contact with such a space when it bought the wealth management business MLC from Lend Lease in 2000. MLC's North Sydney headquarters, Campus MLC, is a striking departure from a traditional financial services environment, one of the first colonists of the new terrain of workplaces.

MLC chief executive Peter Scott, now executive general manager wealth management at the National, is well known for working from the same style open plan workstation on Campus as all employees. He moves around the building regularly, clearly visible and approachable to all of MLC's people.

"I queue up for coffee in the café and people can just talk to me in the line. I hear different ideas and problems everyday – you just can't overestimate the power of that sort of transparency," he says.

Campus MLC was the inspiration for N@D but the full extension of the idea was limited by the traditional building it occupied. At N@D, the design and interaction of the workplaces has flowed out from the idea. This project is very different in a property sense from any other project that has been done in Australia," says principal architect James Grose of Bligh Voller Nield. "It was initiated as a people project, not a property project. Usually with property developments, the owner buys a box, finds a tenant, then they try to work together on a fitout – but it's always a compromise." He says rather than take any traditional developer box and try to fit a work environment into it, for the first time the architects were able to holistically reconceive a new building.

"It was refreshing to put the needs of the employees first when designing the building," he says. "If you are designing a workplace then you need to consider what will drive the productivity of the people that will work in it. Our design includes activity-based work settings where we have studied a person's work function and determined what they need in their space to be effective.

"This is more efficient than just allocating a set space based on an employee's place in an organisational hierarchy."

Grose says the design of the interior space also had to support and nurture human activity. "Instead of designing a building based entirely around cost, we were able to incorporate employee needs," he says. "However, the design also had to be commercially sound. Every decision had to be proven on a value-for-money basis. Having the same team continue to work together, building on the knowledge developed over the previous three years, was instrumental in making this happen."

The project manager Bovis Lend Lease was also involved in a number of firsts in the project, including what amounted to a total redesign of the base building to reflect the cultural aspirations of the National, all within a fixed cost cap and just in time for construction. It is believed no other builder would expose themselves to this risk and deliver on the promise without one dispute. Bovis also played its part in the employee engagement process by hosting some 2000 staff on the site during 2003.

Significantly, the success of the partnership approach to the project has already demonstrated the value of one of the National's

New approach for traditional industry

The National's new building at Docklands has no offices, is decked in primary colours and has an interior which places a premium on light, freedom and informal communication. It's hard to believe that this will be the home to staff of a traditionally conservative financial services organisation.

The overall aim of the design of the National's new corporate headquarters at Victoria Harbour was to discourage hierarchy and encourage informal discussion. "It's about building better relationships with employees," explains Rosemary Kirkby, joint project director of National@Docklands (N@D). "To do this we went to our employees and asked them what they wanted in an office space."

Convincing commitment

She says it was difficult at first to convince staff that they had a say in the design of the building. However, after the governance structure was put in place – that is, after directors made a commitment to review the project every six weeks – it gathered momentum.

"The new building is part of a business re-engineering process taking place at the National," she says. "To encourage feedback from staff about what they wanted from their new office, each division put in place teams of people who worked at multiple levels to ensure feedback about the work environment was gathered. We also introduced newsletters to keep staff informed about what we were doing. It has been a partnership process at every level."

When it is finished in 2004, the building will become home to 3500 people, about two-thirds of them from the banking side.

While spectacular on the outside, the innovation continues inside as well. Its south-eastern public entrance houses a call centre – the most obvious contact for most customers with the National. At ground level is a coffee bar and the internal public space can host community events – a symbol of the National's desire to be an active member of the Docklands community.

The building itself consists of two eight-storey buildings, linked by bridges which also house meeting and common spaces. Other features include gardens, verandahs, connecting decks criss-crossing the atrium and windows that open to allow for sunlight and natural ventilation funnels.

Common space in the light-filled northern zone is naturally ventilated, the air being drawn through the adjacent funnels. For staff, the view across N@D's large floor plates – known as "piers" in homage to Docklands – is open and mobile. On the north-eastern corner, a tower of "project rooms" has been added to the original design. The rooms also open to fresh air.

The project has been a hotbed of building innovation and includes:

- Increased connectivity through the use of criss-crossing bridges and stairs.
- Increased daylight penetration into the workplace using a new sun screen design.
- Fresh air infiltration to winter gardens, sunrooms and project towers.
- Human needs embodied in the work setting. The teapoints include a large kitchen table similar to those many people have at home and staff are encouraged to cook their meals at work. Many floor surfaces are timber – there is no granite or marble as you traditionally find in large corporate offices.
- Building community. Significant amounts of space within the building have been dedicated to create community facilities, for example, a cafe has been built in a park-like setting for 100 people
- Humanity is built into the design through the use of colour, the diversity of spaces, materials and furniture.

Most of these attributes have been reviewed by the end users – National's employees.

Peter Affleck, joint project director for the N@D, says while the building will be a great place to work, what excites him the most is that the environment is really just the conduit to facilitate major organisational and cultural change. He adds that, while instigating cultural change and creating a sense of community is frequently talked about in various dimensions, it is not often appreciated that the physical environment can be a catalyst for that change.

"Changing the physical environment that somebody works in can promote a whole new way of thinking and working collaboratively and effectively," he says.

Fresh air and natural light flow throughout the building.

mission statements: "Growth through excellent relationships". Lisa Gray, the National's chief executive of wealth protection, will be the most senior wealth management executive on campus. She too is excited. For the first time, her team of bankers and wealth managers will be together under a single roof.

"Instead of many different people being located on many different floors, where it's hard to stay in touch, you can just bump into someone and fix something on the spot," she says. "Things get solved because you are together."

Peter Scott adds the campus environment promotes a more mature workplace which also adds to the sense of responsibility. "You do have to be more aware of one another; there's a stronger community sense."

Unique beings

Gray adds: "First and foremost this building is about strategy, about offering 'growth through excellent relationships'. If that is going to be more than just a slogan we need to provide an environment that facilitates it for our staff and customers. Each customer is unique and we need to work differently. Part of that is allowing different ways to work."

N@D marks the first time GPT has invested in an office development and not a finished building. Martin Ritchie, GPT's office investment manager, says buying into the building and the process was a risk for GPT but one with which he is comfortable.

"When choosing a building to buy, we look for a secure and growing income stream," he says. "In 2001, we were looking at Docklands and at the potential of working with the National. The bank was proposing to lease the space for a 12-year term and as the

New paradigm for urbanism

If we restrict our consideration of the impact of the new workplace to the office building we will miss the deeper significance of the silent but powerful social revolution that is occurring around us. The same forces that underpin the arguments for a new workplace are reshaping our personal and collective expectations of appropriate living environments.

The challenges in the workplace to be innovative in the search for enhanced corporate performance have spilled into our whole lives and raised the expectations of what is possible for a better quality of life.

In all, this is creating quite different expectations of what makes for a good city. The emerging criteria for satisfying work and living emphasise that which creates and supports our sense of personal significance coming from what we do, where we do it and with whom we do it

The conditions have become variety and innovation, responding to a society no longer structured into singular models of acceptability but open like a big database, into which we plunge to find what we need. This is multiplied to respond to the needs of each of us, and of the groupings of which we are part, resulting in a demand for a diverse array of activities and places.

Neighbourhood work

Taking our work into neighbourhood places as much as we are now doing requires changes to the form and function of eating places, parks and other amenities. Those of us who want to work and live contiguously, and those who work extended hours, or different shifts, expect opportunities to relax and socialise when not working.

These and other implications of the new lifestyles require new paradigms to guide the re-formation of our urban areas especially in those cities not endowed with natural amenities for relaxation and re-creation. Otherwise we will not be able to support the corporations – global and local – that follow the new sensibilities to the workplace, nor the employees that find the new workplace to be such a rewarding innovation.

The National's building at Docklands has successfully applied many of these human needs criteria to a life/work environment which will inevitably establish a new standard for the long-term.

By Professor Graham Brawn – Architecture, Urban Design and Strategic Planning, University of Melbourne.

amount of rent was calculated from the cost of the building, we knew that the initial yield would not be eroded as the partnership designed extra enhancements.

The long-term lease and the high quality of the National's covenant meant that we had security of income."

Ritchie also says he felt the Victoria Harbour location was the best block of land in Docklands and the rental value in five years will have grown enormously. "Buying a new generation, campus-style development as opposed to a traditional CBD tower building also adds diversification to our portfolio," he says.

Meanwhile, Cull says one thing the project has shown is that the future of property developments is all about strategic alliances.

"What we have achieved is an excellent example of this," he says. "If we weren't all seeking the same value-based outcomes, it would not have worked. There has been a huge investment in building intellectual capital and each party took considerable risk.

"The result is the most innovative property to have been built in Australia in recent years, one that has been accomplished through a partnership process while ensuring the needs of the end users are met." ■

By Kevin Bailey

Keeping Australians wealthy and wise

Measures to make people more responsible for their financial well-being are being emulated around the world.

Financial planning is one of the fastest growing industries in the Western world. An ageing population and the need for self-provision in retirement has led to more people turning to financial planners for advice on how to make their money last.

Protection against any potential issues associated with people's life savings – such as selling products that benefit the planner more than the customer – requires a strong and effective professional body. Such a body provides investors with the confidence that their needs are being overseen by an independent authority. Australians are fortunate that the need for a single voice in relation to planning was recognised early on in this country. The Financial Planning Association of Australia (FPA) was formed by merging previously disparate groups and given a strong representative mandate for professional standards.

Establishing such a body was an innovative move in an environment where commercial imperatives often dictated an approach that served members rather than the public. The vision of the FPA and its members is to strive to improve the financial well-being of all Australians.

It aims to be the pre-eminent body representing the financial planning profession, to be greatly valued by members, highly respected by regulators and consumers and listened to by government. This approach initially brought it into conflict with some of the more short-sighted of its membership base who believed the FPA made life difficult for them as they were forced to raise their standards. However, the benefits of a highly-respected and relevant group quickly became apparent.

Expanding industry, growing wealth

Today the FPA represents more than 80 per cent of the financial services advisory industry. The growth in financial services is reflected in the increased capitalisation of finance and insurance companies listed on the Australian Stock Exchange. These companies now comprise more than a third of total market capitalisation, a rise of 132 per cent in the past decade. Australia is the sixth largest managed investments market in the world and employs more than 354,000 people in the finance and insurance sector.

The *IMD World Competitiveness Yearbook 2001* ranks Australia first in the world for the availability of finance skills. As an example of Australia's sophistication in this area, former FPA chairman Ray Griffin became the first non-American financial planner to be elected chairman of the International Certified Financial Planner (CFP) Council. Australia was the first country outside of the United States to adopt the CFP mark. This mark is recognised around the world as representative of the highest level of excellence in financial planning expertise. Australia has built a financial sector regulatory regime that is viewed by many countries as a model for their own regulatory systems. Recent reforms have created a framework that is streamlined and efficient while maintaining the highest standards of market integrity and consumer protection.

The landmark Financial Services Reform Act 2001 (FSRA), which commenced in March 2002, introduced uniform licensing, disclosure and conduct provisions for all financial services providers. The International Monetary Fund has hailed Australia's regulatory changes as "path-breaking reforms which put Australia at the forefront of international practice".

The innovative submission to the 1996 Wallis Inquiry by the FPA laid out the roadmap for financial services reform that was substantially adopted by the FSRA regulations and is now being emulated around the world. The FPA submission argued that its

The International Monetary Fund has hailed Australia's regulatory changes as "path-breaking reforms which put Australia at the forefront of international practice"

vision for a new regulatory system was one that was functionally-based. This involved looking across the financial services industry at its various activities, how these activities are regulated and grouping them into more coherent and compatible function sets.

Regulation of advice should be separate from regulation of product manufacturer and market trading and intermediation activities. The regime should ensure the competence of those giving advice through an appropriate licensing system that should impose high standards of conduct and disclosure and be geared to the provision of advice appropriate to the needs of consumers, not the demands of product providers or sales targets.

In 1995, the FPA introduced the first external dispute resolution scheme for financial planners called the Financial Complaints Resolution Scheme. This was ultimately merged with the life insurance scheme and run independently under the name of the Financial Industry Complaints Scheme.

In this respect, the FPA was well ahead of government and regulators in ensuring that the public was able to have total confidence in its dealings with planners and, if any problems emerged, that they had access to proper recourse.

Following this theme, the FPA introduced the National Quality Assessment Program in 2000. This program was designed to assist

The FPA works with its members to help Australians build a healthy nest egg for retirement.

licensed planners with compliance assurance to ensure consumers receive the highest standard of professional financial advice.

Raising awareness

The consequences of the innovative and proactive approach of the FPA has helped raise awareness of the need for proper planning in the Australian community and ensured that our economy is better positioned than most with a greater emphasis on self-reliance and savings.

Combined with one of the most successful superannuation systems anywhere in the world, Australians are uniquely positioned to embrace the challenges ahead with a better understanding of the issues and responsibilities that face us all. Of the 9 million working Australians participating in the $A500 billion superannuation pool, a large proportion rely on the retirement and estate planning skills of the FPA's members to achieve retirement independent of government support.

The FPA is dedicated to promoting and protecting the financial interests of Australians. These interests are promoted through an efficient and competitive market for financial services.

The path that the FPA took to ensure consumers are adequately protected against the wrongdoing of service providers has ensured that the marketplace has grown and confidence is justified.

A healthy, vibrant financial services sector served by truly competent practitioners is to the benefit of all concerned. Nowhere in the world has the industry advanced to the degree that it has in Australia, a situation largely due to the pioneers in the FPA. ■

Kevin Bailey is the managing director of The Money Managers.

By Gayle Bryant

Keeping watch on business behaviour

Serving as a financial conscience to the business community is a complex role for the Institute of Chartered Accountants.

Keeping a steady eye on the financial markets is a key role of the ICAA.

In recent years, public confidence in the financial markets has taken a beating. One of the key players in the continued debate over corporate governance, accounting standards and financial responsibility is the Institute of Chartered Accountants in Australia (ICAA). As a professional association, its main objective is to enhance and promote the reputation and role of chartered accountants. However, much of its work benefits the business community and the public as a whole.

Employing an innovative approach to its role, the ICAA has taken a leadership position in rebuilding public confidence in the financial markets and their institutions by focusing on three strategic imperatives.

Such initiatives have the potential to assist with microeconomic, macroeconomic and socio-economic integration and efficiency

First, standards and integrity are high on the ICAA's agenda. To develop and maintain a skilled and respected membership base, it is continuously "lifting the bar" to ensure that members comply with the highest educational, professional and ethical standards. Second, leadership – that is, serving as a financial conscience for the business community – means it also ensures appropriate decisions are made and appropriate legislation is enacted. Third, it is also charged with building a strong chartered accountant brand and restoring the image of the accounting profession.

The importance of education

These strategic imperatives are complemented by an ongoing focus on education, relationship building, membership growth and satisfaction, and effective governance. Education is a key feature of the ICAA. Apart from running seminars and conferences to ensure members are kept up to date with legislation and topical issues, it is committed to developing thought leadership papers on a number of topics, often in association with external organisations. One recent topic, which is of particular interest in today's current business climate, concerns the responsibilities of auditors.

Stephen Harrison, chief executive officer of the ICAA, says the role of an auditor is one of the most widely misunderstood roles in capital markets. "Most people think an auditor is engaged to examine all opportunities for fraud to be perpetrated against a company and this is not true," Harrison says. "Through the release of the thought leadership paper, we are looking at ways at overcoming this expectations gap of what an auditor is contracted to do and what the community in general might expect of them. However, it is not only about educating the consumer but also expanding the scope of audit to better meet expectations."

Harrison says the ICAA is committed to generating the best outcomes for the public, not just for the profession. "With our thought leadership papers we have the opportunity to promote and lead comprehensive and integrated business performance reporting, communications and assurance development initiatives," he says. "Such initiatives have the potential to assist with microeconomic, macroeconomic and socio-economic integration and efficiency."

Shaping legislation and regulations

In addition to these papers, the ICAA plays an active role in shaping legislation and creating and maintaining a set of high standards. "One of our roles is to lead the debate on issues affecting the profession and influencing regulators and policy-makers through representation and advocacy," Harrison says.

It has played a key role in lobbying for personal tax reform. The ICAA has called on the Federal Government to implement a package of reforms for personal tax to address the current weaknesses in the personal tax system.

"The role of our body is about establishing our members' reputations not only to meet their own personal needs but also the public interest," Harrison says. "Attached to this we look to lobby government and other relevant organisations to ensure legislation that is introduced meets the community's needs." ■

It was when I found out I could make mistakes that I knew I was on to something.

ORNETTE COLEMAN

By Paul Wright

Leaping forward

→ Sometimes all a company needs is the right advice and Invetech is one company that specialises in providing it.

Most of Australia's Top 100 companies are instantly recognisable by their own trademark product or service. However, one of the country's top technology firms has built a strong business out of helping today's Fortune 500 and tomorrow's top 100, create their own market-leading products and services.

Invetech, the innovation engine room of publicly listed Vision Systems – a technology manufacturing company – is located in the outer eastern Melbourne suburb of Mount Waverley. It has grown annually by 35 per cent since 1999 with exports to international market leaders accounting for a significant proportion of this growth.

For more than 25 years, Invetech has provided leading-edge, technology-based innovation services, creating more than $A1 billion of value from the design of new products and manufacturing systems for its clients. Almost 5000 projects have been conducted on everything from the world's first drip-free golden syrup containers, crash-proof bicycle helmets, cancer detection systems and breakthrough wool processing systems to the most sensitive smoke alarms installed in over 80,000 sites.

Client collaboration

Client names read like an international Who's Who of business: Abbott Laboratories, Ansell International, Bayer Diagnostics, Braun, Coca-Cola, Cochlear, Formica, Leica, Mattel, National Foods, Phillips, Roche, Shell, Siemens, Telstra, Uncle Tobys, Whirlpool and The Woolmark Corporation.

Recently, the company has made dramatic inroads into the burgeoning biotechnology/healthcare sectors. Invetech, in collaboration with its sister company Vision BioSystems, applies an equally high level of ingenuity and rigour for world-class suppliers and manufacturers of clinical and research laboratory instrumentation. The two organisations have developed and supplied more than 5500 instruments and continue to win major development projects from global leaders.

Several major projects have involved the development of sophisticated testing and analysis systems used in the diagnosis of cancer, anaemia or leukaemia. Scientists use instruments produced by Invetech and Vision BioSystems to process tissue specimens or prepare stained blood smears from blood samples prior to microscopic examination.

What was once a dangerous manual process has now been fully automated and programmed, with significant benefits in operator safety, increased efficiency and quality and substantially reduced labour costs and time scales.

One of these projects involved Silicon Valley-based Applied Imaging Corporation. Working in conjunction with Vision BioSystems, Invetech's SL50 took out top prize in the Australian Design Awards. The SL50 is used by Applied Imaging's customers, in conjunction with specialist image analysis cancer screening systems, to detect rare secondary cancer cells.

Pathologists usually record the identity of each slide prior to manual loading for automated scanning by image analysis systems. The SL50 enables pathologists to load 50 slides for automated scanning, allowing the instrument to run overnight, reducing the turnaround time for biopsy results.

In an effort to fast track projects with other US-based biotechnology firms, Invetech has opened a beachhead office in San Francisco. Within 12 months, more than $8 million in new projects had been secured, including one with Streck Laboratories, a Nebraska-based innovator of haematology control products.

Streck Laboratories turned to Invetech to enhance reliability levels and provide manufacturing expertise for its blood analysis devices. An innovator of haematology control products, Streck manufactures 25 per cent of the haematology controls used in the US and 35 per cent of the controls worldwide.

"We were extremely impressed with the way Invetech handled this multi-faceted project," Brad Hunsley, Streck's research and development manager says. "We had spoken with several other companies, but were unable to find any that could provide the one-stop-shop approach that Invetech could."

While maintaining its involvement with blue chip companies, Invetech is turning its expertise towards helping start-ups become the Fortune 500 companies of tomorrow.

One of these ambitious start-up companies – Opdicom – is launching a new computer-based peripheral stacking system that aims to revolutionise disc storage.

The system is an automated carousel that stores, protects and retrieves CDs, DVDs, music and game discs. It is aimed at business, studio and home markets and effectively gives users instant and efficient access to their disc libraries.

Within 12 months, more than $8 million in new projects had been secured

Invetech played a key role in developing the product, having been called in not long after Opdicom first opened. Lindsay Lyon, Opdicom's president and chief operating officer, says while the company's core expertise lay in sales, marketing and software development, he realised it needed to establish strong partnerships with other companies to get its product ready quickly.

"We knew from the outset that to get this product to market fast, we'd need to partner with other professionals," he says. "Invetech provided us with industrial, electronic and robotic design, effectively delivering us a functioning prototype in 90 days."

He says the benefits from working with Invetech far outweighed the costs of outsourcing. "At day's end, it's an investment," he says. "You only get what you pay for in product development."

Invetech helps companies face the unknown and move to new levels of commercialisation and profit.

Looking before you leap

The decision by any company to take a leap forward with innovative products can propel profits to new levels, but it is not without risk. Invetech also assists clients in managing risks associated with new product commercialisation by providing support in innovation management, research and development project and portfolio management and commercialisation process management.

For example, Mildura-based Tandou Ltd came to Invetech for help in assessing various technology offerings. Invetech's commercialisation strategy provides robust, objective advice from initial product/technology search and assessment and commercial valuation of intellectual property through to determining the most effective value proposition and route to market.

Tandou Ltd had grown from a privately held company with $45,000 of issued capital to an ASX-listed company with a market capitalisation of $40 million.

It sought commercialisation counsel on how best to fast track development of an Australian technology that gives wineries an environmentally sensitive way to boost the recovery rates of wine available in grapes from 88 per cent to 98 per cent.

"This project would not have been possible without Invetech," says Bob Smith, 's founder and managing director. "The company not only provided invaluable due diligence on the relevant technology to be adapted, but developed a realistic financial model, represented our interests in business dealings with other partners, and prepared a submission for a major Commonwealth Government grant."

Invetech's practical application of technology reduces the risk in new product and process innovation through the experience of its integrated skill base. It can also either work closely with the clients at their own premises or deliver a world-class product quickly and efficiently from its Mt Waverley facility. ■

Paul Wright is chief executive officer of Invetech.

By Jane Breusch

Beyond borders

→ Austrade is successfully taking up the challenge of putting Australia's products and services onto the world stage.

Turning sport into good business: former Wallabies coach Rod Macqueen at one of Austrade's Rugby Business Club Australia events.

Australia's geographic isolation provides constant challenges to exporters seeking new markets. Attracting foreign interest in local goods and services is one of the aims of the Federal Government's trade body, Austrade. The organisation is always seeking new ways to showcase the country and secure important deals.

Austrade's chief economist Tim Harcourt says the organisation is the eyes and ears of Australian exporters and an innovative approach is required to showcase Australia to the world.

"Australia's distance from major markets provides a challenge but one that we are meeting with a range of initiatives," he says. "Some of the ideas have taken advantage of major world events such as the 2000 Olympics and the 2003 Rugby World Cup. These events provide prime promotion opportunities for exporters."

Two initiatives that were set up to attract foreign business people during these events were the Business Club Australia (BCA) during the Olympics and the Rugby Business Club Australia (RBCA) during the Rugby World Cup.

Both clubs aimed to connect Australian exporters with overseas buyers and investors through a global networking club. According to Austrade research, around 50 per cent of all new exporters enter the global marketplace "by accident" through a random event or a chance meeting. The business club concept was developed to increase the likelihood of such a meeting.

Mixing business and pleasure

"With the Olympics, we decided to have a business club so that business people who were travelling to the Olympics could be matched up with locals interested in joint ventures," says Harcourt. "We also felt that visitors who were here for the rugby would be interested in opportunities to forge stronger ties in trade and investment with Australian businesses."

BCA was a huge success for Austrade and Australian exporters. One member, architecture company Woodhead International, secured a contract worth around $A1.35 billion to design the interiors at Singapore's Changi Airport Terminal 3 after networking during the Olympics.

Thinking outside the square is also required to help Australian exporters maintain contact with clients in countries affected by violence and terrorism such as Indonesia and Iraq and to deal with unexpected situations such as the SARS (severe acute respiratory syndrome) outbreak.

The situation in Indonesia following the Bali and Jakarta bombings, when the Australian government warned all Australians against non-essential travel, meant Austrade had to develop new ways of keeping the lines of communication open. Teleconferencing became a popular way of doing business with Austrade representatives in the affected country, helping to tie up the final parts of any deals.

"They basically had to do every part of the deal online and then we've had to help with the face-to-face consultations in the end," Harcourt says. "It was a new way of carrying out the deals and people aren't used to that approach. It certainly helps to take an innovative approach when it comes to assisting our exporters." ■

Showcasing strategies

Austrade showcases Australian exports by developing innovative strategies to promote the country. Some strategies include:

- E-commerce workshops
- Business forums for new exporters
- Networking opportunities to allow exporters to meet with officials from foreign governments
- Trade fairs for different industries and products
- Trade missions to explore opportunities in various countries
- Seminars, conferences and exhibitions to showcase Australian exporters and teach them new skills.

By Gayle Bryant

A winning institute

→ A new Australian organisation is fostering innovation and helping companies to commercialise their intellectual property.

Getty Images

> Commercialising innovation keeps Australian companies ahead of the pack.

Australia is one of the world's most competitive economies – a situation that has come about through implementing initiatives in both the private and public sectors.

However, to remain competitive, the country needs to encourage innovation and commercialise worthwhile initiatives as they arise. One organisation that is dedicated to carrying out these tasks is the Australian Institute of Commercialisation.

In a report entitled "Assessing Australia's Innovative Capacity in the 21st Century", authors Professor Joshua Gans and Professor Scott Stern state that Australia cannot continue to live on the fruits of its natural endowments. While this is not a new thought, it has been masked by the stellar growth rates of the past two decades and driven by the success of micro- and macro-economic reform.

According to Dr Rowan Gilmore, chief executive officer of the Australian Institute of Commercialisation (AIC), as productivity improvements from these reforms begin to slow, Australia's ongoing prosperity and competitiveness will be affected by its ability to create new products and services. It needs to continually adapt to remain ahead of its competitors.

Commercialising innovation

"Establishing Australia as a 'first-tier innovator nation' requires a systematic upgrade to our innovation and commercialisation environment," Gilmore says. "Such an environment is a result of a combination of factors, including well-trained people, an innovation-oriented corporate investment climate and greatly improved collaboration between research organisations and businesses."

He says Australia's research capability is often world class and needs to ensure this capability is converted into social and economic capital, new businesses and new jobs.

"The AIC's role is to improve the skills required to better commercialise intellectual property. It also develops programs and policy recommendations to change the Australian culture to embrace outcome-based research, collaboration with business and sensible, calculated, risk-taking behaviour."

Three programs to provide assistance to people and industry looking for commercialisation opportunities have been set up by the AIC.

"First, our AIC Connect program provides a platform to establish national and global networks to better coordinate and leverage existing commercialisation activities," Gilmore explains. "For example, as part of this program, the AIC has launched Aussie Opportunities – a web-enabled program and database to showcase Australia's research and technology projects. It is being delivered through a joint venture comprising the AIC and the Pacific Capital Corporation with support from Knowledge Commercialisation Australia and InnovationXchange."

The second program, called AIC Know-How, assists in expanding Australia's depth of commercialisation management experience. The program partners with universities and industry to bring together professional development courses and create a pool of knowledge on commercialisation practices.

The final program – AIC Access – assists in better measuring Australia's commercialisation activities.

"What cannot be measured cannot be managed," Gilmore says. "Armed with improved information and assessment tools, industry can better allocate resources and government can better formulate public policy."

The AIC and the Victorian Government commissioned a report entitled "The Economic Impact of the Commercialisation of Publicly Funded R&D in Australia" by the Allen Consulting Group. It found the value of continued improvement in the commercialisation of Australia's research, compared to a levelling off in performance, could be worth about $A20 billion a year to the economy and $18 billion in additional exports by technology-based Australian companies by 2020.

"Hand in hand with the allocation of funds to research should be a greater allocation to the skills for commercialisation of this research base," Gilmore says. "The business environment must also become conducive to fostering the conversion of small technology-based businesses into emerging global competitors in new markets." ■

By Jane Breusch

Taking care of business

→ Australian companies can find help in reinventing themselves, achieving export success and identifying their unique strengths.

Australian Business Limited (ABL) knows an innovative approach is required to assist its members in managing their operations. It has introduced several programs including a knowledge management system and a customer relationship management database to help with their processes and expansion plans.

According to Mark Bethwaite, managing director of ABL, innovation is not something you do once.

"It must be done systematically and continually," he says. "This is a major aim of ABL."

The organisation, formerly the New South Wales Chamber of Manufacturers, provides advice, training and guidance on business processes such as exporting, but it also helps companies to reinvent themselves to remain current and to identify their unique strengths.

ABL is one of Australia's oldest and largest industry and business

support organisations with 5000 full members and another 14,000 associate members mainly in New South Wales.

"Our mission is to assist our members to manage their businesses and to grow," says Bethwaite. "Our first strategic goal is to be member- and customer-driven, and to meet this goal we have introduced the knowledge management program and customer relationship management database." The knowledge management system seeks to capitalise on the knowledge that ABL has amassed in its 120-year history, particularly in industrial relations, trade and State and Federal Government policies.

The system will centrally store all of this information in one database rather than in multiple databases and it will "tag and flag" every document that goes into that central repository with key words, subject and relevant content. This will enable the information to be more easily retrieved.

"This will help us to avoid the 'reinvention of the wheel' syndrome," Bethwaite says. "It is currently being developed and should be rolled out in early 2004. We expect it will make quite a significant change to our internal processes. I would hope that we will become much more efficient and responsive."

Navigating members through innovation and growth strategies is a key role of Australian Business Limited.

Getty Images

Satisfied customers

ABL's customer relationship management database aims to streamline processes for its members and clients, and to provide a portal through which they can directly access the information they need.

The database is also used to facilitate services such as ABL's apprentice placement program. Potential apprentices can log onto the database and view details of the apprenticeships that are on offer.

> We also help companies to approach the market in a more innovative way because innovation isn't just about new technology

Express Coach Building, an ABL member, has sourced all of its trainees and apprentices through ABL's Australian Business Employment Network. Another company, Clements Medical Equipment, hopes to use ABL's network to establish a union of Australian medical equipment manufacturers to encourage product development and increase the industry's competitiveness internationally.

Boosting the chances of international success for Australian companies is what Chris Gibbs Stewart's job entails. As ABL general manager, International Trade & Business Solutions, Stewart

Helping hand

Australian Business Limited (ABL) is an independent member-based company, founded in 1885. Since that time, it has worked with thousands of businesses, from owner-operators to large corporations, connecting them with the information, people and advice they need to make sound business decisions.

Some of the services that ABL provides include:

- Workplace relations advice, including changes to awards and implications for particular businesses. Industrial relations lawyers are also available to advise ABL clients.
- Occupational health and safety guidance including diagnosis of potential hazards and unsafe work practices, policy development and training programs.
- Apprenticeship and traineeship advice for businesses and assistance in completing paperwork and arranging access to government incentives.
- Resources and expertise to enable companies to develop their businesses in international markets and access capital for growth.
- Help to raise capital for projects.
- Lobbying of State and Federal Government policy makers to ensure the best possible environment for business growth.
- Events, involving key industry speakers, to give its clients opportunities to learn about important issues and changes facing their businesses.
- Defence business intelligence to provide members with a source of information on business opportunities, networking activities, policy and member success stories.

and her team help 400 to 500 companies a year with their international endeavours, including preparation for an initial foray into export markets, assessing their product or service, and building strategies for their moves.

"We work with many innovative companies that have great products and services that are world-firsts," she says. "We also help companies to approach the market in a more innovative way because innovation isn't just about new technology, but it can be innovation in processes and strategies that help to drive the business forward and make it more competitive globally."

She says that ABL can add value to the companies by helping them to think of different ways to crack particular markets. ABL's International Trade & Business Solutions unit has 20 staff including country specialists. These are people who have lived and worked overseas and have intimate knowledge of particular markets.

The team also has one person based offshore in India. Raj Kaul recently helped Australian company Hydramatic Engineering to enter the Indian market through organising meetings and making visits as the company's representative.

Hydramatic Engineering was established in the early 1970s when it began manufacturing a unique roof-bolting machine used for bolting the roof and sidewalls of underground mines. It moved from retro-fitting roof bolters to continuous miners to building dedicated bolting machines on caterpillar tracks.

The company, already a successful exporter to various parts of the world, realised that local knowledge and language was needed to break into the Indian market.

Kaul, along with the Hydramatic sales and marketing director, approached Coal India with a unique product – the world's first long wall face bolter – and now the company is confident of winning a tender.

When helping a company to successfully export, ABL also looks at aspects such as new distribution channels, product repackaging, different strategies to develop networks internationally and involving key opinion leaders involved in what a company is doing.

Improvements in these areas can be particularly important in offsetting obstacles such as Australia's geographic isolation, higher transportation costs and high labour costs.

"Australian companies need to be head and shoulders above the rest and the only way they can do that is by coming up with unique approaches to the market, and that's where we come in," Stewart says. "What Australian manufacturers are going to do, particularly as China comes online and more manufacturing is carried out in lower labour cost markets, is to really look at where they're adding value and what their uniqueness is in the Australian market so they can translate these individual strengths overseas." ■

Foundation for growth

Australian Business Limited's (ABL) decision to set up an independent private sector research think tank in 1997 was in itself an innovation, according to ABL Foundation managing director Narelle Kennedy. In doing so, ABL took the chance that the foundation might actually come up with "heretical thoughts" about business policies and needs.

"We sometimes come up with answers or solutions that might be anathema to ABL's business, so it was quite courageous in its approach," Kennedy says. "Having both the vision and the financial wherewithal to support a research think tank of this kind is part of its special difference."

The Foundation, in partnership with universities and outside experts, researches strategic priorities, which it identifies at the start of each year. The research is aimed at helping three key groups: businesses, public policy makers and opinion leaders, analysts and academics.

It has produced several landmark research papers, including Alternative Futures, which details four potential scenarios for Australian business in 2015, and The High Road or the Low Road, which pioneered the idea that innovation rather than cost-cutting is crucial for future growth of business.

"Alternative Futures defines the philosophical and practical methods we work from in forecasting outcomes and business needs," Kennedy says.

Future evaluations

"While not all our work is based on alternative scenarios, the reverberation of the intelligence from that project has influenced many aspects of our research. We continually evaluate the future from the point of view of what is going to make Australia competitive."

According to Kennedy, fresh competitive strategies based on innovation and knowledge are vital in the face of an increasingly volatile and globalised business environment. She says the Foundation's research has shown the only difference between the old economy and new economy is that business is now faster, more interconnected and knowledge-driven.

"This means there is a pressure on companies to grasp and incorporate change and innovate quickly because everyone can access any unit of knowledge at the click of a mouse. We take that to be the critical, sharp edge of a trend like globalisation."

The Foundation has also found that the ease with which business can be conducted online is pushing a trend for companies to reinvent themselves. "Businesses are competing by adding products and services together; a crucial point worthy of serious consideration," says Kennedy.

She cites the example of a gearbox manufacturer which found that cheaper imported products were cutting it out of the market, so it decided to expand into installation, training and design work.

"They have now moved from being a competent manufacturer of the gearbox to also being a service provider of training, quality accreditation and finance. They are now much less likely to lose a customer because they're entwined and embedded with that customer."

Kennedy says while the Foundation is not a lobby group, its research is used by other groups, such as ABL, to influence government policy.

"We can't rely on the reforms of the past to carry us into the future," she says. "While we have a strong economic performance right now, a lot of our work suggests that this progress might have been generated by the reforms of the past 20 years. Maybe that's a well that runs dry and it's time to take the next quantum leap. We need public policies that help foster a climate where businesses can take risks."

By Gayle Bryant

Start me up

→ ATP Innovations is a business accelerator that helps emerging companies to commercialise their ideas.

Children with diabetes are just one group who will benefit from the initiatives of ATP Innovations (ATPi). Currently under development are tests that monitor hypoglycemia in diabetic children to alert parents when an episode is likely. It is one of the innovations of this business accelerator that supports emerging businesses in the biotechnology, ICT and electronics sectors.

This particular product, from startup company AI Medics, is rapidly moving through the commercialisation phase – a situation that would have proven more difficult if it were not for the support and expertise received from ATPi.

Established in 2002, ATPi is owned by four universities: Australian National University, University of New South Wales, University of Sydney and the University of Technology, Sydney. It is Australia's largest technology business commercialisation precinct. Paul Field, director of ATPi's BioBusiness division, says the organisation focuses on assisting emerging technology companies to establish their businesses.

"We only work with early-stage business opportunities from both the private and public sectors," he says. "We offer them a targeted program of business development activities that gets them to the stage where they can approach outside companies for finance."

ATPi's operations are based out of its premises in the National Innovation Centre, located within the Australian Technology Park in the Sydney suburb of Eveleigh. Already, about 30 companies are based at its site.

"We gain equity in some of the companies that come to our precinct – also some that are located off-site – and we take what we call 'sweat equity' in return for our commercialisation experience, expertise and hands-on assistance," Field says. "We have access to a range of contacts that companies need when first setting up, such as patent attorneys, business angels, accountants and mentors. If we don't know the answer to any of their issues, then we know where to turn to find out."

We take 'sweat equity' in return for our commercialisation experience

The companies locate to ATPi and pay minimal rent for office space. The ground floor of the National Innovation Centre is being converted into a medical precinct with funds it received from a $A2.5 million biotechnology grant awarded by the NSW Government in 2002.

Field says he receives about two proposals a week from start-up companies, with the ICT division receiving about the same. Before engaging the services of ATPi or relocating to the precinct, a company has to show it has a product or service that can be commercialised and an entrepreneur or management team with realistic expectations.

Getty Images

ATP Innovations helps start-ups move to the next level.

"We judge proposals on the likelihood of their being commercialised and this includes judging the skill and enthusiasm of the people starting the company," Field says. "At the end of the day, you are backing people as much as a technology because history shows resourceful people build successful companies even if the technology is not necessarily earth-shattering."

He adds the hot areas in biotechnology are tissue engineering, stem cell technology, biomaterials, artificial bone and muscle replacement, nanotechnology and implanted medical devices. Companies at ATPi include Medsaic, which makes diagnostic tests for leukemia, and cardiac catheter developer CathRX.

The ICT division is supporting a number of companies that are gaining a profile. One is WebMCQ, which was named in *BRW Magazine*'s list of the fastest growing companies. Another is sensor company E-Nose, which is developing a product for the electronic detection of odours. It has carried out tests at an abattoir in Goulburn to detect when odours have reached an unacceptable level.

Field says ATPi aims to be the premier site for emerging technology companies in NSW. "We are also working in regional areas with local universities, including Charles Sturt in Wagga Wagga, Southern Cross at Lismore and ANU in Canberra," he says. "We also have links with many of Sydney's teaching hospitals and research units in health sciences – a mutually beneficial and valuable arrangement that is unique in Australia." ■

By Jane Breusch

Measuring Australia's progress

Encouraging informed decision-making is one of the main innovations taking place at Australia's official statistics organisation.

Does our country measure up?

Innovation is such a way of life for the Australian Bureau of Statistics (ABS) that breakthroughs in the collection and presentation of data and use of technology are expected on a regular basis.

Australian statistician Dennis Trewin says the ABS is constantly searching for new ways of conducting its business and, as a result, efficiency improvements and product development occur almost as a matter of course.

"Innovation is essential," he says. "The nature of the demand for our products and services is changing all the time. We also have to generate productivity improvements in order to fund salary increases and the return to government. Most of the innovation comes from our smart use of technology or from our smart use of statistical methods."

One of the ABS' recent innovations is the groundbreaking publication, *Measuring Australia's Progress*. The publication is widely acknowledged as having set a new standard in measuring a society's progress through its comprehensive compilation of economic, social and environmental indicators.

Smart stats

The ABS plans to showcase its smart use of technology by contributing to a project to develop an integrated system that will allow medical researchers to access through a single portal statistical information currently held in various databases.

This project is being carried out under the leadership of the 2003 Australian of the Year, paediatrician Fiona Stanley. It will provide her with quick and easy access to a wide range of data on children's health held by governments, universities and the ABS.

Trewin says the strong relationship between technical and methodological staff and the business areas is the driving force behind innovation at the ABS. Part of this process is the annual ABS Tech, a week-long event aimed at bringing all 3000 employees in the nine ABS offices up to date with the latest technology and the different ways it can be used.

"The key message is the importance of maintaining the strong alignment between our technical and methodological people and the rest of the organisation," he says. "This has to be two-way. The technical and methodological people have a need to understand what the rest of ABS staff are doing but our other staff need some understanding of what's happening with technology and with methodology."

As part of its ongoing innovation campaign, the ABS maintains contact with statistics bureaux in Canada, the United Kingdom, New Zealand, Sweden, the Netherlands and Finland, sharing ideas and experiences in mutually beneficial relationships. ■

Jane Breusch is a finance writer and former Reuters correspondent.

MAPping Australia's measurements

The Australian Bureau of Statistics (ABS) has developed a new range of indicators to track key aspects of Australia's progress over a decade.

Known as Measuring Australia's Progress (MAP), the report combines economic, environmental and social data and allows readers to decide whether Australia is progressing and at what rate.

The basis of the report is that you need to look at many facets of life, such as health, quality of the environment, incomes and security to understand progress. This is a significant departure from the traditional dependence on economic indicators such as gross domestic product.

"A lot of the information was already in the public domain but it was all over the place and this report just brought it all together in a nice, concise summary," Australian statistician Dennis Trewin says. "We deliberately designed the publication for a reading age of about 15 so it was accessible to a large proportion of the population."

The US General Accounting Office is using MAP as the model for a similar project while Sweden and Ireland are also considering using it.

By Valerie Khoo

Discovering entrepreneurship

Teaching people how to embrace innovation and become entrepreneurs is the challenge for this progressive organisation.

Getty Images

Personal passion drives people on the road to entrepreneurship.

Coming face to face with the fear of doing what you feel like doing, and doing it anyway, is one of the key themes of a new Australian course in entrepreneurship. According to Hans Wijgh, co-founder of Feelike, potential entrepreneurs need to confront some personal truths before they can truly succeed.

"Everybody is born with an entrepreneurial spirit," says Wijgh. "But as you go through life, many layers build up and that spirit becomes buried in the process."

Rediscovering your capacity for entrepreneurship is what Wijgh's Feelike seminars are all about. Held in all states in Australia, Feelike is a four-day seminar that takes participants on the first few steps of their entrepreneurial journey.

Wijgh says entreneurship and the capacity to innovate and succeed are not intangible concepts only espoused in motivational books. "To make the most of an innate entrepreneurial spirit, people need to access their core values and identify what sets them apart from other people," he says. "They need to be given the basic tools to turn their passion into a business. That's what will get you on the road to entrepreneurship."

Passion is a key word in Wijgh's vocabulary. He is a strong advocate for personal fulfilment as a fundamental driver of innovation and success. "I have a distinct dislike for the way society restricts people's opportunities," he says. "Many social structures confine people's talent. When that happens, you simply end up giving in to other people's passions and excitement instead of developing your own."

Wijgh says that although people have something to contribute in life, many don't know what that 'something' is. "When people identify what they want to do in life, they are more fulfilled and society also benefits. Feelike is a seminar program that challenges people and gives them opportunities to design such creative lives."

Unlike seminars that see hundreds of people listening to a speaker for hours on end, Wijgh says that Feelike adopts an innovative role in the learning and discovery process. Participants are not involved in a theatre-style set-up, taking notes and watching Powerpoint presentations. "It's a challenging course – people come face to face with the fear of being an entrepreneur," he says. "The reality is that being an entrepreneur is like being on a rollercoaster ride. Your heart thumps, and you are forever on the edge looking over a precipice, not knowing whether that hole you are looking at really is a hole or merely the sky reflected in a shallow puddle. The seminar is four days of high-energy interactive experience."

Although comparisons with motivational guru Anthony Robbins are inevitable, Wijgh says that participants are presented with practical business tools that are focused on issues such as identifying markets, nurturing start-ups, securing capital and building your team. "Participants learn how to convince someone else that their project is worth funding," he says.

Encouraging entrepreneurship and innovation is one of Wijgh's personal passions. "A society that has a high birth rate of companies is going to be a successful society," he says. "In fact, this is essential because individuals and corporations need to continue reinventing themselves. What exists today will often no longer be relevant in five years."

According to Wijgh, Feelike incorporates an examination of personal values with a firm foundation in practical business sense. Added to this are resources and tools to assist entrepreneurs hoping to take the plunge into a new venture.

"Everybody can make it," he says. "Don't let other people pull you down. Find your sweet spot and go for it." ■

I had to make sure my own management style was constructive

leaders SPEAK

Performance through people

Imagine you are the chief executive of a company that shows signs of business stagnation. Also imagine that the company is facing external pressures such as changes in traditional end-user markets. What do you do?

At Bayer Australia, we implemented a change management program to create a constructive company culture after we measured our culture profile and realised there was evidence of a silo mentality.

Company culture is not a job the chief executive can delegate. It is core to leadership. And the first step I had to take was to "look in the mirror" and make sure my own management style was constructive.

The next step resided with the whole leadership team. Eighty-one managers underwent an in-depth analysis program and received feedback on their behaviour and how it impacted on the company's culture.

Our human resources department rolled out a new competencies framework and much-improved appraisal system, all of which focused on rewarding performance.

The outcomes have been enormously positive. Over the past two years, we have sustained and exceeded our financial targets. We have created a constructive culture, helping us to retain and attract good staff.

Our communication has become more transparent and everyone in the company knows the direction in which we are heading. Perhaps more importantly, we are engaging our employees to shape the future direction of the company.

ALAN MCGILVRAY
SENIOR COUNTRY REPRESENTATIVE
BAYER AUSTRALIA AND NEW ZEALAND

Improving profit margins

There are two conflicting methods to drive improvements in a company's profit margin. Cost-cutting has been easier and a primary focus in past years but can limit a company's ability to grow long term. The greater challenge is to demonstrate tangible value delivery to your customers while leaving your company poised for the future.

The key is to design and implement a strong customer-facing corporate culture through measures such as:

- Developing staff capable of operating across all lines of business;
- Ensuring you differentiate your products or services;
- Sharing and disseminating information effectively;
- Moving up and across the value chain by leveraging strategic partners; and
- Planning based upon customer needs.

ANDREA GALLOWAY
PRESIDENT AND GROUP MANAGING DIRECTOR
SPHERION GROUP

By Mark Hollands

The business of people

→ Drake International helps companies find the right person for the job – all at the touch of a button.

Getty Images

Yes, no, maybe: software solutions take the paperwork out of recruitment.

When it comes to innovative ideas and best practice, recruitment leader Drake International likes to think long-term. Its Australian operation has spent four years and more than $A2 million building software to provide clients with technology to improve candidate selections and drive supply-chain efficiencies.

Taming technology will be the most influential factor for success in the recruitment industry for the next 50 years, the chairman of Drake International's Australia board, Ron Urwin, believes.

The company already boasts arguably the most sophisticated web-based staff hiring and management system in the Australian market, providing companies and candidates with the opportunities to leverage the power of the internet.

Some 30 per cent of its Australian client base have already made the Drake "Online Ordering" solution an integral part of their recruitment process, using it not only for candidate selection but as a portal through which they manage multiple recruitment suppliers.

It has been designed to reduce invoicing costs, assist the batching of internal documents and integrate contractor details regardless of agency loyalty.

"If you cast a long-term view," says Urwin, "then you cannot deny the importance of technology and innovation in business practice."

Enhancing efficiency

Urwin cites one of Australia's four largest banks as a significant beneficiary of Drake's web technology, achieving more than $7 million in cost efficiencies.

"These are the bank's own figures and the savings have nothing to do with the calibre of selection," continues Urwin. "Some $4 million of efficiency came from integrating vendor and staffing contracts into one system, making the bank's recruitment system easier to manage. Another $2.6 million came from simplifying the invoicing process."

An important facet of the system is the ability for clients to monitor recruitment projects. "For organisations who seek 150 nurses, or 20 data processors, it is important to be able to see progress in real time," says Urwin.

The system has been such a success that the Australian operation is now beginning to seek export opportunities, looking initially at sharing the innovation across Drake's global customer base and group companies.

Innovation in the recruitment sector stretches beyond cost efficiency, Urwin says.

Drake is one of the first agencies to introduce the interactive resume in which a candidate presents themselves on a pre-recorded video that is streamed across the internet. The webcast features details of the individual's resume, which appear on-screen in synchronisation with their presentation.

Urwin says: "The technology helps you shorten the recruitment cycle. The normal process sees a recruiter get 85 resumes for a position and then separate them into three piles – yes, no and maybe." Using Drake Express, the recruiter can make instant judgements on who should be shortlisted. Drake recently placed a doctor using the interactive resume. It was a special appointment because the individual was a refugee and in detention.

The use of technology will continue to be a strong feature of the recruitment process. Another prevailing trend will be shortage of labour "across the board", Urwin says.

He has already seen the stresses created in countries with an insufficient number of nurses. Drake has become the largest importer of nurses in the United Kingdom, placing more than 150,000 after permission from the British Government to accept nurses from Asian nations such as the Philippines.

Other areas of the economy and society can expect similar stresses and will need to be equally flexible and open to new solutions, he says.

The increasing demand for a work/life balance will be another significant factor for the recruitment industry over the next decade and beyond, Urwin predicts.

In an attempt to help companies cope with this challenge, Drake has built a software application called Cost Controller. It aims to help measure the optimum staffing levels of a company or business unit, basing its calculations on workload and time pressures.

"This is another example of our commitment to using innovative technology to help our clients come to terms with the changing demands of staff and society so they can build stronger and more flexible organisations," Urwin says.

By Jane Breusch

Tell me what you want

→ Making employees' needs a top priority has increased staff retention rates for consumer goods company SC Johnson.

Listen up. That is what fast-moving consumer goods firm SC Johnson is doing – and it is making sure that it hears its employees loud and clear. As a result of an innovative human resources program, retention rates have dramatically improved. One particular issue which the program highlighted was that if employees find their needs are not being met, they will switch to another employer.

Since 2000, when its "Enjoy the Difference" program was introduced, staff retention rates have increased to more than 90 per cent, up from 75 per cent in 1999. Organisational health – a measure of recruitment and the value of the lost employee which is costed at 12 months' salary and benefits – has also improved as a result of the program.

Andrew Miedler, human resources director for SC Johnson Australia and New Zealand, says the program is based on the premise that employees are increasingly switching employers to satisfy unmet needs. For example, Generation X employees are expected to have three careers and up to eight different jobs in their working lives, while Generation Y is likely to have five careers and 15 different jobs.

Changing expectations

"Generation X and Y are not demonstrating the same employer loyalty as the baby boomers did because they have a different mindset," Miedler says. "They're looking for an employer that offers something different, such as a value proposition which meets their individual needs while providing a vibrant work atmosphere. If they find their employee needs are not satisfied, they move to another employer who promises to satisfy this need."

Using external market research and internal surveys, SC Johnson identified what current and future employees wanted. From this, it prioritised the top five needs as leadership, career, training, recognition and work-life balance.

The company then developed initiatives aimed at meeting the needs of its employees. These include a wide range of leadership development, competency-based training and career plans; a three-tier recognition plan; flexible hours and a compressed work week; preferred meeting times and onsite services such as pick-up and drop-off of dry-cleaning.

The program has been successful, not only with increasing employee engagement but also in winning a number of industry awards. These include the 2002 Best Retention Strategy at the Australian HR Awards and recognition as the Employer of Choice for Women; 2003 Best Employer Brand at the Australian HR Awards, and Leading Organisation for the Advancement of Women (under 500 employees) at the EOWA Business Achievement Awards for Women. Miedler says external recognition of its HR strategy confirms to current and potential employees that SC Johnson offers leading-edge practices. ■

Catering to the diverse needs of Generation X and Y is vital for effective staff retention.

A model leader

■ **Leadership:** use of leadership models for different levels within the company. SC Johnson has also introduced a uniform leadership vocabulary so that feedback is delivered consistently throughout the organisation.

■ **Careers:** a key aim is to internally fill vacancies through cross-functional moves, vertical moves and international opportunities.

■ **Training:** competency gaps are analysed within each of the company's functions, and employees' skills are developed for their current and future roles.

■ **Recognition:** a three-tier program is established that begins with rewarding immediate behaviour and has additional layers of recognition for higher levels of achievement.

■ **Work-life balance:** programs include flexible hours, compressed work week, preferred meeting times (between 9am and 4pm), onsite services such as dry-cleaning, tennis courts, gym and cafeteria, and casual dress. SC Johnson also offers paid parental leave, paid carers' leave and six months unpaid sabbatical leave after five years' service.

Intelligence recognises what has happened.
Genius recognises what will happen.

JOHN CIARDI

By Dr Robin Batterham

All together now

→ Australia's chief scientist says the country needs to direct more funding into collaborative ventures to encourage innovation.

Australia faces many challenges when it comes to raising its scientific profile but there are a number of programs in place to see that we are able to meet them.

For example, the Government's five-year, $A3 billion science and innovation program, Backing Australia's Ability, has been in place for over two years and is clear recognition of the importance of long-term funding to sustain a culture of innovation in the Australian community.

There is no doubt about the excellence of our science base – the advances in quantum computing, astronomy, neuroscience down to the cellular level and the nano-sized membranes that will help bring in the hydrogen economy. These are snippets of an exciting range of world-class science that has been a long tradition in Australia and one that has given us our fair share of Nobel Prize winners.

Collaboration is vital if Australia is able to face the next chapter of innovation.

Critics criticised

Australia is often criticised about how its business expenditure on research and development is much lower than the OECD average. Critics should look first at the economic base of Australia. The agriculture and mining sectors, and some aspects of the manufacturing sector, compete well on the world stage partly because they perform the right level of research and development.

That said, the knowledge economy is with us, along with global markets, high values placed on intellectual capital and ever-changing alliances and markets. To compete, we must generate more opportunities as well as maintain our competitive base. Australia's commercialisation of government-funded research is improving and we are familiar with the success stories of Cochlear and ResMed but we need 10 times as many successes. So how do we see more entrepreneurial activity and innovation in the market place?

One way is to target the country's small to medium-sized enterprises. They are characterised by tight funding, an inability to cope with the turbulence that the marketplace can generate and lack the time to develop promising ideas and deliver them to the marketplace. CSIRO and the Australian Institute for Commercialisation join many initiatives targeting this area.

A second way is to realise that innovation is more than just new technology being brought to the marketplace. A winner of a competition for a title to describe the capabilities of Australia to innovate came up with the line that Australia is a "triple A nation: we adopt, adapt and advance". This emphasises the key areas that contribute to innovation.

Finally, the nature of science and innovation is changing. The opportunities are coming from the cross-cutting intersection of disciplines or fields of endeavour. The insect vision work from Professor Mandyam Srinivasan at the Australian National University and its application to flight control is an example.

Collaborative research and development and innovation are new beasts compared to what we have been used to. While our Cooperative Research Centres program is one example of bridge-building between universities and industry, we need many more initiatives. We must direct a significant amount of our funding and thinking into collaborative ventures. This is one of our biggest challenges. ■

Dr Robin Batterham is Australia's chief scientist.

By David Salt

Scientific wisdom

→ A leading scientific institution, the Australian Academy of Science, has been honouring and promoting good science for 50 years.

What are the major issues in science and technology that will impact on our nation over the coming decades? How should we respond to emerging concerns about genetic engineering, population growth, national security and environmental degradation? Such issues are being dealt with by one of Australia's leading scientific institutions – the Australian Academy of Science.

Regarded by some as a national treasure, the Academy, which turns 50 in 2004, was set up to deal with all types of scientific concerns. The Academy is an independent, non-profit organisation comprising about 350 of Australia's leading scientists. These scientists are known as Fellows of the Academy and they occupy a range of senior positions throughout Australia in universities, the CSIRO and industry.

Dr Jim Peacock, president of the Academy, says Fellows are not paid for being a member; instead they participate for the privilege, opportunity and responsibility that comes with membership.

"The responsibility lies in ensuring the Academy fulfils its charter which is to honour and promote good science," he says. "Good, strong science is vital to the economic, social and environmental wellbeing of the nation. The Academy does this by recognising excellence in research, advising government and organising scientific conferences."

The Academy is also involved in science publishing, fostering international scientific relations and contributing to science education and public awareness. When Australia's top scientists pool their collective wisdom on an issue, people listen.

"You might not see the Academy striding the public stage and engaging the government in highly-charged, confrontational debates," Peacock says. "However, it is working behind the scenes ensuring our nation's top scientists have the opportunity to make a positive contribution to important debates."

The Academy is on the Prime Minister's Science, Engineering and Innovation Council, which advises the Prime Minister on important scientific issues. It also publishes a range of reports on public issues such as stem cell research, pesticides and human cloning. These reports have a big influence on government policy. The Academy also makes direct submissions to government ministers and inquiries.

"Fellows of the Academy are in a position to make a real difference," says Peacock. "For scientists, making a difference through their science is usually a driving ambition."

Science stars shine

It is not just the Fellows that are making this contribution. The Academy actively encourages emerging career scientists to join the debate. One way it does this is through its annual Frontiers of Science conference in which rising scientific stars are invited to speak on areas of emerging interest.

The Academy is also engaging teachers, students, policy-makers and members of the general public to be part of important scientific debates. Current issues being explored include mapping salinity, nanotechnology and the role science can play in national security.

[The Academy is a leading player in science education.

Getty Images

The Academy is a leading player in science education with its highly successful "Nova: science in the news" website and its award-winning Primary Investigations program which encourages science at primary school level.

The Australian Academy of Science is turning 50 at a time when science and technology is changing the world faster than ever before. Australia's brains trust of eminent scientists will play a key role in navigating the challenges thrown up over the next 50 years and beyond. ■

By Kristin Austin-Bird

Reading, writing & research

→ By developing industry partnerships, one Australian university has pushed innovation beyond its previous boundaries.

Thomas Edison once said, "If there's a way to do it better … find it." Therein lies the heart of innovation and often it runs contrary to established customs. This is exactly what the University of Technology, Sydney (UTS) is doing with its research – finding a way of not only doing it better, but differently. How? Through partnerships.

"Globalisation has encouraged UTS to develop true partnerships both internally and externally," says Stephan Wellink, director of UTS Research and Commercialisation. "We're lucky we are a modern university, which gives us a more flexible approach and a less limiting outlook. However, we've had to build different work models, and review and focus on our areas of strength. This is integral to any true innovation process. We've also had to think through how we ensure partnership success."

The best way to get a great idea is to get lots of them

Wellink quotes Linus Pauling, a two-time winner of the Nobel Prize, who said: "The best way to get a great idea is to get lots of them". Wellink says this is the essence of UTS Research's innovative model. Not only do its partnerships pool the wisdom of many academic and corporate minds, which in turn leads to the gathering of many ideas, but they combine the minds from different schools of thought, areas of specialty and disparate disciplines.

UTS is ambitious in its plans for research and development leadership. Its most fundamental value is the capability to attract a critical mass of scientists and researchers with different skills and backgrounds. "We want to be a key university research centre with worldwide standards and achievements," says Wellink.

A meeting of minds

An example of a meeting of the (different) minds is UTS's partnership with Alcatel, the global telecommunications giant.

The UTS/Alcatel research partnership brings together people from the Engineering and IT faculties and the Institute for Information,

Strong partnerships are key to success at UTS.

Getty Images

Communication and Technology (IICT), which has input from across the university. The breadth of expertise on offer was just one of the reasons Alcatel chose to work with UTS.

"We see our global universities' alliance as the way of the future and therefore it was critical we got the 'fit' right," says David Hudson, partnering director for Alcatel Australia. "There is such value in this long-term partnership, not only for extending Alcatel's knowledge but in encouraging young Australians to play a part in developing some of the best R&D in the world."

But Alcatel is not a one-off for UTS. UTS has already tested the theories of cross-disciplinary input and partnering and knows they produce results. MindSwitch technology and AI Medics Pty Ltd, just two of UTS's innovations, saw engineering and science researchers developing concepts (and working models) of mind-driven wheelchairs and non-invasive blood glucose monitors for people with diabetes, among other things.

MindSwitch is a technology that allows anyone to turn an electrical appliance on or off with the power of thought. This incredible-sounding invention can be used by anyone but is designed for the disabled. You close your eyes and focus your thoughts to a specified device. A burst of alpha waves from the brain does the rest.

AI Medics uses breakthrough technology to help those who can go into seizure while asleep and are unaware of their blood-sugar levels. It is a particularly powerful solution for babies who suffer from the condition and require constant parental observation.

AI Medics has been bolstered by a $A200,000 Federal Government grant and a further $80,000 from the NSW Government.

Another UTS venture is PacMab Pty Ltd, which received $250,000 from the Biotechnology Industry Fund to continue its work in the field of genetically-engineered cancer therapeutics. This is a global market worth more than $1.5 billion a year.

A joint venture has also been established for IT benchmarking software that allows computer systems engineers to measure and evaluate performance. UTS set up its company, Avolution, to work with a Federal Government start-up, Item 3, to commercialise the technology, called Architecture Based Analysis of Complex Systems (ABACUS). These products and services will ultimately be of great benefit to society. UTS views this as being core to its business.

Cross-disciplinary partnerships

UTS is also clear that its research and development innovation does not just extend to "saleable" objects. "The cross-disciplinary partnership model extends across the whole university," says Wellink. "We've also got great partnerships taking place through the Faculty of Nursing, Midwifery & Health with hospitals both locally and internationally. We are also probably leading the world in the new discipline of social capital (the study of the collective value of social networks), which combines researchers from across the Business Education and Humanities and Social Science faculties."

> The cross-disciplinary partnership model extends across the whole university

What seems special about the UTS model is that the research and development knowledge gained from its partnerships allows innovation to be passed on in a teaching and learning capacity. That can only build better outcomes for the future. ■

Kristin Austin-Bird is manager of marketing and communications, Research and Commercialisation, UTS.

Patience pays off

The University of Technology, Sydney (UTS) has formed relationships with many industry associations and organisations to further its innovative research projects. It has a number of institutes that operate under the UTS banner, such as ones specialising in biotechnology to combat infectious disease, information technology and communication, nanotechnology, water usage and environmental solutions.

UTS also boasts the Institute for Sustainable Futures, which formulates ideas and policies for multiple aspects of the environment, ranging from energy-saving practices for the building industry to sustainable forest management techniques.

A number of specialised units are also allied to UTS, such as centres for Corporate Governance, Health Economic Research and Evaluation and Health Technologies. These groups also include those looking at health technology, innovation collaboration and vocational adult training.

Stephan Wellink, director of UTS Research and Commercialisation, says UTS has helped establish a world-class Centre for Quantitative Finance Research, which provides research and information to brokerages and other financial institutions. This structure ensures UTS is more than simply a dream factory for innovators and scientists.

"You must be patient to be successful in research," says Wellink. "We adopt the philosophy that you have to invest over longer periods of time. Not only must you take a technology from an idea and design to proof of concept, but also understand whether it has true value either to society or commercially."

The preoccupation for Wellink is the sourcing and application of funds that UTS receives to allow its innovation work to continue. A major source of this lifeblood comes from the Australian Research Council. Philanthropic organisations and Australia's corporations are important contributors, too.

Wellink says his university will often seek out finance from companies, offering either equity in a joint venture or the opportunity to license the intellectual property.

"If we were working on an application for a mature technology, then we would attempt to establish a linkage program with a third party through a licensing arrangement," says Wellink. "However, the situation can change when we are dealing with a disruptive technology, in which case we might set up our own company to take it to market."

Meanwhile, UTS continues to keep its eyes well and truly open as it plays its part in driving innovation and creating a future for society and economic benefit for Australia.

By Edward O'Daly

Taking a bee's eye view

→ Using bees as the basis of a new system for aircraft navigation is just one project being undertaken by the Australian National University.

At the Australian National University (ANU), research is about making a tangible difference to tomorrow. Its latest research covers areas as diverse as aircraft navigation systems modelled on bees, a new form of justice and computer chips that operate at the speed of light.

While man achieved flight a century ago, even the most sophisticated fighter jets can be significantly improved with lessons from nature. ANU Professor Mandyam Srinivasan, head of the Visual Sciences Group in the Research School of Biological Sciences, is working with bees to help develop a new generation of navigation systems.

Impressed by bees' ability to fly through narrow openings and land on unstable surfaces such as flowers without the means to triangulate distance with their eyes the way humans can, Srinivasan and his team began investigating bee navigation.

"We were intrigued," he says. "When you see a bee flying into a small hole it doesn't bang into the side walls. We asked ourselves 'How does it do that?'"

Srinivasan found bees regulated their velocity and distance from objects by balancing the speed with which the surrounding environment passed their left and right retina.

He has helped transfer the principles of bee navigation into algorithms for the guidance of pilotless model aircraft. The same principles that put honey on the breakfast table could soon guide the next generation of artificially intelligent military aircraft (unmanned aerial vehicles) into battle.

A project from within the social sciences examines new ways of delivering justice. A judicial system based on punishment has been the accepted model for centuries, yet crime and violence remain a part of our lives. This situation prompted ANU Professor John Braithwaite to ask whether there is a better way. He believes an approach that moves away from the punitive to empower victims could be the solution. Braithwaite and colleagues at the ANU Centre for Restorative Justice are pioneering research that factors compassion into the legal system.

"It may well be our retributive traditions were historically valuable, that tribes which didn't have the will to fight back were wiped out by competing tribes for scarce agricultural land," he says. "But the usefulness of striking out in a retributive way is rather doubtful in contemporary conditions, as it is much more likely to engender endless cycles of violence like we see in the Middle East and the Balkans, for example."

Crime hurts, justice heals

Restorative justice is based on the philosophy that because crime hurts, justice should heal. At its most basic level, it brings victims of crime and perpetrators together to discuss impacts, reasons and just outcomes, agreed upon by both parties.

The most obvious context for the restorative justice approach is in street crime situations but it has also been applied by Braithwaite and his research team to nursing home regulation, white-collar crime and armed conflict resolution.

The humble bee could shape the future of artificially intelligent military aircraft systems.

Meanwhile, photonic research could soon outshine the silicon chip in revolutionising computing.

"Governments and corporations across the globe are placing an increasing reliance on computer-based information but their work is currently impeded by silicon-based technology," says ANU Professor Yuri Kivshar, who is spearheading the research. "Photonic crystal devices are a new generation of tiny, nanoscale optical devices operating entirely with light, which can be controlled far more precisely than silicon-based technology."

He adds that photonic crystals promise a wealth of new devices that could satisfy the need for much faster computers and optical communications.

The ANU is a unique national university inspired by the legacy of its own Nobel laureates whose work has changed the world. Today it is home to a generation of leading researchers who are striving to broaden our understanding of the world to improve our lives for tomorrow. ■

Edward O'Daly is the marketing writer at the Australian National University.

Innovation is the transformation of knowledge into money. Research is the transformation of money into knowledge.

BAYER
Germany

By Gayle Bryant

Learning about education

→ International students are set to make a significant impact on Australia's economy and social fabric in the years to come.

Best and brightest: education could soon rival tourism as an Australian money-spinner.

International education is central to Australia's future economic sustainability, providing a rich source of innovation and growth for the labour market as the nation embraces the knowledge economy while bracing for the impact of an ageing population.

IDP Education Australia – the largest marketing, research and support organisation for international students studying at Australian institutions – has released research that is likely to spark a fundamental rethink of the role of international education in Australia.

About 200,000 international students are enrolled in Australian universities, with up to 300,000 across all education sectors, making international education the country's third-largest services export. The IDP report "Global Student Mobility 2025: Forecasts of the Global Demand for International Higher Education" predicts that by 2025 some 996,000 international students will be studying in Australia. Further, education exports will have jumped to more than \$A38 billion by 2025, up from \$A4 billion in 2000.

One of the report's authors and head of IDP Education Australia's Planning and Research branch, Anthony Böhm, says the implications of the research are significant and far-reaching.

"With international education becoming increasingly more

important to the future growth and development of Australia, it is essential that we more clearly understand the future of this important industry," he says.

Multiple effects

IDP's innovative and advanced research on international education has had far-reaching effects on a number of stakeholders including institutions, governments and society and, according to Böhm, it has sparked a lot of debate both within Australia and overseas.

Getty Images

"The debate revolves around the realisation that the global demand for Australian education may exceed the number of Australian students by 2025," he explains. "This research questions the fundamental role of international education in Australia's future economic development. These questions are all the more important in a rapidly changing global environment – particularly given the growing drive toward knowledge economies and the concerns over economic sustainability in industrialised countries facing ageing populations."

The research has stirred great debate within Australia's universities as well as in universities around the world. "Some universities believe adopting strategies to proactively meet the global demand for Australian education is key to their survival," Böhm says. "Other universities have put in place strategies to control demand. These universities believe obtaining a balance between international and local students is essential."

He adds other institutions are dealing with the expected increase in other ways. For example, the University of New South Wales has stated that it wants to have a clear balance between the two types of students and that a natural limit of about 25 per cent would be imposed on international student numbers.

Böhm says applying a limit is a short-sighted approach. "I think the critical issue is whether or not you have the best students within your institution," he says. "Some of the best business schools in the world have up to 90 per cent international students."

The discussion about international students has now been taken up by governments and tied into debates about migration, the ageing population and the knowledge economy. According to Böhm, international education is going to play a much more fundamental role in the sustainability of industrialised economies.

"When you look at ageing populations, no one quite knows how we are going to deal with the situation of having more pension recipients than tax-paying contributors," he says. "This situation is already been forecast for Italy because they have a negative birthrate. No one has worked out the solution to this problem. It creates a productivity challenge if a smaller group of people are to maintain living standards at the level at which they are now."

Böhm says Germany is a good example of this as well because it has a declining labour force and it is seeing international education as one part of the solution.

Looking for long-term solutions

"Migration acts as a short-term increase in the fertility rate," Böhm says. "It is not a long-term solution but it provides short-term relief. The migration of the past was low-skill migrants but the migration of the future has to be high-skilled migration or what

Australian international student forecast

In 2002, IDP Education Australia issued the report "Global Student Mobility 2025: Forecasts of the Global Demand for International Higher Education". This was updated in 2003 to provide an increased understanding of global demand for international education.

The updated report aims to give Australian governments, the international education industry and other peak bodies a sound basis on which to underpin their strategies and shape policy for international education. The results show that international education is an industry that could soon rival tourism.

Some of the forecasts in the report include:

- Asia will be the dominant source region of Australia's international students.
- The relative importance of Europe as a source region of international students will decline.
- All top 10 source countries will be Asian with the top two being India and China.

Drivers of international education in Australia

The international perception of the quality of Australian higher education is the key factor in deciding how fast education exports will grow, according to the "Global Student Mobility" report produced by IDP Education Australia. While the results also indicate that countries need to attract international students in order to address the economic challenges associated with declining birth rates and labour force participation rates, it is the quality of the programs that will be the key determinant in attracting students to our shores.

The study uses a statistical model, which it calls a discrete choice model, to analyse the importance of various factors in the decision students make about their destination country. Some of these factors include quality of education, lifestyle, cost, employment prospects and personal security.

However, the key attribute was found to be the perceived quality of education. The report devised a "quality shock scenario" in which a "plagiarism and corruption scandal" in the overseas campus of an Australian university erupts and eventually leads to international professional associations deregistering Australian qualifications. Such a scenario was enough to make international students choose another destination for their studies and within two years numbers had halved. A strengthening Australia dollar and the higher living costs of Australia also contributed to this situation.

According to Anthony Böhm of IDP Education Australia, if Australia wants to be successful in attracting students, it needs to further enhance its high-quality programs. "International students are buying a pathway to success," he says. "Therefore the focus needs to be on delivering the best outcome for its students."

Böhm says the market is responsive to changes in the perception of quality. "Our findings show that, in south-east Asia, 1 per cent change in the positive perception of quality of Australia's courses would lead to an increase in Australia's market share of 2.7 per cent."

some people call 'smart' migration. International education is one way of identifying the innovative people from around the world, educating them and having them move seamlessly into your labour force.

"Migration through international students partly offsets the ageing population and also drives innovation in the economy," Böhm says.

Some of the best business schools in the world have up to 90 per cent international students

"The forecasts presented in our report have got people thinking about international education in a much different light; it has become almost a national imperative that countries have international students. Countries have a number of strategies to attract them. For example, Germany provides free education. It realised international students were a solution to its declining labour force. It is also increasingly delivering its programs in order to attract international students."

He adds that another spin-off is the greater involvement of German industry in sponsoring students because they realise that the lifeblood of their organisations is their ability to attract high-skilled intelligent staff.

Böhm says the Australian Government has not been as proactive in the international student market, but he believes it is now becoming an important part of its agenda.

Another debate arising from the research centres on how sustainable Australia's own universities will be. "There has been debate over the impact of increased numbers of international students on Australian students – that they will be squeezed out of the places," he says.

"This is one of the myths of international education as there are polices that prevent crowding out. Instead, there are significant benefits to local students of having more students from overseas." Böhm says these include local students benefitting from the resources the students inject as well as the expanded range of quality courses on offer.

"By making the curriculum more international, Australian students benefit by getting a global perspective, especially as the programs need to be at the cutting-edge of any subject covered," he says. "There are also social and networking benefits of having international students on campus."

While Böhm says it is difficult to work out the effects of the increase in international student numbers on gross domestic product, indicative forecasts were made in the report.

"In 2002 dollar terms, we worked out that international education would be worth $38 billion by 2025," he says. "At the moment it is around $4.2 billion. International education is one of Australia's most sustainable green exports and will eventually challenge tourism."

However, while the increase in international students appears to paint a rosy picture of advantages and benefits, Böhm says there are always negatives.

By making the curriculum more international, Australian students benefit by getting a global perspective

"Some commentators are asking about levels of social tolerance," he says. "Some say having more international students here will create tension and be an issue but I don't think it will be a big problem purely by virtue of the fact Australia is so multicultural. Overall, Australia can only benefit from an increase in overseas students and we need to start seriously thinking about innovative ways of attracting international students to our shores." ■

By Kathy Graham

Planning for the future

→ Foresight planning is becoming big business for governments, corporations and academic institutions.

Getty Images

Strategic planning for the future is far more scientific than looking into a crystal ball.

Imagine that you are sitting on top of a hill and that it is night. You are scanning the horizon when all of a sudden you see a light way off in the distance. You turn around and say to your senior management team behind you: "I've seen a faint light. It is either a bushfire or someone has just lit a candle, and it could go out before we get to it. We need to be prepared for all eventualities."

Marcus Barber, a strategic foresight analyst with the Foresight Planning & Review Unit (FPR) at Swinburne University of Technology (SUT) is using this metaphor to explain "foresight planning", a process which is facilitated using various techniques including visioning, environmental or "advanced warning" scanning, and scenarios.

Visioning involves formulating goals that then become an organisation's core driver. Advanced warning scanning involves looking for "weak signals" or changes that may occur for which the organisation can then prepare. The scenarios process involves posing a range of "what ifs" and then devising strategies accordingly. All aim to assist organisations to enhance the way they approach the future by being more innovative in their thinking.

Exploring all possibilities

"Such techniques work by literally releasing the organisation from the constraints of short-term, reactive management styles," explains Barber. "They also inject a freedom to think of what could be possible with a proactive forward view that incorporates a strategic and deliberate approach to innovation."

SUT is the only academic institution in Australia to have extended the concept of innovation through two foresight units: FPR and the Australian Foresight Institute, a research and teaching school headed by Professor Richard Slaughter. Both were established in 1999 by SUT's then vice-chancellor, Professor Iain Wallace. SUT was making the transition from institute to university at the time and he saw a strategic foresight and planning division as a vital addition.

FPR was originally set up to inform the wider SUT community and its senior managers of potential changes in its operating environment, leading to a better-prepared organisation that was more able to embrace innovative thinking.

However, it now also consults on foresight planning outside the university. "At first we had to go out and really sell the concept to people," recalls FPR head Maree Conway. "Now people are coming to us and asking, 'How can we use foresight?'"

According to Barber, notwithstanding the external demand for FPR's consulting services, it still goes against the grain for most organisations to think creatively about the future.

"Innovation demands upsetting the status quo," he says. "Yet big business and government generally want to maintain it. They want to hang onto their market share or win the next election. This is the type of culture that is ingrained."

He adds that strategic foresight encourages taking a long-term perspective. "Unless such organisations are given the leeway to do this, they don't," he says. "From a risk management view, that exposes them because they're not thinking widely enough, and from an innovation view, it limits them because they're not looking at what potential growth opportunities there may be." ■

FPR has consulted to:

- The core management team in a large State government department. When taken through a visioning process, this generated more effective awareness of possible impacts of changes arising from shifting demographic demands and increasing public pressure.
- A major federal department utilised the scenario methodology to flesh out what services it would need to supply in differing economic conditions and to identify new service opportunities.
- A major financial services company used the advanced warning scanning process to help its risk managers and strategic planners avoid the future "quicksand" and to be more readily able to capitalise on opportunities for future growth.

By Tim Treadgold

Virus vigilantes

→ Finding solutions for the spread of viruses is the focus of a new Cooperative Research Centre at Curtin University of Technology.

Getty Images

With the acceleration of travel and trade, viruses can move between countries in a matter of hours.

Mention viruses these days and the chances are people will think you're talking about something nasty being delivered via the internet. In fact, there are much nastier viruses being cooked up in the human and animal populations of the world and, given the right conditions, they could be arriving soon, courtesy of the next Boeing or Airbus from Africa, Asia or South America.

The main difference between the two types, apart from one virus being electronic and the other organic, is that no one has yet invented the animal equivalent of Norton, McAfee or Sophos anti-virus software. In this regard, Australia and every other country, is exposed to a viral outbreak which has the potential to be deadly, or economically devastating – or both.

Disease detectives

Britain's experience with foot and mouth disease and the United State's exposure to anthrax as a terrorist instrument are examples of what may lie ahead. This also explains why Australian governments, corporations and academic institutions such as Perth's Curtin University of Technology have joined to create the Australian Biosecurity Cooperative Research Centre (CRC).

Three of Australia's leading researchers in the field of biosecurity will form the backbone of Curtin's contribution to the CRC. Professor Aileen Plant has won international acclaim for her work investigating the SARS (severe acute respiratory syndrome) outbreak in Vietnam – a classic example of the rapid and global spread of an infectious disease. Professor Michael Alpers is a leading human biologist with a background in anthropology, medicine and virology. In Papua New Guinea he worked with Carleton Gadjusek who won a Nobel Prize for his work in demonstrating the transmission of diseases via cannibalism. The third researcher is Professor John McKenzie, a world expert on arbo-viruses, such as Japanese encephalitis.

Meeting the urgency

The team will be part of a much wider CRC structure that will have its head office at the University of Queensland and will involve researchers at nodes in Sydney, Geelong, CSIRO and a number of Federal and State government agencies that have a role in disease detection, prevention and eradication.

The urgency and size of the task facing the biosecurity teams is best illustrated in the decision by the Federal Government to commit $A65 million over seven years and to engage 60 scientists on the project.

Curtin's pro vice-chancellor of Research and Development Dr Barney Glover, agrees with the urgency. "We are ensuring that our research programs are responding to key issues of national interest," he says. "Tangible outcomes that are beneficial to industry, business or the community are essential. Curtin's strategic partnerships with government and relevant industry groups ensure that our research outcomes achieve these aims."

Lisa Adams, the Perth-based former executive director of the CRC and now a consultant, says the primary focus of the biosecurity teams is to protect public health from the entry, establishment and spread of unwanted pests and diseases.

"The CRC brings together the best in human and animal expertise, it is multi-disciplinary and has a range of objectives," she says. "With the acceleration of the movement of people and their ability to travel quickly between countries, a disease such as West Nile Virus could be in Australia in just 12 hours."

Meanwhile, Glover says the university is harnessing its research capabilities to create real solutions for real problems. "Curtin's world leading in biosecurity will enhance Australia's ability to be prepared for potential threats."

There is a need for more effective communication of commercialisation opportunities between the research sector and industry

leaders SPEAK

Commercialising innovation

One of the obstacles to commercialising innovation in our universities is that business does not always have access to relevant descriptions of the research being undertaken.

There is a need for more effective communication of commercialisation opportunities between the research sector and industry.

Knowledge Commercialisation Australasia (KCA) helps commercialise research by providing effective showcasing opportunities. It is the peak body for people involved in commercialising innovation in our public research institutions.

A number of technologies were showcased by KCA at the Commercialisation Forum and Fair of Ideas in 2003.

The technologies competed for the prestigious Peter Doherty Prize for Innovation, awarded to the technology with the most outstanding commercial potential.

KCA provides a stimulating forum for members to share their experiences and to interact with government, industry and investors, facilitating commercialisation of taxpayer-funded research in Australia.

PAUL FIELD
VICE-CHAIR
KNOWLEDGE COMMERCIALISATION AUSTRALASIA

By Katrina Lobley

Improving the flavour of alternative research

From Chinese medicine to goldsmithing using nanotechnology, RMIT's innovative research programs showcase a range of diverse pursuits.

For thousands of years people have sworn by the healing powers of Chinese medicine. Yet major stumbling blocks have stopped the alternative therapy taking off in Western countries such as Australia.

First of all, there's the taste – Chinese herbal remedies usually taste terrible. Then there is the inconvenient preparation of potions and powders. Patients may be given ingredients that need to be cooked at home for up to an hour before they can be taken as medicine. Because people use different utensils and cook in various ways, there is little hope of exercising any quality control over the finished product.

Enter Dr Charlie Xue, the head of RMIT University's Chinese Medicine Research Group. The associate professor established the research group in 2001 to not only overcome Western scepticism of Chinese medicine by testing the efficacy and safety of the remedies but to produce the medicine in a more palatable form.

The researchers recognised that Westerners prefer to take their medicine as tablets – there's no preparation time, no bitter taste and a reliable, quantifiable dosage can be offered. As a result, the group approached a Taiwanese manufacturer about turning the bitter herbal brews into tablet form.

Herbal hiccups

"There was a number of technical difficulties to overcome," Xue says. "The first one was that most of the tableting machines require a minimum of three kilograms of powder to start tableting."

The massive quantities did not sit well with the individualised nature of Chinese medicine so the researchers and the Taiwanese manufacturer developed a new machine. They came up with a design that could produce tablets with as little as three grams of powder. The revolutionary machine has since been introduced into clinical practice – a development that prompted a report in the *New Scientist* journal.

With a new mode of delivery taken care of, Xue turned his mind to the content of the tablets as the researchers realised the quality of herbal medicine from China was a concern. The solution was to develop a herbal farming industry in Melbourne's north-east, around RMIT's Bundoora campus where Xue works. He initiated discussions with the local council, the Victorian Farmers' Federation and the Northern Melbourne Institute of TAFE about the project, while also drawing upon the knowledge of herbal farming experts from China and Hong Kong.

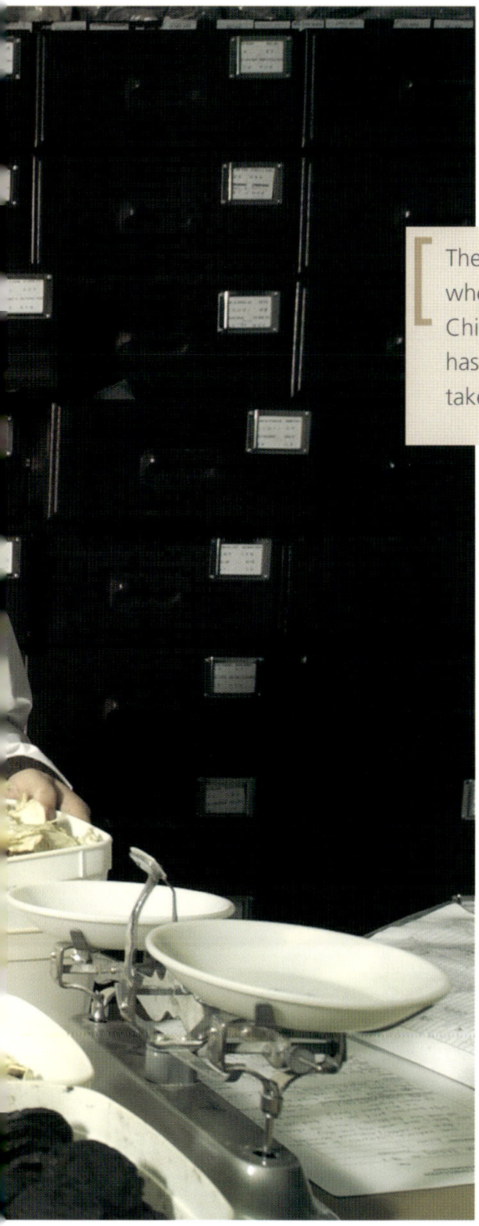

The patience required when preparing Chinese remedies has hindered its take-up in the West.

"I was wondering whether farmers would have any idea about this or any interest but I was amazed by the reaction," Xue says.

The group drew upon DNA fingerprinting to choose the right species of the right plants that would be in demand for herbal remedies both in Australia and internationally

"We conducted a farmers' meeting and were expecting a handful of people to attend but we had more than 50. Some of the farmers were farming Western herbs and not making a profit, so they also came to our meeting."

The group drew upon DNA fingerprinting to choose the right species of the right plants that would be in demand for herbal remedies both in Australia and internationally. The varieties also had to be compatible with local climatic, soil and water conditions.

"We were able to identify a fairly long list and we now have more than 40 varieties of plants growing in our lab in the Department of Biotechnology," Xue says. "In 2004, we'll put some of them into a pilot farming study."

Xue also has to contend with the scepticism concerning Chinese medicine, so he has adopted an evidence-based approach to establish the effectiveness of traditional remedies. This approach has been so successful that several of the Chinese Medicine Research Group's papers have been accepted by mainstream medical journals.

"We reported on the case of a patient who was suffering from very late-stage lung cancer," Xue explains. "Normally, her life span would be less than five years. She decided not to use Western medicine but to try herbal medicine. We conducted careful monitoring at the hospital – she had regular CT scans and photographs. She's survived for eight years and is still very healthy."

Biotechnology is one of four strategic areas that RMIT is looking to "grow" during the next five years (the others are globalisation, global sustainability and interactive communication).

RMIT's pro vice-chancellor (Research and Innovation) Professor Neil Furlong says across the university the focus will be on those four areas. "We expect a range of different disciplines to be able to contribute into those areas and, ultimately, if people can point their skills towards those four areas, they just might find that someone else in the university is pointing in the same direction," he says. "Then we start to get crossing boundaries and interdisciplinary work. If we're all interested in making an impact, then that clearly comes from disciplines working together."

One innovative area of research is the work being carried out by RMIT gold and silversmithing lecturer Robert Baines. Using a scanning electron microscope, Baines has been able to identify the techniques used in Etruscan and Greek gold jewellery. His research has since been used to establish the authenticity or otherwise of what is believed to be ancient jewellery.

Fusing research with the creative arts might sound like a crazy idea, yet an important tenet of any research university is remaining open to such ideas. Furlong says that 20 per cent of RMIT's research reserves are put aside for "creating new areas and emerging research".

"It's unstructured, so we hope we don't miss too many of the passionate researchers along the way," he says.

One of those passionate "outside the square" researchers is Dr Roslyn Russell, who works out of RMIT's Faculty of Business although she doesn't belong to any traditional school or department. Russell works across several disciplines as she ponders the complexities of employers managing two age groups: those known as Generation X (born between 1961 and 1980) and Generation Y (1981–1995). While Generation Xers are known for valuing flexibility, family and leisure time, Generation Y is shaping up to be every employer's nightmare – they're extremely confident, easily bored and unwilling to work for work's sake.

"It's a matter of understanding that research can have many different dimensions," Furlong says. "It's not about just sitting in the lab and making new chemicals or designing bridges – it's about the uptake of ideas. That's what RMIT is about." ■

Katrina Lobley is a journalist who contributes to The Sydney Morning Herald.

A question of identity: RMIT's barcode encryption project helps identify pathology samples.

RMIT's initiatives

Interactive Information Institute

When Australian Rail wanted to train its staff in handling subway emergencies, it went to RMIT's Interactive Information Institute (also know as I-Cubed). Its Virtual Reality Centre allowed the organisation to re-create scenarios using 3D simulation. The cutting-edge training bypassed the expense and danger of Australian Rail using real-life scenarios. The British company Tickford has also used the centre to visualise several vehicle designs. I Cubed contains an ideas incubator to help take ideas to the proof-of-concept stage.

The ID Group

The ID Group comprises seven RMIT staff and several graduate students. It looks at ways to enhance Australia's security systems. The projects include developing identification systems such as fingerprint and iris scanners, and the encryption of barcodes that identify pathology samples and forensic evidence.

Tethered Systems

More precise manoeuvring of water-bombing planes could help Australia's fire-fighting efforts. This is one of the goals of RMIT's Dynamics and Control of Tethered Systems Group, which is working on controlling an aircraft's flight path via a cable that draws upon preset manoeuvres. The group doesn't confine its work to fixed-wing aircraft – it is also developing intelligent tethered systems for ships, submarines and spacecraft. The research could one day improve sea rescue missions and the delivery of supplies to remote communities.

By Kathy Graham

Tomorrow's Nobel laureates

→ Fellowships from the Australian Research Council foster innovation and provide a bright future for our country's leading scientists.

The ARC is nurturing scientific talent within Australia.

The Australian Research Council (ARC) is committed to attracting and retaining excellent researchers through its competitive grants program. Its recommendations and initiatives create an environment that encourages innovation and drives ground-breaking research within the country, thus reversing the "brain drain" of Australian scientists heading to overseas research facilities.

The ARC is an independent body that advises the Government on investment for national research. Through the Federal Government's innovation action plan for the future – Backing Australia's Ability – up to 25 Federation Fellowships, worth $A230,000 a year for five years, are offered by the ARC.

Federation Fellowship recipient Professor Mark Von Itzstein, from the Institute for Glycomics at Griffith University, was one of the leading scientists involved in the development of the anti-influenza drug Relenza. His fellowship has allowed him to continue his work in the area of carbohydrate science.

Von Itzstein says the funding has enabled him to intensify his search for new glycotherapeutic drugs to treat certain types of cancer.

"The potential of carbohydrates has only recently been realised," he says. "Just as bacteria and viruses use carbohydrates in their life cycle, many cancers use carbohydrates in the events that spread the disease throughout the body. If we can better understand how this happens, we can start to target those processes."

According to ARC chief executive officer Professor Vicki Sara, grants and funding are vital for continued innovation within Australia. "Competitive funding provides support for the highest quality research," she says. "This leads to the discovery of new ideas and the advancement of knowledge, training and skills for the development of our next generation of researchers – tomorrow's innovators and potential Nobel laureates."

Professor Marcela Bilek from the University of Sydney's School of Physics was one of four women who received a Federation Fellowship in 2003. She says that the fellowships encourage younger women who might be contemplating a career in science.

"A perception still exists that careers in science aren't feasible for women," she says. "I think that our visibility to younger women has a significant impact."

Basic research contributes to innovation in a fundamental and long-term way. There is a direct link between research and economic growth, jobs and the social and cultural benefits that research brings to our society. Bilek says: "Funding research that makes an impact in these areas is vital. This also applies to those projects that are a platform for Australia's sustained competitiveness in the global knowledge economy."

Von Itzstein agrees: "The people who are driving new technologies need to be free to do so in order to expedite the outcomes for the nation." ■

Diverse research

The ARC funds a wide range of research topics through its Federation Fellowships. Some examples include:

- Quantum and atom optics.
- Cellular plasticity in the brain – discovering the mechanisms controlling the production of brain cells during brain development, function, ageing and disease.
- The manipulation of nanoscale assembly to create new therapeutic and electronic products.
- New methods helping the detection, monitoring and attribution of changes in the greenhouse effect.
- How organisms coordinate energy supply and demand.
- Precision pulsar timing and its applications.
- Atomic-scale devices in silicon – the ultimate limit of microelectronics.
- Salinity tolerance and long-distance transport in cereals.
- The physics of self-organisation, from space plasmas to brain dynamics.
- Sustainable reform of the Murray-Darling system – property rights, uncertainty and institutions.

By Shirley Glaister

Intellectual engine room reverses brain drain

An expanding hub of research centres is attracting many of Australia's best scientific brains to The University of Queensland.

Constructing materials atom by atom, a process known as nanotechnology, may seem like something out of a science fiction movie but researchers at one of The University of Queensland's research centres are bringing this possibility into the real world.

The Australian Institute of Bioengineering and Nanotechnology (AIBN) is part of the University's expanding hub of centres and institutes that is uniting some of the world's best researchers.

The Institute for Molecular Bioscience at The University of Queensland.

The Institute is leading a two-pronged drive into both nanotechnology and bioengineering. Nanotechnology is the ability to systematically build structures atom by atom at the molecular level, such as miniature machines that could travel in blood systems to help fight disease.

Bioengineering involves the design of artificial human organs and tissues, and the development of new drugs. For example, artificial bones and joints developed by the $A60 million AIBN will one day replace the steel and plastic implants used to fix broken or decayed joints in injured or aged patients.

Bioengineering and nanotechnology are important new areas of research with potential benefits for fields such as health, manufacturing, information technology and the environment.

The university's vice-chancellor, Professor John Hay, says the AIBN is at the leading-edge of discovery, especially in the field of organ replacement. Here, developing artificial organs that are not rejected by the body is becoming a critical issue.

"This is driving the university's goal of building research areas in which it is a leading player both in Australia and overseas, and reflects Queensland's growing status as a region for research excellence," he says. "The university has provided the intellectual engine room for this change, playing a major role in reversing the so-called 'brain drain' from the State."

Solutions from snails

Meanwhile, another centre, the university's Institute for Molecular Bioscience (IMB), is developing powerful new painkillers derived from the venom of marine snails. These drugs could be used to treat Parkinson's disease, nervous disorders and strokes. IMB research has also uncovered an enzyme that prevents the establishment of bacterial infection in the lungs of people with the debilitating condition cystic fibrosis.

In early 2003, the $A105 million Queensland Bioscience Precinct was opened. Located at the University's St Lucia campus, it brings together 700 scientists from the IMB and CSIRO.

The development of new insights into the molecular, genetic and physiological regulation of brain development and function to provide a deeper understanding of ageing, neurological disease, cognition and mental illness, will be the role of the university's Queensland Brain Institute (QBI). QBI researchers are working towards stimulating nerve-cell production in the brain. The objective is to replace those cells lost due to strokes, trauma or degenerative diseases such as Alzheimer's, Parkinson's or motor neurone disease.

With such an impressive range of centres and institutions, the University is in a unique position to make groundbreaking discoveries in biology, medicine and agriculture. ■

Shirley Glaister is a senior communications officer at The University of Queensland.

University of Queensland Centres of Excellence:

■ **Australian Research Council (ARC) Centre of Excellence for Integrative Legume Research:** Concentrates on agricultural development as legumes continue to improve soils by introducing nitrogen.

■ **ARC Centre for Functional Nanomaterials:** This centre is focusing on new forms of sustainable ener[gy] as well as the potential of biomaterials and engineering tissue for organ replacements.

■ **ARC Centre for Complex Systems:** Applications being examined include air-traffic control, genetic regulatory networks and the evolution of economic systems.

■ **ARC Centre for Bioinformatics:** Advanced computer-modelling techniques will be used to examine how the genome is activated in the mammalian cell.

By Tim Treadgold

Earn as you learn

→ A world first in the way a global company works with a local community is taking place in the progressive regional city of Ballarat.

In the historic Victorian goldmining town of Ballarat, one of the world's biggest corporations is finding it has a lot in common with one of Australia's newest universities.

IBM and the University of Ballarat are working together to create a research and development centre. While the company and the university both benefit, there is a third winner in the equation. These are the students of regional Australia who are being exposed to some of the world's best thinking without having to move to a capital city. The regional city of Ballarat is located 100 kilometres from Melbourne.

IBM's Regional Software Solutions Centre is housed inside the university with company staff working alongside academics. A centrepiece of the arrangement is a program IBM has dubbed "earn as you learn" which sees students working on IBM commercial projects as part of their schooling.

"People ask me what's our model, how do we make this work," says Werner Hulsmann, manager of IBM's Regional Software Solutions Centre, which was established at the University of Ballarat with support from the Victorian Government. "The simple answer is that both parties were completely committed to our shared success."

The Regional Software Solutions Centre was created in 2002 and followed IBM's initial move to Ballarat in 1995 with the establishment of its Southern Region Data Centre, a customer service hub that was developed after IBM won a major contract to service the information technology requirements of a number of Victorian government agencies.

Getty Images

Looking at IT education from another angle: Ballarat partners with IBM.

Those who made the move are shaping a model which is a world first in the way a company works with a local community and its primary academic institution

Hard work is required to make all relationships work and this was the case with IBM and the university. The initial move to the city was not an easy process and occurred largely because IBM was successful in winning a 1994 tender to manage the Victorian Government's public transport information technology requirements. Part of the contract specified the establishment of an information technology centre in Ballarat.

Adapting to change

"There were issues with getting people to move from Melbourne," says Hulsmann. "The people who did move generally agree it's the best thing they ever did."

Those who made the move are shaping a model that is a world first in the way a company works with a local community and its primary academic institution. Three years ago, Hulsmann and John Harvey of IBM and university leaders Kerry Cox (now vice-chancellor) and Robert Hook jointly approached the Federal Government to secure funding for the course, which IBM calls "earn as you learn" and the university calls a Bachelor of Information Technology (Professional Practice).

Pick of the crop

"The course is all about exposing the students to corporate life," Hulsmann says. "Students study for four years but, in the second year, they are asked to work with us for three months. By the fourth year they are working with us for six months."

He adds that IBM gets a lot from the relationship, as do the other participants.

"Our intention is that after their four-year course we get to pick the best students," says Hulsmann. "If we don't choose them, then there are many other companies looking for students with corporate experience." ■

By Kathy Graham

Dirty, dull and dangerous

→ A bold and innovative approach to research is paying off and resulting in successful advances for Australia's oldest university.

Replacing human beings with robots might sound like a plot line out of Hollywood. In fact, robots or autonomous systems are already replacing people in a broad and diverse range of applications, including mining, cargo handling and transport.

But these machines have little in common with the humanistic robots of Arnold Schwarzenegger's *Terminator* or *The Stepford Wives* model in the 1970s cult classic film of the same name.

So why has the Australian Centre for Field Robotics (ACFR) at the University of Sydney been established, if not to improve on imperfect human beings?

"I have three words for you," says centre director Professor Hugh Durrant-Whyte. "Dirty, dull and dangerous. Essentially, we want to make robots that are able to do those jobs that are really too dirty, too dull or too dangerous for people to do. These are jobs in bushfire-fighting, deep sea and space exploration and defence, to name a few."

Hence the term "field robotics" is applied to this particular area of research. It refers to the use of robots in outdoor, harsh and unstructured environments.

Getty Images

Robots could soon take over the dirty work.

"Autonomous systems represent the next great step in the fusion of machines, computing, sensing and software to create intelligent systems capable of interacting with the complexities of the real world," says Durrant-Whyte. "They're the physical embodiment of machine intelligence."

> One of the biggest systemic problems in Australian research is the low investment by business in research and development

Meanwhile, there are still a number of obstacles to overcome. According to Durrant-Whyte, one of these is the failure by current models to adequately interpret the information-rich representations of complex environments.

"Perception by autonomous systems, in unstructured dynamic worlds, is one of the outstanding and significant research challenges in building intelligent systems," he says.

Together with the Department of Artificial Intelligence at the University of NSW and the Mechatronics and Intelligent Systems Groups at University of Technology, Sydney, the ACFR is a partner in the ARC Centre of Excellence for Autonomous Systems, currently the second-largest centre of its kind in the world. According to Professor Gavin Brown, vice-chancellor and principal, University of Sydney, this is in line with the university's policy to be "a leader in productive innovation in science, technology and the arts". Indeed, recent independent figures from IBISWorld show the University of Sydney is Australia's second-highest spender in research and development, investing hundreds of millions of dollars more than some of the nation's major companies.

Investing in research

One of the biggest systemic problems in Australian research is the low investment by business in research and development. In fact, as a nation, Australia is way behind OECD norms and the gap is widening.

Brown says the good news is that the Australian Government is encouraging universities to help improve research and development performance, although he says he is happy if incentives go directly to industry, provided these are used to develop strategic and creative partnerships with universities.

The ACFR is a case in point, having attracted more than $A5 million in industry funding. Similarly, the Brain and Mind Research Institute (BMRI), another centre for discovery within the university, houses 65 independently-funded research laboratories.

The long-term goal of the BMRI is that it will be of direct benefit to people whose lives have been affected by common and debilitating forms of psychiatric and neurological illness.

"At Sydney, we believe in being bold, innovative and successful," says Brown. "The result is that our applied research is paying dividends across a range of endeavours."

By Kathy Graham

Walking on water

→ Innovative, self-funding, regionally-focused projects are a key factor in this university's progressive research strategy.

When the Brisbane City Council wanted to encourage people to leave their car at home and walk or cycle through the surrounds of the Brisbane River, it embarked on its ambitious River Walk project. As part of the 34-kilometre stretch, a floating walkway was needed to enable pedestrians and cyclists to commute to the city. The council turned to the University of Southern Queensland (USQ) for a solution and the resulting walkway was made possible with the clever use of fibre composites.

Floating feet: Brisbane residents can now float to work on the River Walk.

Initially, these materials were used in the aerospace industry, such as in the construction of stealth bombers, but these days fibre composites are used in a wide range of applications. As Professor Gerard Van Erp explains, they have almost all the attributes sought after by designers and engineers. "We now see everything from cars and boats to prosthetic devices and wheelchairs manufactured with fibre components," he says. "This is because they offer unique performance through the combination of polymers with carbon, glass and/or Kevlar fibres.

"They are extremely strong yet a fraction of the weight of steel or concrete. They're also non-corroding, non-magnetic and can be ingeniously designed to locate strength and stiffness where needed."

Our research activities as a regional university need to be locally important and nationally significant

Van Erp heads USQ's Fibre Composites Design and Development Group. When it came to designing the floating walkway, the challenge for his team was how to secure the floats to each other safely and securely with a material that was not just strong – but also non-corrosive and malleable enough to allow for the twists and turns of the walkway as it followed the river.

According to Van Erp, the walkway paves the way for the next generation of fibre composite products made and designed specifically for civil engineering applications.

Regional priorities

Deputy vice-chancellor (Research and Enterprise) Professor Malcolm McKay says, USQ has different priorities to many of the metropolitan universities which can engage in pure research for the sake of it. "At USQ, we believe that our research activities as a regional university need to be locally important and nationally significant." To this end, academics compete for internal university funding and their research projects are chosen by a panel of their peers through a competitive internal grants program.

As McKay points out, Australia's "sandstone" universities gain more in research funding from the Federal Government than do the regional ones. "It's imperative that the regionals 'kick start' innovative projects that offer potential for industry support," he says. "Especially ones that have direct outcomes for their particular region."

This is reflected strongly in research activities carried out at USQ's main research centres: the National Centre for Engineering in Agriculture; the Centre for Rural and Remote Area Health; the Landuse Research Centre; and the Centre for Rural and Environmental Biotechnology.

For example, USQ scientists working at the latter institution are looking for molecular markers and "DNA fingerprints" to identify "good genes" in wheat, a globally important food crop.

No more rot

"We are looking for genetic markers in wheat for economically important traits," says Associate Professor Grant Daggard, a scientist working on the project. "These include resistance to diseases like crown rot, absence of grain defects and desirable quality traits like high milling yield. By identifying markers linked to genes for such traits, the long-term goal is to take the range of disease fighting and other characteristics found in several wheat species and to combine them to develop new and better wheat varieties."

This progressive research is made possible through the regionally-focused approach of USQ's research and is set to develop a new generation of agricultural products, civil engineering strategies and biotechnology solutions. ■

By Julie Hare

It's only natural

→ Innovative plant technology from the University of Western Sydney is reaping rewards and proving to be a commercial success.

Getty Images

Coming to fruition: enabling cold-climate fruits to grow in warm weather.

Back in 1972, horticulturalist Dr Graeme Richards was wondering what he could do to help supplement the income on his family's dairy and banana farm when he stumbled across some American literature about growing cold-climate plants in warm climates.

It sparked his interest and he embarked on a cross-breeding program of stone fruits which now, 32 years later, is about to come to fruition with the development of a "spin-out" company to commercialise his intellectual property called Phytonova (literally 'new plant').

"I see the opportunities that exist both here and overseas and I'm getting very excited," says Richards. "In fact, there seem to be too many opportunities."

Richards' research was conducted through the University of Western Sydney's (UWS) Centre for Horticulture and Plant Sciences (CHAPS). In early 2003, UWS brought on board commercialisation expert Hans Wijgh to be the general manager of Phytonova and to ensure the commercial application of Richards' ideas. Through its Office of Business Development, the university secured seed money of $A500,000 from Melbourne-based venture capital company SciVentures which was used to prove up the market in the US and Europe and to protect the intellectual property through process patents and plant breeders rights (PBRs).

In layman's terms, the main application of Richards' invention to date has been the development of stone fruit trees which can grow in warm climates where traditionally they only grow successfully in cold climates, all without the help of genetic modification.

It's only natural

"Genetic modification only existed in people's imaginations when I started and because of the consumer backlash about GM foods, we've decided to keep it GM-free," says Richards. "Stone fruits need around 800 to 1200 hours of chill – temperatures less than 6C – to produce good fruit," says Wijgh. "Graeme's trees need only 100 hours."

What this means is that high-quality fruit will be available on the market much earlier in the season. It also means quality fruit can be grown in traditional high-chill areas that might be affected by global warming or other climate changes.

"The invention, is without doubt, well and truly ahead of the market," says Wijgh. "It's how Graeme has applied the cross-breeding and the technology he has developed based on normally traditional techniques, where the true invention lies.

"It's tough to make projections about its potential dollar value but I would imagine it could be worth tens of millions within Australia alone."

Working mainly with plums, nectarines, cherries and peaches, Graeme is also applying his methodology to producing faster-growing walnut trees for their wood. "It's a really wonderful thing to be able to help someone like Graeme bring to fruition a lifetime's work," says Wijgh. "Often these projects fall flat because the innovator or inventor are not entrepreneurial. So I'm in a really privileged position to help take the project forward."

Harris Madden from the Office of Business Development at UWS says: "Phytonova is a great example of the university's ability to create technology that has commercial potential and broader benefits. We're a highly 'applied' university with some very talented people in our 14 major research centres, and we have some emerging projects that could have worldwide impact." Watch this space. ■

Julie Hare is a journalist based in Sydney.

Discovery consists of seeing what everybody has seen and thinking what nobody has thought.

ALBERT VON SZENT-GYORGI

By David Woolbank

Driven by design

→ Electrolux's commitment to new product innovation has put the customer firmly in the driving seat.

In the historic inner-Sydney suburb of Pyrmont, the past and the future come together with new residential housing and workplace developments nestled among some of the city's oldest buildings. It is here that Electrolux has chosen to locate its ultra-modern Design and Innovation Centre.

High-profile: Electrolux's Design and Innovation Centre at Pyrmont.

These days, innovative designs with practical features are made with the consumer firmly in mind and drive all aspects of Electrolux's business in Australia. The new centre has quickly become a creative incubator for new technology and products and emphasises the importance the company places on the process of design for its Electrolux (Gallery, Chef, Dishlex and Kelvinator), Westinghouse, Simpson and Hoover products.

Relocating the 15 designers from their workplace at the back of Electrolux headquarters at Riverwood (in Sydney's south-west) to the Design and Innovation Centre 40 minutes away has significantly lifted the division's profile within the organisation.

"Everyone from the customer service representatives to technicians are now asking about the new designs and even offering their own suggestions," says Electrolux Home Products Design director Lars Erikson.

Swedish-born Erikson is committed to incorporating innovative European design and styling into the company's Australian-made products. As a result of his arts-based education, his approach to functional industrial products represents a shift in thinking about how appliances are both designed and displayed.

Electrolux has always been at the forefront of design but there is now a strong commitment to produce innovative products that people take pleasure in using.

According to Erikson, the company is listening closely to what consumers need and rejuvenating its entire range to reflect future desires and trends.

"Before, we used to say to our customers, 'Here is the product, do you want to buy it?' Now we ask the customers what their needs are."

Recent examples of consumer-driven innovation include the repositional deli compartment, the quick-chill feature and the store-safe utility drawer in Westinghouse's new Ophir range of refrigerators. These fridges are the most energy efficient on the market.

Innovation is a word Electrolux has embraced in all aspects of its business since it manufactured the world's first modern vacuum cleaner, the Lux 1, in 1912. Not only has the company led the world in developing appliances such as the Assistant food mixer (1940s), its first washing machine (1951) and the first compressor-driven washing machine (1956), it also pioneered door-to-door selling.

Before, we used to say to our customers, 'Here is the product, do you want to buy it?' Now we ask the customers what their needs are

As the world's largest producer of powered appliances for kitchen, cleaning and outdoor use, the Electrolux Group has made its name with products that were cutting edge for their time. These have included refrigerators, washing machines, cookers, vacuum cleaners, chainsaws, lawn mowers and garden tractors.

Each year the global Electrolux design team produces what are known as "concept" products. These are appliance ideas that are conceived in order to gauge both consumer confidence and the viability of a particular product before it is mass produced.

Smart cookie

The Automower is an appliance that demonstrates the Electrolux philosophy, which is ease of use, ease of mind. It is a robotic

Integrated design process

All new appliances at Electrolux Home Products have their origins in a strict product development program known as the Integrated Product Design Process (IPDP). The process involves all the designers, sales, development and marketing teams working together to meet specific goals. This collaboration allows for refinements and adaptations along the way.

Consumer insights

When designing new products, Electrolux Home Products takes seriously any direct feedback it receives from consumers, retailers and staff. The "consumer insights" process involves interviewing satisfied and dissatisfied customers on a regular basis, conducting focus groups and analysing and responding to

Getty Images

Looking good: design becomes a priority.

lawnmower which, when the battery runs low, navigates its way back to the charging station. After its energy source is replenished, it resumes mowing the lawn. The Electrolux Robot vacuum cleaner or Trilobite works in a similar way. It is even listed in the Guinness Book of Records as the world's most advanced domestic robot.

The Electrolux Screenfridge is another such product. A showcase of so-called "intelligent technology", it incorporates features that enable it to display dinner recipes and record shopping lists and messages courtesy of an in-built video camera. It can be linked to the internet and even used as a television or radio receiver.

Meanwhile, Electrolux India has introduced the world's first talking washing machine. The Washy Talky actually tells the user how to get the perfect wash.

Another concept product is the Bess refrigerator, the appearance of which has been inspired by a mobile phone. The cabinet is made of plastic and the shelves rotate to allow easy access to all stored items. A quick-chill unit is located under the clear hood on top and can chill water, an aluminium can or even a bottle of wine, in less than a minute.

More than 55 million Electrolux Group products are sold every year to customers based in about 150 countries. The Swedish-based Electrolux Group includes world-famous brands such as Electrolux, Kelvinator, Westinghouse, Simpson, Chef, Dishlex, Zanussi and Husqvarna. Electrolux Home Products is the largest household appliance maker in Australia, recording sales in 2002 of more than $A1 billion with a major share of the Australian whitegoods market. Continuing with its innovative strategies while listening to customers will ensure the company stays one of Australia's leading brand names. ■

David Woolbank is director of Brand Marketing, Electrolux Home Products Australasia.

problem detection surveys filled in by customers whenever an appliance is installed.

This process provides Electrolux with a greater understanding of how it can improve on its products. For instance, it may mean installing specific new features such as a bottle rack or a child-proof lockable compartment for storing make-up and medicine in fridges. Other improvements may be to existing features, such as tall bottle storage in fridges.

Meeting new government safety and energy efficiency regulations is one reason for product modifications. Project Ophir, being carried out for the Westinghouse brand, is one successful example of this, delivering fridges that are 33 per cent more efficient than any other in the marketplace. Outdated fridge models were redesigned with increased capacity, modern styling and a higher energy rating. The project was named after a small town called Ophir, located about 30 kilometres north-east of Orange in central-western New South Wales. It is also the site of Australia's first significant gold discovery.

History dictates that a project like this can take up to four years. But in the case of Ophir, more than 40 new fridge models were on shop floors within two years of senior designer Justin Formica submitting his first sketch.

This was mainly due to improved manufacturing techniques as well as involving more people at each stage of the process. Even in global terms, the team's output, given its small size and the short time frame, was exceptional.

By Jan McGuinness

Haven from home

→ Creating loyalty among hotel guests is a challenging task, but the Como Melbourne quickly found its own niche.

Hotel guests are traditionally fickle and any new hotel has to compete with others already established in the market. One way to attract guests is to find an innovative niche offering that will entice them inside and keep them coming back.

The emphasis at the Como Melbourne is on clean lines.

The Como Melbourne – which opened in 1989 – came up with its now renowned gimmick when a guest noted that the hotel lacked for nothing in the way of comfort and detail except for a rubber duck in the bath. Soon, takeaway Como ducks appeared in every bathroom, instantly becoming the hotel's trademark and a status symbol among those who stay there.

This story illustrates two facets of the Como Melbourne's enduring success: Australia's most innovative hotel derives its edge from listening to its customers and a willingness to be a bit "offbeat" to attract them.

The 107-room hotel was Melbourne's first boutique hotel but it differentiated itself with its modern design. Its décor was "of the moment" and the antithesis of the prevailing sumptuous décor and more suffocating atmosphere associated with luxury hotels. The décor continues to be contemporary, understated and uncluttered, and is based on straight lines and clear surfaces. Another difference is the staff uniforms, which are smart but casual and provide a stark contrast to the traditional hotel look.

The design of the hotel was modelled on New York's Morgan Hotel. Like the Como, it was an ultra-modern boutique hotel that targeted chief executive officers and representatives from the entertainment world. The Morgan Hotel offered a quiet, discreet and elegant service, believing that by the end of the ostentatious 1980s the pampered, high net worth traveller market was ready for something different.

This difference – something that offered both style and substance in a more streamlined, understated package – can be found in the Como Melbourne. Instead of luxury and opulence, management listened to the needs of the modern traveller and found they were craving home-away-from-home comforts, simplicity and up-to-the-minute facilities, including the latest technology offerings. For the latter, Como introduced in-room business facilities.

From day one, the aim of the hotel was to provide a level of service that not only met but exceeded guest expectations. This was the vision of Patrick Griffin, the Como Melbourne's first general manager and a driving force whose firm ideas of what the hotel and its culture should be have prevailed through several successors to the present day.

Going the extra mile

The result is a high level of return business that current general manager, Paul Bidgood, says he's never encountered in his 30 years as a hotelier. "Staff are prepared to go above and beyond their normal duties to meet the needs of guests and this dedication is highly appreciated," he says.

Tennis great and regular Como guest, Steffi Graf, says of the

The Como Melbourne's decor is contemporary rather than sumptuous.

hotel: "If my service was as good as yours, it would be easier to reach the finals of the Australian Open."

The days of always going the extra mile to fulfil a guest's request continue. A Hollywood producer staying at the hotel told a local radio station he didn't get fries with his room-service order. When hotel staff were told, the chef whipped up a batch of fries and tossed them into a hot box, which was delivered by limousine to the radio station before the interview had finished.

Management listened to the needs of the modern traveller and found they were craving home-away-from-home comforts

Other instances of providing outstanding services to guests include taking out a full-page advertisement in a metropolitan newspaper to wish radio announcer, Derryn Hinch, a happy 50th birthday from the Como Melbourne.

"Such is our hotel's reputation for personal service that staff are motivated to maintain it as a source of pride and point of difference," says Bidgood. "A culture of openness and discussion encourages them to be ever alert to ideas they can improve on and introduce to the hotel."

These ideas range from the simple to the complex. One example revolved around changing the room-service menu. As many guests, particularly film crews, stay for months, the menu was redesigned to feature weekly specials that included comfort foods.

Another innovative idea targeted rewards for regular guests. Where other hotels might reward frequent patronage with upgrades, the Como has guest amenities graded according to the number of return visits. These amenities include chocolate-covered nuts, duck cuff links, champagne and theatre tickets.

Other individual touches include having each room fitted with CD players – guests can pre-arrange a personal selection of music. In-room traveller's survival kits exclusive to the Como are replete with everything from toothbrushes to mouthwash, shavers and condoms.

"Such personal touches make a huge difference and because the hotel is small they can be implemented quickly and without fuss," Bidgood says. "Behind the scenes, the special needs of guests are anticipated at daily staff meetings, with the aim being to get as close as possible to the guest before arrival through personal assistants and agents."

The Como Melbourne has an initial staff training program that follows a 90-day plan, during which staff are versed in every facet of the hotel's operation. An ongoing program provides a focus on different skills and performance is monitored throughout the year. Senior staff, including accountants and the technical services team, stay over regularly to experience the hotel first-hand.

Como's latest innovation is to position itself as a resource for the local social scene. Its three duplex penthouse suites are hired out for catered dinner parties at which the hosts can stay over and the hotel has converted the neighbouring Maxim's restaurant into function rooms – all named after species of ducks.

"Local" means South Yarra, the chic inner-city suburb where the Como Melbourne is located. The hotel began as a feature of the vast Como apartment, office and retail development planned by Singaporean developer Jack Chia. When he was no longer involved, his finance company completed the project along more modest lines.

Intended as the jewel in Chia's dream, the Como instead became a focus of South Yarra's fast-growing Chapel Street retail, café and bar strip. In the years since the hotel opened, what was already a smart precinct has become the place to shop and socialise.

Inner Melbourne has become a city of apartments and the Como Melbourne is a place where guests don't just stay, but live. This quintessentially modern hotel reflects the city's reputation for good design and liveability. ■

Fashionable features

Services and features introduced into the Australian hotel scene by the Como Melbourne include:

- In-room spa baths
- Leading hotel outside CBD
- Signage-free foyer
- Seated check-in
- Key card access to all floors
- Modern "designer" décor
- Business facilities (fax, computers) in rooms
- Duplex penthouses

By Julie Hare

Recognising achievement

→ Driving and promoting excellence in innovation is a key role of the Australian Design Awards.

Motoring ahead: Award-winning design from the Hunwick Harrop team.

While many associate innovation with more technical aspects of science and technology, new ideas and products are often driven by the aesthetics and functionality of design. The Australian Design Awards, run by Standards Australia, is a program that recognises the role of the design process in the development of innovative products.

As group chief executive Ross Wraight points out, the organisation's background in standards position it perfectly to promote, endorse and foster excellence in innovation.

"The aim of the Australian Design Awards is to present to the Australian public and the world the best examples of Australian design and innovation and the high quality of design expertise available to manufacturers both here and overseas," Wraight says.

"Standards Australia has a heritage that comes out of the development of standards and benchmarks that affect pretty much everything that is built or designed.

"Good design tends to stretch the imagination and move the benchmark higher."

For example, a small, steamlined, easy-to-read, simple-to-use, hand-held, ergonomic blood glucose metre for monitoring diabetes took out the 2003 Australian Design Award.

The Betachek G5 is the embodiment of innovative design – fabulous looking, technically smart, low cost and with an almost unlimited market potential both in Australia and overseas.

It is in good company and the kudos attached to the Australian Design Awards is growing with some of the country's biggest brand names – Qantas, Holden, Ford, Victa, Telstra, Electrolux, Cochlear, Sunbeam and Breville – competing against smaller inventors. Redesigns of existing successful products, such as the Victor mower, are among entries, while others, such as the Betachek G5 are without precedent.

Lifting the bar

Wraight says that Standards Australia supports and promotes innovation in Australian business and industries not only by running the design awards, but by actively promoting best practice through a range of activities including holding the annual, equally prestigious Business Excellence Awards and being the largest management systems auditor in Australia, specifically of ISO 9000 companies.

"Driving innovation is the next big wave after driving business excellence," says Wraight. "Continuous improvement has been around for a long time and the ISO 9000 and business excellence frameworks are about that."

However, he says organisations need to go a step further to be truly innovative.

"Companies that want to foster innovation have to create a culture of creativity within," says Wraight. "They need to go beyond best practice."

Wraight says one of the most exciting aspects of the Australian Design Awards is the sheer range of products entered each year, of which there are about 250. "Whether its fabric for car seating, a superior supermarket trolley, motorcycle or robotic vacuum cleaner, the commitment of Australian companies to good design, breakthrough thinking and commercial applicability is undeniable."

A new category was introduced in 2003 to specifically foster innovation. "We call the new category 'Invention' and it allows us to recognise great ideas rather than just the application of brilliant design to improving existing products," says Wraight.

And beyond that, Wraight is adamant that some of the most inspirational designs come in the highly competitive student category, which is sponsored by Dyson to the tune of $10,000 each year.

Standards Australia is, says Wraight, a company which practises what it preaches, a company which aspires not only to best practice but to new and innovative ways of doing business.

"We have made about $100 million in sales this year and about 25 per cent of that comes from offshore. We are leading the charge for some pretty innovative Australian thinking.

"This is a company that has seen a massive reformation. It is an organisation that is 81 years old and some might think of us as sturdy and uneventful, but in fact we have reformed the entire business and the whole industry."

Spare the innovation and
ruin the company.

ROBERT HELLER

By Phil Robertson

Protecting intellectual property

→ Australia is lagging behind similar countries in the business of generating patents to protect its innovations.

Compared with other countries on a per capita basis, Australia is well below par in generating patents to protect its technology inventions in the leading world market. In the box "How does Australia rate?" on the facing page, the figures show Australia lagging behind other countries with similar standards of education and technology infrastructure.

If this trend continues, it is cause for concern as a strong research and development (R&D) capability in Australia and success in generating commercial benefit is vital for the future of R&D in this country.

Of course all commercialisation mechanisms are important but unless we have something tangible to commercialise, we're not likely to make much progress. What clearly stands out from the country figures is that there is a high level of correlation between the per capita numbers of patents and levels of export earnings from technology-based industries.

What are patents?

Patents are the most tangible form of intellectual property protection. A granted patent gives the owner a monopoly to exploit the invention commercially in the region of jurisdiction of the patent in return for the disclosure of the invention. Most organisations retain ownership of patents invented by their staff through their employment contracts.

Why is this important?

Technology is a global business, dominated by large companies whose businesses rely on technology-based products and related services, with smaller companies feeding into the "food-chain". At a globally competitive scale, companies have to protect their products with patents, otherwise someone else can reproduce the same products with lower R&D amortisation costs.

Of course patents are only one area of intellectual property protection. But other aspects such as copyright and trade marks are generally secondary factors because they are much easier to circumvent or rely on achieving an initial market position.

Is patenting only for companies who can afford it?

It's often said that patents are only important for large companies who can afford the significant costs involved. Small companies generally don't have the resources to maintain extensive patent portfolios. Instead, they often rely on early sale of their technologies, strategic patents or the whole company to large companies. What in tangible terms do they have to sell or license if they don't have patents? If small companies don't have patents and want to grow their businesses, the risk occurs when they start generating significant revenues. They may then attract the attention of a company whose patents they may already be infringing or of a company that decides to move into the market created by the smaller companies.

> We cannot ignore the US, or major European and Asian markets, if we are to grow our technology based exports

How much does it cost to file and maintain patents?

It is important to look at the costs objectively. The cost of developing and maintaining a patent portfolio can be extensive overall but mechanisms such as the Patent Cooperation Treaty (PCT), which simplifies and reduces the cost of international patent protection by allowing a single filing to cover a wide range of countries, allows many of the costs to be deferred until licensing or other commercial arrangements can be made.

Australia has a provisional filing system which allows an inventor to file a patent at relatively low cost, with a one-year window before the complete filing is required. At this point the decision about other country filings has to be made.

The cost of drafting by a patent attorney may range between $A5,000 and $10,000. That is probably less that 5 per cent of the overall annual cost of employing a technology specialist. Maintaining a patent may cost a similar amount every year, because of the need to respond to "office actions" from patent offices in the various regions in which it is filed as it undergoes the process of examination prior to granting. This may take several years. So the cost of maintaining a patent portfolio may rise to around 10 per cent of the overall cost of employing technology specialists, recognising that the average number of patents filed per person per year may be around 0.25 for a productive technology company.

How important are US patents to companies in Australia?

The reality is that the US is one of the major leading markets in the world. In some areas, countries such as Japan or Korea may be leading markets, but overall technology-based products and services are likely to find markets first in the US or Japan. Patent protection in the US is important for global products. In general, the Australian market is not large enough to support the levels of R&D and market development required for leading products,

Getty Images

Keep prying eyes at bay: patents protect your innovations from exploitation by other companies.

so companies have to target other larger markets to build successful businesses. We cannot ignore the US or major European and Asian markets if we are to grow our technology-based exports.

How do we improve the situation in Australia?

We can start with university education. Very few if any universities in Australia have intellectual property protection as an embedded part of their courses. Very few students have ever read a patent by the time they leave university or written one. That is slowly changing with the help of IP Australia's recent activities and also through companies such as Canon Information Systems Research Australia (CISRA) working with universities.

We can also look at our professional practices and strengthen the skills of researchers and patent attorneys to improve the quality and reduce the drafting costs of our patents. We need to look at the cultural factors that inhibit a focus on patents and recognise that better commercial outcomes will increasingly be required if we want to justify continued public funding for R&D.

Increasing the overall amount of R&D being undertaken in Australia, particularly by global companies and their subsidiaries, would have a significant impact on building our abilities to patent. This is because multinational subsidiaries have access to their parent company IP strategies and practices and can develop their staff skills in patenting. The more we do, the more experienced people there will be and the more we will improve not only the amount of patenting we do in Australia but also how well the patents we produce can be commercially exploited.

So then, patents are important?

Yes, patents are important. They are critical for global exploitation of R&D. In fact, they are probably the single best indicator of gaining commercial benefit from R&D. As a nation, we should be targeting a profile of patents more in tune with other countries that generate significant export revenues from technology R&D. That means there should be a five-fold or greater increase in the numbers of patents we generate in Australia.

Numbers are not everything of course but without something to exploit, it's hard to exploit it. We also have to develop significantly better targeting strategies for the commercial exploitation of patents – in reality the costs help focus this. The barriers are as much cultural as lack of skills and we have to tackle both with urgency. ■

Phil Robertson is a director of CISRA.

How does Australia rate?

The United States Patent and Trademark Office publishes lists of the numbers of patents granted in the US, organised by country of origin, which is determined as the place of residence of the first inventor. The following number of US patents were released for 2001, listed by country of origin.

Finland	769	Canada	4063
Australia	1031	United Kingdom	4356
Israel	1031	France	4456
Netherlands	1494	Taiwan	6545
Switzerland	1557	Germany	11895
Sweden	1935	Japan	34891
Italy	1978	United States	98666
South Korea	3763		

In the US, the state of California generated 20863 US patents. Twenty-two other states generated more than 1000 patents and five of those generated more than 4000.

By Gayle Bryant

Patently protected

Safeguarding innovations is a key role for IP Australia and the Institute of Patent and Trade Mark Attorneys in Australia.

A good number of the world's brightest ideas originated in Australia. From the Stump Jump Plough invented back in the 1870s through to modern marvels such as the cochlear ear implant and the Black Box flight recorder, Australian innovations have placed the country as one of the cleverest in the world.

Creativity and ingenuity are critical factors in maintaining a competitive edge in today's global economy and vital to Australia's future prosperity. A healthy flow of groundbreaking inventions, leading-edge designs and imaginative innovations are the key to maintaining and building a prosperous country.

Getty Images

Protect your assets to avoid losing out to competitors.

Intellectual property (IP) can be described in simple terms as the product of your mind or intellect. In the commercial world, IP can often be viewed as a financial asset and, for this reason, it requires protection.

Forms of IP protection include: patents, for new or improved products or processes; trade marks, for words, symbols and pictures, such as brands, logos and names; designs, for the shape or appearance of manufactured goods; copyright, for original material in literary, artistic, dramatic or musical works, films, broadcasts, multimedia and computer programs; circuit layout rights, for the three-dimensional configuration of electronic circuits in integrated circuit products or layout designs; plant breeder's rights, for new plant varieties, and confidentiality/trade secrets including know-how and other confidential or proprietary information.

An organisation or individual who comes up with an original idea, whether it be a scientist working on a new cure for cancer or a fashion designer creating distinctive garments, should be aware of his or her IP rights. The consequences of not doing so can lead to confusion, dispute or potential litigation which will rely on common law to prove ownership.

As the Australian government organisation responsible for administering patents, trade marks and designs, IP Australia is celebrating its centenary year in 2004. To ensure information is accessible to all, its website contains a wealth of information about the IP process. It also works closely with the Institute of Patent and Trade Mark Attorneys of Australia (IPTA), which represents patent and trade mark attorneys registered in Australia, both in private and corporate practice.

Although experts in the laws of intellectual property, Australian patent attorneys must have a tertiary science or engineering degree and often have had a career in scientific or engineering circles before joining the profession.

With such a background, Australian patent attorneys play a critical role in protecting Australian innovation in Australia and, importantly, in key overseas markets.

President of the IPTA Colin Macauley says the role of a patent attorney lies primarily in the registrable intellectual property rights of patents, trade marks and industrial designs.

"Patent attorneys assist clients with the protection and exploitation of intellectual property rights," he says. "They comprise a distinct profession as they are concerned with assisting the owner of registrable forms of intellectual property to secure and maintain registration of their property as well as with advising clients generally in matters arising in connection with these forms of property."

Almost all national governments grant to the owners of inventions an exclusive right of limited duration to prevent unauthorised use of an invention, known as a patent. Such protection serves to encourage a national innovative culture by encouraging and rewarding innovators who bring their inventions to the attention of the public.

A key aim of the IPTA is the maintenance and improvement of Australia's intellectual property framework for the benefit of inventors, trade mark owners, designers, business and the community as a whole. It is set to create an environment that will produce the next generation of Cochlear implants and Black Box recorders. ■

By Jane Breusch

Claiming credit

→ Managing intellectual property and protecting innovation is not only good business sense, it's vital for Australia's economic health.

Few Australian businesses are aware of the lost opportunities or even the disasters that can befall them by not managing and protecting their innovations – whether they be products or processes – and how easily these can be avoided.

"The majority of Australian companies don't appreciate the need for intellectual property (IP) management," says Griffith Hack principal Rob Wulff. "IP is often given low or no priority."

Wulff says it is something that all businesses, including individuals and small-to-medium enterprises, must quickly start to recognise.

"Australia's economic success and livelihood depends on us competing effectively in the global market place because the Australian domestic market is, for most products, just insufficient. Strong IP enables this," Wulff says.

"The only way we can do this is if we effectively manage the whole process of innovation, including the management of a firm's intellectual assets. If we don't do that, then the Chinese, the Americans, the Japanese and the Europeans will leave us for dead."

The first, and most important, step in the IP management process is identification of IP. It is then critical to keep innovations secret, as disclosure in many countries can invalidate the IP. Innovations should then be evaluated by a qualified patent and trademark attorney to see whether they are protectable. At the same time, they should be assessed for market viability.

"There's no use getting protection and spending the money on protection if there isn't any market or opportunities," says Wulff.

Costly consequences

He cites the instance of an Australian radio station that spent $A150,000 on re-badging and renaming only to discover the name was already registered as a trade mark to another radio company.

"The $150,000, not to mention the time they had spent on corporate re-imaging, was a complete waste, whereas all they had to do was a $500 trademark search," Wulff says.

"We find businesses are making these mistakes all the time and not doing the proper checks. It shows that many have no concept of what intellectual property is or what the ramifications of it are."

In another case, a large, publicly listed company engaged an external research organisation to do some contract research without discussing ownership of the intellectual property. The external research organisation decided it would not sign over the intellectual property, which later proved to be a commercially significant development in the agricultural area.

"It took two years of protracted negotiations and considerable legal expenses to sort out the issue and finally reach a settlement," Wulff says. "Once they did get it right, the returns because of the IP were significant."

Wulff says the establishment of an in-house IP management position and a management strategy can prevent such incidents from occurring, adding that good communication is vital.

Getty Images

It's mine! Don't give up your intellectual property rights.

"All key employees need to be educated as to what the intellectual assets are," he says. "In addition, the processes for identifying, documenting, protecting and managing those assets must be clearly outlined and you need to have somebody on staff that is directly responsible for these tasks". ■

A history of IP

Griffith Hack Patent and Trade Mark Attorneys and Lawyers has a long association with intellectual property in Australia. It can trace its roots back more than a century to two of the oldest patent attorney firms in Australia: Griffith Hassel & Frazer in Sydney and Melbourne's Clement Hack & Co.

Griffith Hassel & Frazer was started by Irish immigrant and patent attorney Arthur Griffith, who became a member of the New South Wales Parliament and was Minister of Public Works from 1910. The New South Wales town of Griffith was named in his honour. He was also president of the Institute of Patent Attorneys of Australia in 1929-1930.

Innovation is the ability to change the way people do something by providing an alternative approach to achieving the desired outcome

leaders SPEAK

Continuing economic reform

Maintaining Australia's relatively high living standards in the face of an ageing population will require, among other things, achieving rapid productivity growth. It is therefore timely (after two decades of economic reforms) to consider the next steps in further enhancing the global competitiveness of Australian firms.

The development of the knowledge economy as a general concept, and of science-based industries and the innovation system as particular concepts, introduces new perspectives and public policy needs – and the need for thinking about the links between these new perspectives and the enhancement of firm-level capabilities.

The Committee for Economic Development of Australia is keen to promote discussion of the key issues involved in innovation. We need to explore the empirical developments and the conceptual frameworks and instruments through which the next phase in thinking and public policy development might be approached.

Other issues would cover the creation of capabilities, the abilities needed to respond to technological opportunities, incentive structures and the management of risk and uncertainty. The key objective is to leverage the capabilities of Australian firms to enhance national economic welfare and living standards.

BOB LIM
CHIEF OPERATING OFFICER & NSW EXECUTIVE DIRECTOR
COMMITTEE FOR ECONOMIC DEVELOPMENT OF AUSTRALIA

What is innovation?

For many people, innovation is coming up with a clever idea. For others, it is the opportunity to create some new technology. In truth, however, innovation is the ability to change the way people do something by providing an alternative, non-obvious approach to achieving the desired outcome. This implies that, while the idea and the technology may be important in achieving the result, they are irrelevant if the idea does not manifest itself in the marketplace.

For Telstra Research Laboratories (TRL), this drive to achieve a real outcome is the raison d'etre of its research work. In the telecommunications industry there are many great ideas about what we can do with the technology in the future, but our focus is on "what can we create next that will make a meaningful difference?"

Consider, for example, the notion that mobile phones will become handheld computing and information devices. While much has been promised, the reality is that little has been accepted by users other than SMS – short messages or text messages.

TRL has found, however, that innovative solutions in this regard can be created using available technology. For example, querying a directory using SMS is often simpler and more effective than making a phone call. The number that is returned can be dialled directly with one click, making it convenient for the user, who does not have to write it down and dial it – particularly useful when you are on the move!

HUGH BRADLOW
MANAGING DIRECTOR
TELSTRA RESEARCH LABORATORIES

Going against the grain

Throughout the 1990s and into the 21st century, the Australian grains industry has been one of the nation's success stories. While on a global scale Australia is a small producer of grains, it is a major exporter, not least of wheat for making bread and a range of other products, as well as barley for beer.

Australia's reputation as an exporter is based on quality, underpinned by strategic planning and an industry commitment to research and development. It is this commitment that has sustained the continued productivity growth of grains, while other major industries have faltered.

Canola, for example, grew from a gleam in the scientist's eye to become the golden grain for the growers of the 1990s. Farming systems changed as new crops were introduced into rotations and reduced cultivation improved soil conditions and water use.

Make no mistake, the Australian grains industry is as "high-tech" as any "dotcom" – and much more durable. The grains industry is sustained by biotechnology at many levels, including the adoption of leading-edge gene technologies by plant breeders.

Innovation in information technology has offered growers the opportunity to gain knowledge while deploying geographic positioning systems in routine farm operations.

The value of grains has bounced from $A3 billion to $8 billion in a decade. More importantly, the grains industry is repositioning itself as a vital and diversified sector within the $75 billion Australian food industry.

And there is to be no resting on laurels. The industry is in the midst of a strategic planning exercise with a horizon of, at least, 2020.

JOHN LOVETT
MANAGING DIRECTOR
GRAINS RESEARCH AND DEVELOPMENT CORPORATION

What can we create next that will make a meaningful difference?

ABIE

By Gayle Bryant

Commercialising new ideas

→ Nurturing start-ups, solving information and communications problems and commercialising technology is all in a day's work at NICTA.

Most of us know the frustrations of waiting while an internet site takes minutes to download when only the day before it downloaded within seconds. While there are any number of reasons to explain why this happens – server issues or poor telecommunications connections – what people want is greater surety that the internet sites they are accessing will be available within a set amount of time.

Solutions to issues such as these are now being developed by a new Australian company known as NICTA – the National Information and Communications Technology Centre Australia. In 2003, as a result of the Federal Government's Backing Australia's Ability program, $A130 million was committed to building an Information and Communications Technology (ICT) Centre of Excellence and NICTA was the result.

NICTA is a limited liability company whose shareholders are the four founding partners of the bid to build the centre. They are: the Australian National University; the University of New South Wales; the NSW Government; and the ACT Government. It is also supported by the University of Sydney.

NICTA's aim is to become the ICT equivalent of the Australian Institute of Sport

A sporting chance

NICTA's aim is to become the ICT equivalent of the Australian Institute of Sport. According to chairman Neville Roach, its contribution will be measured "not in gold medals and world cups but in the transformation of Australia into a significant producer and world's best practice user of ICT".

NICTA has been charged with taking on a leadership role in Australia's ICT research community by drawing together in a single institution, research, training, commercialisation and collaboration.

The focus of research to be undertaken by the Centre follows five broad themes:

- Infrastructure technologies;
- Intelligent systems;
- Software engineering;
- Human-machine interaction; and
- Foundations (the basic theories and algorithms on which ICT technologies are built).

Within each theme there are several programs which all the researchers in the centre are involved with. A number of initial programs are already underway, including the application that looks into guaranteeing quality of service in the internet environment. This program is part of the networks and pervasive computing program and runs under the infrastructure technologies theme.

Program leader Aruna Seneviratne says one of the main thrusts of the research is mobile internet applications. "We are especially interested in applications such as guaranteeing the quality of service in the internet environment," he says. "What users want is to be able to send a request and know that there will be a response in a predictable time. This is especially important if you are moving into real-time data such as telephone calls, video or audio."

Access all areas

The basis behind the networks and pervasive computing program is the fact that today's laptops and handheld computers are being equipped with hardware that provides access to different wireless networks. To take advantage of the developments in devices, different wireless networks are being deployed which provide services within local as well as wide area networks. These developments give rise to pervasive computing environments that make it possible to provide access to any of the available services from anywhere at any time.

Providing access to users anywhere at any time is the focus behind another project within the group, that of being able to seamlessly move communications from one type of phone to another, or to a laptop.

"At the moment you can't move from a GSM environment to a LAN without breaking the connection," Seneviratne says. "We have carried out some trials that allow a person to seamlessly move between networks. The results of these trials have led to us filing patents. In addition, we have formed a spin-off company that focuses on this area."

Seneviratne hopes that the projects his group is involved with will lead to the creation of pervasive environments.

"This means we are always connected to some type of network," he says. "We want to build applications that allow users to always be connected, although there are social consequences of this that need to be considered. For example, if a user has a number of different devices, such as a mobile phone and a laptop, then being connected all the time could mean that they are constantly available to people and this could be stressful. There would be no downtime."

Remote care

Another area the group is involved with and one that has major implications for Australia's aging population is home telecare. One application of this is placing monitors on people in their homes and monitoring their movements remotely.

"If a patient is taking high-blood-pressure tablets and the doctor is worried that they may experience side effects, then the telecare monitoring system can provide information that tells doctors whether their patient is sitting up or walking," Seneviratne explains. "We have carried out some trials already at a hospital in Wagga Wagga which have gone well. Such a system also allows people to remain in their own homes rather than having to go to hospital."

Getty Images

Raising the bar – NICTA aims to be the ICT equivalent of the Australian Institute of Sport.

Seneviratne's group was previously based within the University of New South Wales. He views the move to NICTA as a positive one.

"The way we were structured at the university meant the research we carried out was done in a fragmented way," he explains. "We can now bring our expertise together as a team and work in a more coherent fashion."

Through external arrangements, NICTA will provide training in general management areas as well as specific subjects designed to promote the successful launch and growth of new start-ups.

NICTA is being developed in two phases. Initially, the organisation has established its executive offices at Sydney's Australian Technology Park. Research, research training and commercialisation are being undertaken here. This phase has also seen two research laboratories established at two of Australia's leading ICT research universities: the University of New South Wales and the Australian National University.

In the second phase, NICTA executive offices and the Sydney research laboratory will move to a new building within the Park. Each of the sites will accommodate more than 100 PhD-qualified researchers, plus PhD students, technical and support staff.

NICTA will use whatever commercialising approach a particular technology requires to achieve its mission, including licensing usage rights, sale of technology, joint ventures, start-ups and high-end consulting. However, to maximise the Centre's impact on Australia's economy, NICTA will have a primary emphasis on creating start-ups and growing small-to-medium enterprises.

Australia has not exploited the opportunities in the marketplace

Seneviratne says Australia has not traditionally been efficient at exploiting the opportunities that exist in the marketplace. He says that patents filed relating to his work have been taken up by multi-national companies rather than Australian ones, which is a situation he hopes will change.

"What we hope is that NICTA will facilitate exploiting the commercialisation of the new ideas that people generate," he says. "This is one of the major advantages of NICTA being set up." ■

Gayle Bryant is a finance and technology writer based in Sydney.

By Mark Hollands

The next big thing

→ Industry analysts have the power to influence the success or failure of a product or service. But how do they decide what will work?

The information, communications and technology (ICT) industry is littered with the quick and the dead. Executives thrive if they are smart enough to read the future and stay ahead of the curve, while those who are inflexible who or procrastinate about changing are unceremoniously dumped on to the streets of history.

Business is an unforgiving world, and the technology sector is more severe than most. That is why it is energised by visionaries such as Microsoft's Bill Gates, Dell's Michael Dell or one of the internet's creators, Vincent Cerf.

Our thirst to know the future has spawned a worldwide industry of soothsayers. Predicting the "next big thing" in information technology (IT) is a multi-million-dollar industry – especially if you are right more times than not.

Predicting trends is the role of the industry analyst – an influential figure who can make or break a company, even an industry, with their insight and foresight. Often supremely confident, the provocative analyst can divide followers into camps of love and hate. Cynics decry their predictions, believing them to be supported by little more science than holding up a damp finger in a wind of change.

They are in the minority, however. The analyst, like the visionary, plays an important part in helping to shape the future of technology, its industry and society as a whole. This role gives analysts a unique perspective from which they can see and articulate the value of innovation and its ramifications.

The name of the game for all analyst companies is recognising and comprehending innovation of value

Gartner analysts have been predicting an internet-driven revolution of the music industry since the mid-1990s. Almost a decade later, it is clear the digitisation of music will change how we listen to and how we buy our favourite songs with the plethora of online music sites now being launched on the web.

Analyst companies were among the first to understand the commercial power of the internet. Some were more aggressive in their optimism than others. Gartner continued its self-confessed conservatism in the face of enormous hype, but its attitude ultimately paid off.

The name of the game for all analyst companies is recognising and comprehending innovation of value. Without it, there is nothing to predict, nothing to analyse, nothing on which to advise.

Innovation is the lifeblood for the industry. Articulating the trends is a vital task which helps executives understand where they should invest, deploy resources or whether to pull out and move on.

Every industry faces its own issues with technology adoption but they have become most intense for the finance institutions, telecommunication carriers and manufacturers who face an ongoing battle to drive inefficiency out of the supply chain in an increasingly commoditised world.

When it is possible for an unknown company such as Smart to design and badge its own fuel-efficient cars globally without so much as a spanner, let alone an assembly line, then it is clear how quickly the world can change through innovation of business principles and technology.

The former head of BT Labs in Britain, Peter Cochrane, loves to tell how there are more scientists and researchers in the world today than at any time in history – "so it is no surprise the world is changing so quickly".

This is not a journey any of us can stop, of course. The key is to embrace and understand what is happening around us and maximise the opportunities to create stronger companies and better living standards for everyone.

For executives in positions of power and influence, this is an obligation not an option. But one of their challenges is to know where to turn for the right advice.

Providing hype

For IT advice, many companies seek out Gartner, a 25-year-old company with more than 800 analysts and consultants. Gartner dominates the IT advisory sector with a global client base of 10,000-plus companies.

Its most popular technique for articulating innovation value and predicting the future is the Hype Cycle – a snake-looking device of peaks and troughs that illustrates what is hot and what is not (see illustration below).

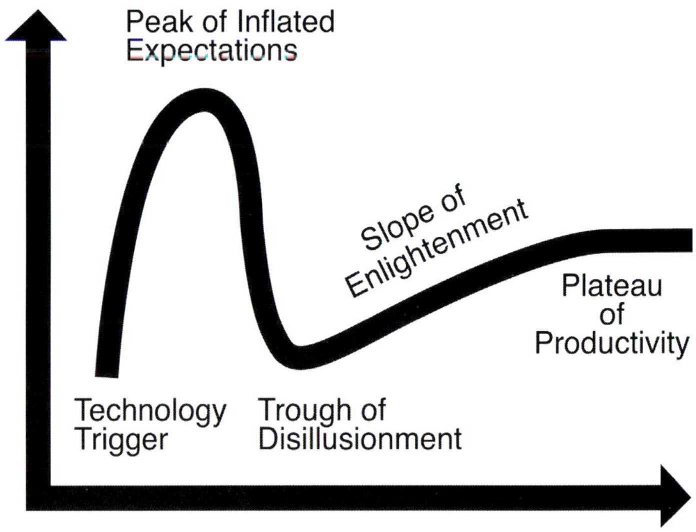

Finding a way to manage the hype is all in a day's work for industry analysts.

Developing a Hype Cycle at Gartner is a complex task. It cannot be done alone, requiring experts from around the world to provide their say on the correct positioning of a technology.

Inside the Stamford-based company, it is the most intellectually intensive and rigorous endeavour any analyst can undertake. Each project requires a significant amount of research which is subject to constant peer review and online collaboration. Few are finished in less than six months. The Hype Cycle has two axes, one signifying industry awareness of the innovation and the other determining the technology's maturity.

The wavy-line is divided into five sections, each one signifying a stage in a technology's life that all modern-day invention experiences. These are called:

- Technology trigger;
- Peak of inflated expectation;
- Trough of disillusionment;
- Slope of enlightenment; and
- Plateau of productivity.

It is difficult to argue against the relevance of these quaint sounding categories. There has not been an innovation in the ICT industry that has not been hyped to the heavens, or one that has not felt the blade of the tall poppy syndrome.

The latest Hype Cycle for emerging technologies covers a wide variety of topics, from smart dust – airborne particles released at high altitude to create a mesh network – to micro-fuel cells and light-emitting polymers that will power the screens of phones, handhelds and notebooks and save battery power.

Two technologies get the order of the boot: holographic storage and wireless networking using Bluetooth technology. Others, such as speech recognition for call centres, are positioned along the "Slope of enlightenment" and will shortly appear on the "Plateau of productivity" before disappearing off the chart because they have become an established part of commercial life.

Colour-coded life cycles

Each technology entry on the Hype Cycle is colour-coded to signify the length of time it will take before it matures sufficiently for established business usage. For example, smart phones that handle voice and data equally efficiently will take 12 months to get through the cycle, while micro-fuel cells will take at least five years.

The Hype Cycle is not used just for emerging big-picture innovations. Each industry sector, such as supply chain management or networks, has a sufficient number of niche technologies to make the technique valuable.

It is just one method deployed by Gartner to help clients understand the power of innovation, the length of time before it matures and the impact it will have on an industry, or even a society.

It is a lot more accurate and powerful than holding up a damp finger into the wind. And its popularity is driven by a simple motivation, which is: everyone in this industry wants to be quick, not dead. ■

Mark Hollands is principal of Hollands Media.

By Gayle Bryant

Talking naturally

→ The frustration of ordering over the telephone may soon be over as natural language speech systems become more prevalent.

Getty Images

Telstra is developing new solutions for telecommunications.

Natural speech language systems are the latest examples of innovation to come out of Telstra Research Laboratories. These systems replace the frustrating procedures we go through when ordering products or services over a telephone via an interactive voice response system (IVR).

At the moment, when ordering movie tickets, for example, we are given a menu of choices from which we have to choose. A natural speech language system allows us to just say what we want and the system automatically understands the request.

It is an amazingly innovative system because it has to recognise the context of the different words

Dr Hugh Bradlow, chief technology officer for Telstra, says how much easier it would be to just say, "I want two adult tickets to the new James Bond movie at the Tivoli for 7.30pm tonight" and have that request recognised immediately.

"While IVR systems are used by many organisations to increase productivity, they make customers frustrated," he says. "Our system automatically learns the language and the grammar that people use. There is also no need for people to learn how to program these skills into the system as it is done semi-automatically."

The idea for the system came about when a Telstra researcher thought that, since humans could be taught grammar, why couldn't machines? The natural speech language system is known as Lyrebird. "Lyrebirds are known for mimicking other birds in the bush, and our Lyrebird system has the capability to mimic humans," Bradlow says. "It is an amazingly innovative system because it has to recognise the context of the different words. For example, it needs to know that 7.30pm is a time, that James Bond is a type of movie and that the Tivoli is a venue."

Bradlow expects that over the next few years, the system will become prevalent in a number of commercial applications.

SMS tracking

Another innovation from Telstra Research Laboratories is a corporate directory that enables employees to look up the telephone numbers of fellow staff using SMS.

Bradlow says Telstra SMS Directory Search – designed and developed at Telstra – was a first for Australia.

"If I want to get the number for someone in Telstra, I SMS their name to a special number and within a few seconds I get an SMS back with the person's name and their fixed and mobile numbers," Bradlow says.

He adds that Telstra is involved in a number of other innovative products and services, especially in the telecommunications sphere focusing on wireless broadband technologies.

"Also important are future broadband technologies such as fibre-to-the-premises and technologies that improve the performance, security and management of internet protocol (IP) networks," he says. "We're looking at technologies that will drive future applications in the IP environment – so-called 'web services'."

Telecoms trends

Telstra spends between $A160 million and $240 million each year on research and development. Innovation is occurring in a range of IP-based services and technologies. Bradlow says mobile devices, for example, are changing rapidly because of the two simultaneous trends.

"There is a rapid uptake of laptops, which now outsell desktops," he says. "Also, mobile devices are becoming richer and more complex with more facilities and categories. That means the potential for applications in the mobile space is very high." ■

By David Salt

Run RABiT run

→ A Melbourne-based research centre's foray into by-wire technology will ensure Australia stays in the fast lane of this new industry.

Forget being able to fly your car around in space. While it may be nice to imagine what the car of the future will be capable of, in reality it will look much like the one you are driving now, with some noticeable differences.

Tomorrow's car is likely to have fewer parts such as brake fluid, steering pumps or fan belts. It may have a joystick instead of a wheel and a button instead of a brake pedal. Manufacturers should also be able to customise it to meet preferences for suspension, braking and steering "feel".

All these changes are possible with "by-wire technology". By-wire technologies replace many of the mechanical systems that have traditionally made up a car with electronic controls. The transition to by-wire technology is being driven by the need to reduce weight and production costs.

A new research centre – the Research Centre for Advanced By-Wire Technologies (RABiT) – has been established in Melbourne to ensure Australia is at the forefront of the new technology.

Alan Stevens, vice-president of Pacifica Group Technologies (PGT), the research arm of the company Pacifica Group, Australia's largest automotive brake manufacturer, says by-wire technologies are poised to transform the way vehicles are designed, manufactured and driven.

Customised cars

"These technologies will make cars safer, lighter, more economical, easier to use and capable of being customised," he says. "By-wire technologies are already being designed into today's cars. Automotive analysts estimate that within 10 years, 40 per cent of all cars produced in Europe will feature almost all by-wire technologies such as steer-by-wire, clutch-by-wire and brake-by-wire."

The adoption of by-wire technology is already well advanced and growing at an exponential rate. The establishment of RABiT ensures Australia will not be left behind.

"RABiT is a $A10.3 million collaborative venture between industry, research and government," says Stevens.

"PGT is the leading partner and we're teaming with the University of Melbourne and Swinburne University of Technology to accelerate by-wire development. A wide range of new skills is required which need to be integrated with existing mechanical and manufacturing expertise."

Stevens says RABiT needs to build competencies in mechatronics, software development, electronics and simulation. "We also need an education program which can gear up a whole industry to the new requirements of by-wire technology," he says. "It's a classic paradigm shift that will affect a wide range of industries such as marine, defence and transport."

The Victorian Government is contributing $4.7 million for infrastructure to ensure that Victoria has a sustainable by-wire industry.

Getty Images

RABiT's mission is to make Victoria a regional leader in by-wire design and central to achieving this aim is the creation of two research facilities: a vehicle dynamics laboratory at PGT's engineering centre in East Bentleigh, Victoria, and an education/simulation laboratory based at Melbourne University.

Future focus

These state-of-the-art facilities will offer RABiT partners and collaborators access to dynamometers, environmental chambers, vehicle testing facilities, advanced electronics and a virtual reality capability for simulating vehicle dynamics and process development.

"Companies such as Pacifica need to invest in by-wire control," Stevens says. "It's the future of the automotive industry and we plan to take a lead role in it." ■

By David Salt

Innovators never sleep

→ Somewhere over the horizon is the next big thing and this is where Bishop Technology Group is firmly focusing its sights.

What makes a company a successful innovator? Is it simply a matter of good luck or is innovation a goal that can be actively grown?

"Luck has nothing to do with it," says Dr John Baxter, joint managing director of Bishop Innovation, part of the Bishop Technology Group (BTG). "Successful innovation is all about looking ahead, focusing on areas where there's a future need, combining your strengths with the strengths of others to meet that need, and making sure you're protecting any intellectual property you create along the way."

While that might be easier said than done, BTG's track record of successful engineering innovation suggests it is an effective formula. BTG holds about 350 patents and patent applications worldwide, earning the group more than $A15 million a year in royalties. Its technology has been licensed to some of the largest automotive manufacturers in the world, and BTG-developed steering technology is used in one in five passenger cars globally.

Its ability to innovate certainly impressed DaimlerChrysler, one of the world's leading automotive organisations. In August 2001, it took a 30 per cent stake in the company – a strong endorsement of the potential of BTG's intellectual property that it holds and is developing.

However, instead of resting on its laurels, BTG is always looking ahead for areas that require innovation. The global automotive industry is now undergoing a revolution in which many of a car's mechanical systems, such as steering, are increasingly being controlled electronically. BTG's work in the area of electric power-assisted steering (EPAS) serves as a good example of how innovative the company is.

"The benefits of EPAS include more accurate and reliable vehicle control and improved fuel efficiency," says Baxter. "These changes are going to affect all cars."

The power of torque

Baxter says about eight years ago the company knew that EPAS would be taking off in 10 years' time and decided to develop a better torque sensor that would measure the torque that was being applied to the steering wheel. "Such a device is fundamental to developing a reliable EPAS system," he says. "The sensors being used back then were based on magnetic analog systems. They were complicated, not very reliable and subject to interference by other electromagnetic fields present in the car."

The first thing BTG did was to look around for other possible systems and here Baxter provides another tip for successful innovations: don't limit yourself to what you can do yourself.

"You need to look at what other people are doing and see if you can re-engineer their strengths to solve your problems," he says. "We realised we needed an optically-based torque sensor, so we harnessed the strengths of a Swiss company to develop with us an optical integrated circuit that would robustly 'read' the marks on a disc attached to the input shaft of the steering gear."

This is a similar technology to what a computer mouse uses to measure the changing position as you move it by hand. BTG just applied the technology to a different problem. The end result was a Bishop Torque and Angle Sensor or BTAS system. This has now undergone further refinement and, in partnership with the German automotive giant Bosch, plans are being developed for its worldwide production over the next few years.

"We looked ahead, saw where the opportunities lay, and then sought out partners with the appropriate skills to realise those opportunities," says Baxter. "Critical to our success was our ability to protect the intellectual property (IP) that was being created. This is a core strategy for innovation, and it is something that BTG takes very seriously."

> We looked ahead, saw where the opportunities lay and then sought out partners with the appropriate skills to realise those opportunities

Baxter says prior to committing to developing any aspect of the BTAS system, the company had drawn up development license agreements with its partners that clearly set out who was doing what and who owned what parts of any IP being generated in those developments.

"There were 10 worldwide families of patents created during the development of the BTAS system," Baxter says. "The IP is where the true value in innovation lies so to capture that value you have to protect it."

Worldwide production of EPAS in passenger vehicles is only now ramping up but BTG technology has already secured its place at the centre of the revolution.

"We're now looking over the horizon at the next big thing," says Baxter. "Successful innovation requires you to always be looking ahead and to look in the right areas you need to ask the right questions and you need innovative staff."

Igniting the inventive spark

Baxter praises the BTG staff. "We have a great team but we're constantly on the look out for new people with those core qualities that go hand-in-hand with innovation," he says. "While they have to have the technical skills that many good engineers and scientists possess, they also need that inventive spark and an ability to sell their ideas to different audiences."

Baxter says after the right people are attracted to the organisation, their capacity for innovation is then actively cultivated by teaching them the skills that will allow them to protect the IP that is being generated.

Looking ahead at future needs drives Bishop's innovative culture.

"Our staff are taught things such as how to draft patent applications, basic commercialisation strategies and how to conduct searches on what other technologies are already out there," he explains. "Our experience suggests that there are certain innate qualities in some people that make them naturally good at innovation. However, we've also found that innovation arises from a certain culture and at BTG that culture is the driving force behind our success." ■

The Bishop behind BTG

Bishop Technology Group (BTG) was founded by Dr Arthur Bishop, one of Australia's most prolific inventors. While Bishop had always planned to be an inventor, World War II propelled his career into an unexpected direction. The war disrupted supplies from other countries so Australian companies had to produce things for themselves.

At the time, Bishop was a chief engineer involved in the production of Beaufort bombers. He believed the original designs for the plane's landing gear weren't up to Australian conditions, so he redesigned it and built a plant that could manufacture the parts. Within 12 months, Bishop had established a factory in Sydney that turned out thousands of aircraft landing systems.

In *Driven By Ideas: The Story of Arthur Bishop – A Great Australian Inventor* (University of New South Wales Press), author Clare Brown charts the life and work of this brilliant and multi-talented man. As well as being a clever designer and engineer, Bishop was an astute businessman. He protected his intellectual property by patenting the design of his landing gear. After the war, he sold his designs to the US Navy and Airforce. He then spent the next 30 years adapting his designs for the car industry, developing a more responsive steering system. He licensed his designs for use by both General Motors Holden and the Ford Motor Company.

The potential for innovation exists in any organisation

Crediting ideas

Generating new ideas and bringing them to life through economic activity is vital to our country's future. It will turn great ideas into products and services, and help to create jobs and generate income here in Australia.

Australians are a resourceful, inventive and creative people. But we are too modest in our appreciation of these qualities. As a nation we must commit to fostering a more innovative culture by celebrating our successes. We need to better acknowledge the contributions of our leading innovators and entrepreneurs, and better appreciate the positive role of innovation in every workplace and organisation.

Our work at National Innovation Awareness Strategy (NIAS) is centred on the broadest – and we feel the most widely constructive – definition of innovation. And this we describe as "any new thinking which demonstrably helps an organisation function better, improves a process of discovery or commercialisation, or one which urges staff to partake in making their organisation more stimulating and competitive".

Such a broad focus naturally encourages the idea that innovation is not merely the concern of those in science and technology but individuals and groups engaged in education, human resources, commerce, sport, the arts and many other endeavours.

Our message is that the potential for innovation exists in any organisation.

DAVID MILES
CHAIRMAN
NATIONAL INNOVATION AWARENESS STRATEGY

leaders SPEAK

Fair exchange

The InnovationXchange Network was launched in May 2003 by the Federal Minister for Industry, Tourism and Resources, Ian Macfarlane, to provide Australian companies with access to much-needed innovation resources and connections.

An "open network of networks", the InnovationXchange Network spans industry, research and educational institutions and government. It consists of network leaders (typically larger organisations, such as CSIRO, a large percentage of Australian universities, corporations such as IBM and government agencies), network partners (smaller non-profit associations and networks); network members (small companies and start-ups), and associate members (the members of network leaders or partners).

With the shift from an industrial to a knowledge economy, smart companies are increasingly aware of the importance of networks and "open innovation" – that is, looking outside the organisation for good ideas – using trusted intermediaries and a year-round program of networking opportunities, which include leadership luncheons, innovation briefings, blue-sky forums and a visiting international Fellows program, to provide significant benefits to participants.

The InnovationXchange Network takes a whole-of-business approach to innovation and the website offers innovation resources in new technologies, research and development, education and training, business services, government assistance, finance and funding, and networks, experts and events. While this is a public service, only members can use it to market and promote their capabilities.

GRANT KEARNEY
CHIEF EXECUTIVE
AUSTRALIAN INDUSTRY INNOVATIONXCHANGE NETWORK

Innovations never happen as planned.

GIFFORD PINCHOT III

By Roger Lough

In defence of innovation

→ DSTO has contributed to many scientific achievements that have led to new capabilities for the Australian Defence Force.

The Defence Science and Technology Organisation (DSTO) has been involved with many innovative breakthroughs that have found their way into global applications for both military and non-military uses. These range from the ubiquitous black box flight recorder to the sophisticated Shapes Vector network protection system. However, it is Australia's Defence Force that has largely benefited.

The Jindalee Operational Radar Network.

Science on the seven seas

The Navy has benefited enormously from the innovations developed by DSTO. An active missile decoy, known as NULKA, was developed in partnership with industry and the US Navy from an original DSTO concept. NULKA has revolutionised ship protection with its unique, hovering rocket that entices missiles away from the vessel.

DSTO's history of innovation goes back many decades. In the 1960s, it developed the Barra sonobuoy which remains the most effective directional sonobuoy system in operational use. Its role is to detect, locate and identify quiet submarines.

Defence and industry are using the Barra technology to enhance capability in air-deployable multistatic active sonobuoy systems.

Another innovation that focuses on defence within the Navy is Ikara, a ship-borne long-range anti-submarine guided weapon which achieved international recognition in the 1960s and entered service with the Australian, British and Brazilian navies.

Search and rescue is a key part of naval defence and DSTO has assisted with the development of a marine dye marker, known as Seamark. Now marketed worldwide, Seamark is a safe, effective and longer lasting alternative to flares and smoke signals for search and rescue at sea.

Materials to strengthen Navy vessels have been another innovation. The Collins Class submarines feature a high-strength steel with welding technology to match, as well as specialised anechoic tiles for signature reduction.

DSTO has also been instrumental in developing array streamers for passive maritime surveillance systems. It developed one known as Kariwara, which is a slim-line buoyant-fibre towable and reelable acoustic array streamer.

Innovation takes to the skies

DSTO has also forged a place as a leader in the use of adhesively-bonded fibre composites, such as Bortex, to repair aircraft structures and arrest stress corrosion cracking. The technology has been successfully marketed worldwide and used in commercial airliners as well as military aircraft belonging to the RAAF, US Air Force and the Belgian Air Force.

The organisation has also pioneered full-scale fatigue testing of aircraft structures to predict their safe operating life. DSTO is now a world leader in this area. The science of fatigue life prediction is now practised by aircraft designers all over the world.

Aligning aircraft onto the correct glide path for landing is one of the most important manoeuvres a plane can make. To assist, DSTO invented the 'T' Visual Approach Slope Indicator System which has now become the international standard and won the Prince Philip Prize for Australian design and the *Diplome d'Honneur from the Federation Aeronautique Internationale.*

Commercialising concepts

Not all DSTO innovations are confined to high-profile items of equipment and infrastructure. Every day, DSTO is innovating by writing new software codes, making engineering improvements, developing capability solutions, redesigning business processes and even applying lateral thinking to people problems.

Being able to identify network anomalies is a key requirement of all computer systems. DSTO has developed a network security system called Shapes Vector that uses intelligent agents and 3D visualisation techniques to identify such anomalies. It represents breakthrough research in real-time visualisation of network attacks. The Defense Advanced Research Projects Agency in the United States says Shapes Vector is a generation ahead of current network visualisation technologies.

DSTO proved it was capable of developing world-beating systems

The Acoustic Probe on a model of the Collins Class submarine in the anechoic chamber.

when it produced a system known as Starlight, that allows users of secure computer systems to access insecure networks, for example, the internet, without compromising their own security. Starlight is now being marketed worldwide under the brand name Veto.

DSTO innovations can be found in many everyday products

In collaboration with industry, a leading-edge detection system is being developed to clear anti-vehicle landmines from unsealed roads and landing areas. Known as Rapid Route & Area Mine Neutralisation System (RRAMNS), it is industry-built and incorporates a world-first patented metal detector array and a DSTO-developed imaging system. The detection system will be quicker, safer and more affordable for clearing landmines.

DSTO innovations can also be found in many everyday products. For example, the 'Wet' xerographic process was invented by DSTO in 1959 for reproducing colour.

It was also responsible for establishing the standard for sunglasses, as well as for pioneering the optical glass industry in Australia by producing optical parts for telescopic gunsights and microscopes.

Whatever the problem, DSTO always rises to the challenge and, more often than not, leaves a substantial mark. ■

Dr Roger Lough is the chief defence scientist of DSTO.

Innovation winners become money spinners

DSTO* has a long history of exploiting its intellectual property. This is often done by partnering with industry, which further develops and markets the technologies worldwide. By commercialising innovative technologies, DSTO is playing its part in national wealth creation and supporting the Government's Innovation Action Plan – Backing Australia's Ability. DSTO has helped generate millions of dollars for Australia. One innovation in particular, the Australian Minesweeping System known as AMAS, has earned more than $A40 million in exports to the UK, United States, Indonesia, Japan, Poland and Thailand. It was also used in de-mining the Iraqi port of Umm Qasr in 2003. AMAS incorporates the DSTO-developed compact, self-powered sweep called Dyad which emulates the magnetic signatures of target vessels, causing sea mines to detonate safely.

In the 1980s, DSTO developed the Laser Airborne Depth Sounder, or LADS, to measure the depth of coastal waters with greater efficiency and speed compared with conventional shipborne sonar methods. In the past 10 years, the LADS technology has generated more than $100 million for Australian companies and it continues to find new markets for different applications, especially in the United States.

* *Includes innovations from former laboratories that are now part of DSTO.*

By Lucio Di Bartolomeo

An excellent defence

→ Developments by Australia's leading defence and electronics systems company are attracting global attention and orders.

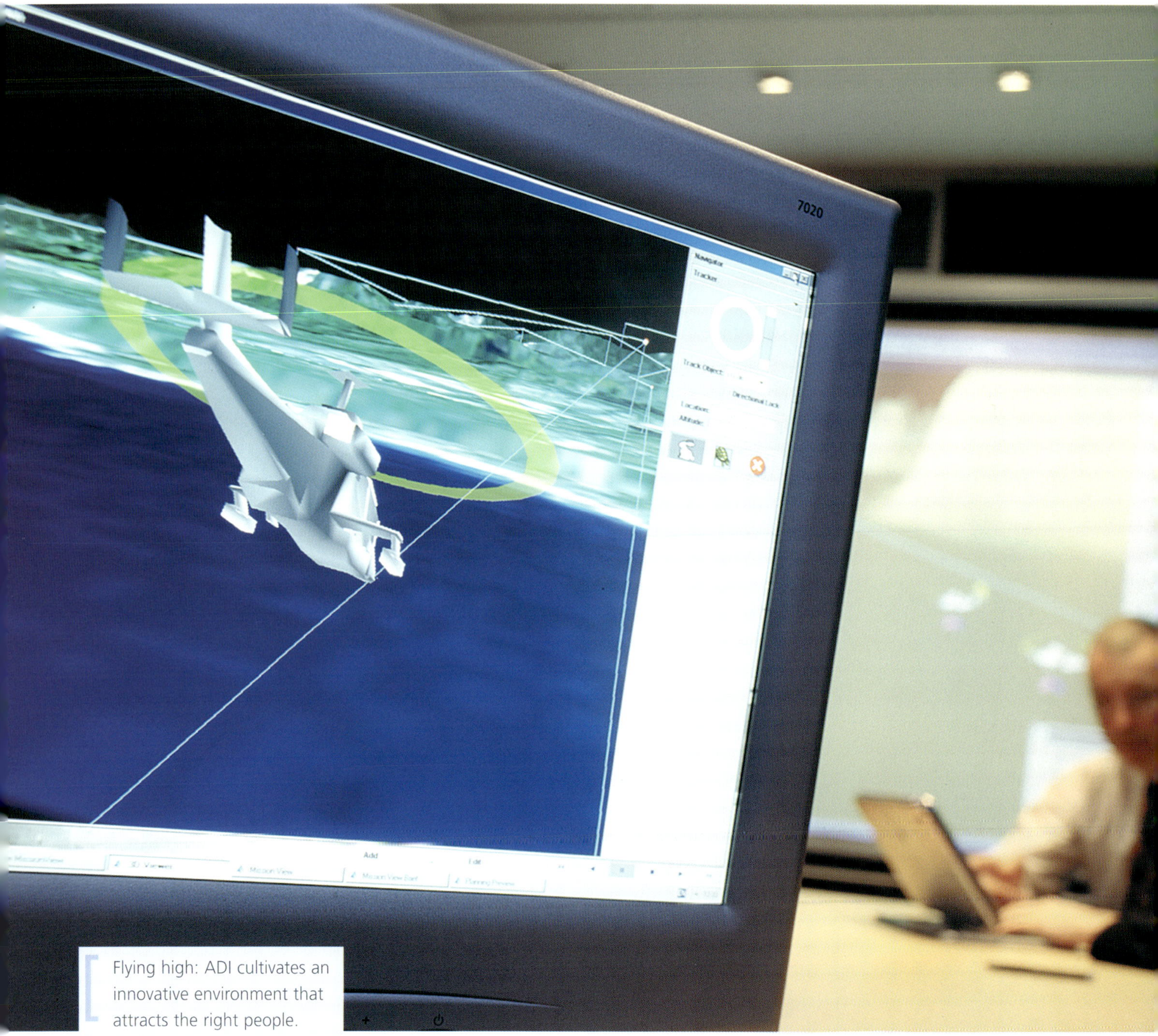

Flying high: ADI cultivates an innovative environment that attracts the right people.

An Australian project that uses research based on insect vision is attracting international interest. The project, being developed by ADI and Defence Science & Technology Organisation (DSTO), hopes to produce a unique tracking and guidance system that uses new techniques for detecting and tracking moving targets against background clutter.

To ensure the technology remains in Australia and is available to the Australian Defence Force (ADF), DSTO and ADI are providing the funding to demonstrate its feasibility. A high-speed, highly manoeuvrable, unmanned aircraft, the turbo-jet-powered CYBIRD 2 has been developed to demonstrate the guidance system's capabilities.

Such a project is typical of the innovations ADI is involved in.

It is one of Australia's leading software development houses and provides large-scale systems integration, project management and software engineering. It operates in five States, with almost half of its workforce employed in regional Australia.

The company's achievements as a leading Australian defence and electronic systems company reflect its commitment to innovation, customer service and personnel selection and development.

It seeks to cultivate a progressive environment that attracts the right people and provides them with the opportunities to develop entrepreneurial skills in research, engineering processes and systems integration.

The contract represents the first time a military vehicle developed and manufactured in Australia has been sold in volume to the United States

ADI supplies the ADF with a wide range of systems, equipment and services. This includes the joint command support system used by all ADF headquarters, the majority of communication systems used by the Royal Australian Navy (RAN), ammunition, weapons systems and military vehicles. ADI's mine countermeasures systems are in service with the RAN and seven international navies.

Australian first

ADI is also supplying 40 high-mobility engineering excavator vehicles (HMEE) to the US Army, which has also shortlisted the company for a 1500-vehicle contract for a new-generation HMEE. The contract represents the first time a military vehicle developed and manufactured in Australia has been sold in volume to the United States.

ADI benefits from access to world-leading research and development in technology, tools and processes via its joint-venture owners, Thales and Transfield Holdings. This is a significant advantage in today's technology-dependent environment, particularly as Thales now spends some $A3 billion a year on research and development.

The company conducts business around the world and teams with major international companies. ADI works closely with research organisations DSTO and CSIRO on joint research and development projects to ensure continuing success in product development and export.

ADI's world-class activities include capability assessment, system evaluation and evolution using virtual environment simulations, system specification, design, implementation, integration verification and validation for complex domains such as C4ISR (Command, Control, Communications, Computers, Intelligence, Surveillance and Reconnaissance).

Key disciplines are project and risk management, software engineering, system development, integration, engineering and logistics support.

Application domains include shipbuilding, upgrade and maintenance; mine countermeasures systems; the manufacture of munitions, explosives, chemicals, propellants and weapons; mechanical, light, medium to heavy-precision engineering and fabrication; design and manufacture of military and specialist vehicles; outsourcing; facilities management; provision of logistic services; electronic warfare support; aerospace; and optronics services.

Caring for customers

Customer commitment and attracting and retaining the best people are ADI priorities. A culture of customer commitment is an essential ingredient of a successful company and ADI is focused on building customer loyalty based on long-term satisfaction with the services and products supplied. Capability assessments to link product development more closely to customer needs are proving beneficial to the company.

ADI assigns mentors and coaches to its specialists to develop their innovative skills, accelerate their careers and to keep skills refreshed by the achievements of other companies and other countries. Best practices are evaluated and institutionalised by common efficiency teams run jointly with partners and in consultation with customers.

Customer commitment and attracting and retaining the best people are ADI priorities

With an annual turnover of more than $700 million and a workforce of about 2500, ADI can apply its size, management and technology to delivering large and complex programs. Through close consultation and cooperation between ADI's scientists and company personnel including former military officers, ADI bridges the gap between the world of emerging technologies and the increasingly complex scope of defence needs. ■

Lucio Di Bartolomeo is managing director of ADI.

Developments in defence equipment

Some of ADI's recent innovations include:

■ Adaptation of techniques in military vehicle design to provide the FirKing bushfire fighting vehicle equipped with a unique crew survival cabin.

■ Acoustic signature generator for sweeping minefields.

■ Development of composite materials and structures for minesweepers.

■ Use of web-based architectures and tools for military command support systems.

■ Dynamic management of bandwidth across a military network to maintain a prescribed level of military-quality service.

■ Use of synthetic environments and operational analysis for refining customer requirements and testing solution effectiveness.

By Gayle Bryant

Sounding off

→ Commercialising defence technology and expanding into other industries has paid off for this creator of underwater sonar systems.

Under the pristine, blue surface of the sea is another world. It is a place where a complex system of marine life co-exists with human activities such as oil and gas exploration, and the quiet movement of underwater craft. It is also sometimes an unwitting stage where battles have been won and lost.

It is in this area of defence that navies have relied on sonar technologies since World War II to counter threats posed by submarines, mines and surface ships. A key player in developing such technologies and products is Thales Underwater Systems (TUS).

TUS is the world leader in the design, manufacture and support

Getty Images

Far-reaching innovations: Thales Underwater Systems is entering into new commercial areas.

of sonar systems. Sonar systems either transmit sound or detect energy transmitted by underwater objects. They detect objects by the sound they make or the way sound reflects from them.

TUS' history of providing sonar systems capability to the Australian Defence Force (ADF) began in the early 1970s and today TUS is its major supplier of anti-submarine sonar systems and mine warfare systems. Its systems are operational today aboard the Collins class submarines, the ANZAC frigates, the HUON class mine hunter and the Royal Australian Air Force (RAAF) Maritime Surveillance P3C aircraft.

TUS supplies its products to more than 50 navies and has developed sonar systems for commercial applications industries such as marine oil and gas exploration. Always at the forefront of innovation, TUS was the first to develop the concept of a forward operating self-propelled sonar, PVDS (propelled variable depth sonar).

This concept is the solution to the new threat that will face surface ships and submarines from mines. The vehicle carries a multi-frequency sonar array which is both effective and compact and the efficiency of this system has been demonstrated at sea.

The right partners

Partnerships are important to TUS and it has long associations with the Royal Australian Navy (RAN), the RAAF and the Defence Science and Technology Organisation (DSTO). The company also has a long-term commitment to the Australian Defence Organisation (ADO) to build strategic capabilities for current and future ADO operational requirements.

"The partnerships we have established, especially with DSTO, have enabled us to collaborate in initiatives that have delivered improved capability to the Navy and Air Force," says Ashley Deacon, marketing and sales director with TUS. "At the same time we are developing a strategic Australian industrial capability providing employment and business growth."

TUS is cooperating with DSTO and the Defence Material Organisation on a number of activities from research and development to sea trials and, ultimately, the volume manufacture of many systems including bi-static and multi-static sonobuoys, towed arrays, signal processing algorithms and systems, open architecture systems studies and sonar displays.

In 1997, DSTO and the ADF signed an alliance agreement with TUS to exchange information on sonar systems technology and trends to maximise the operational effectiveness of sonars. This alliance has benefitted the Collins class submarine, allowing new sonar functions supplied by other companies, including small to medium enterprises, to be implemented onboard.

Today, the ADF, DSTO and TUS are cooperating to develop sonar systems such as PETREL (the three-dimensional mine and obstacle avoidance sonar for surface ships) and the TESS 2 Sonar Predication & Performance Modelling system.

Defending innovation

An important area for TUS is torpedo defence. It is supplying the complete Underwater Warfare System for the RAN's FFG-7 Guided Missile Frigate upgrade program. The upgrade includes the world's first torpedo defence and mine and obstacle avoidance systems.

The torpedo defence system in itself is a fundamental requirement as it is the means by which torpedoes are detected and evaluated to enable the correct countermeasures to be applied. Such a system must provide all-round coverage and be capable of detecting the full range of threat weapons that are likely to be encountered.

Torpedo defence has long been regarded as a significant capability shortfall and it is in this regard that modern towed array-based systems are having considerable success. This is particularly evident in the FFG upgrade project where TUS' torpedo defence system – Sea Defender – will provide the ship with the most advanced system available.

> TUS is second only to the Australian Government in the level of funding it directs to R&D in sonar systems

Sea Defender detects possible torpedoes in the noisy environments around ships in time for the ship to take evasive action. The system detects, classifies and tracks torpedoes and advises the ship's commanding officer on tactics to evade the torpedo, including the use of off-board acoustic decoy systems.

"Sea Defender is optimised for own-ship self-defence but many navies are looking to develop the capability for force protection and/or area defence," Deacon says. "This will be achieved by integrating long-range surveillance passive towed arrays with torpedo defence arrays such as Sea Defender and then fusing the outputs of multiple ships' sensors to produce wider coverage than would be achieved by individual ships."

Towed to the future

Meanwhile, the towed array-based surface ship sonar continues to be the most expedient means of providing the surface ship with an acoustic suite that enables the ship to undertake anti-submarine warfare (ASW) missions effectively. In addition to being effective in combating today's threat, the towed array-based system clearly has a growth path into the future. When installed into a networked tactical ASW system, which might include input from airborne sonar systems such as dipping sonar and sonobuoys, the towed variable depth sonar has the capability to enable the surface combatant to contribute as an effective node in a network-centric ASW system. The towed array-based ASW system is almost mandatory for navies seeking to develop real ASW capability against the growing submarine threat.

TUS is a member of the Thales Group and has more than 50 years experience in undersea warfare. It is a world leader in its field, the major exporter of sonars and allied systems for navies and airforces, and is making major advances for civil applications. More than half of the mine warfare vessels in service throughout the world are fitted with systems designed and produced by TUS.

Innovation is key to TUS remaining number one in its field. To ensure its innovations are leading-edge, it spends a considerable amount on research and development activities.

"We spent $A9 million on R&D in 2003," Deacon says. "We are committed to undertaking major self-funded R&D programs in Australia. In fact, TUS is second only to the Australian Government in the level of funding it directs to R&D in sonar systems."

TUS has also established a significant export business in marine oil and gas seismic survey systems, producing more than 700 kilometres of seismic towed arrays since 1996 and generating sales of more than $350 million. ■

By Anthony Schmidt

Taking the high-tech road

→ Road construction has become a high-tech activity with one Australian company now a world leader in servicing the market.

ARRB is taking a new look at road infrastructure.

Getty Images

The complex task of building and maintaining the nation's road infrastructure network relies on an array of technology – a significant amount of which has been developed in Australia by ARRB Transport Research. It is responsible for a range of world firsts including specialist software packages for both the Australian and international road infrastructure markets.

ARRB is regarded as Australia's leading provider of value-added research and technical services to the road and transport industries and has been helping to shape the future of Australia's road infrastructure network for more than 40 years. Its primary focus is on developing tools that help to improve the quality, safety and efficiency of the road network.

According to Gerard Waldron, managing director of ARRB Transport Research, road authorities are now responsible for much more than maintaining the state of our highways. "They are also involved with building, managing and monitoring them," he says. "Increasingly, they rely on our equipment, which ranges from small hand-held testing units through to state-of-the-art multi-function network survey vehicles."

King of the road

He adds the extensive range of specialist testing and monitoring equipment from ARRB is held in such high regard that in some instances it is recognised as the international benchmark for testing accuracy.

"It has become the standard against which other equipment is tested and calibrated. We have a holistic approach to the design, construction, management and maintenance of roads and road transport infrastructure, which has earned us an enviable reputation as a world leader in the field."

Waldron says the company has also developed a number of innovative world-first software packages. "One of these is a system to monitor pavement deterioration and provide asset management modelling," he says. "Another is a package to assist with the setting of appropriate speed limits, and more recently, we have developed a new system for evaluating and prioritising road safety projects."

ARRB is owned by Australia's eight State and Territory government road authorities, the Federal Government's Department of Transport and Regional Services, and the Australian Local Government Association. The organisation is not only a recognised leader in the development of new equipment and technology, it also plays a major role in setting new Australian and international standards in road engineering, transport technology, evaluation, transport management and road safety.

ARRB is considered to be Australia's most extensive and comprehensive "knowledge base" for roads and transport-related matters. Waldron says the organisation's success highlights the benefits of taking an holistic and cooperative approach to research and development that focuses on the development of practical and cost-effective solutions.

"An efficient road network forms the basis of a strong economy and is a key factor in any trading nation's competitiveness," he says. "Whether for the transport of goods and materials, personal travel or other business, everyone relies on transport infrastructure. He adds that life as we know it today could not have developed, nor could it be sustained without roads. "Our success in this area stands as testament to the quality of Australian engineering and technology and of Australia's ability to compete – and win – in the international marketplace."

Anthony Schmidt is managing editor of Highway Engineering in Australia.

By Jane Breusch

The alternative airport

→ Brisbane Airport Corporation's 20-year master plan will transform a regional city airport into a global airport city.

Getty Images

[Airports play a pivotal role in regional economies.

With aircraft becoming quieter, modes of operation becoming more community-friendly and building technology more advanced, the line is blurring between airports and their surrounding communities.

Brisbane Airport Corporation's (BAC) chief executive officer Koen Rooijmans says airports are no longer places just to take off from and land, but are becoming mini-cities in their own right. They represent opportunities for businesses that need strategic access to transport infrastructure and for aviation-related industries.

BAC's master plan outlines a vision for a global airport city. This includes the leading-edge Number 1 Airport Drive precinct, which includes a factory outlet centre, homemaker centre, golf course, convenience retail stores, childcare, health and fitness facilities, restaurants, cafes, hotels, corporate offices, car parking and possibly even a theme park.

"It is now more important than ever for the modern airport to reinvent itself and to achieve a balance between providing critical economic infrastructure and an efficient, profitable airport for shareholders," Rooijmans says. "Blessed with the sort of land, cost structures and scope for developments that other airports can only dream of, Brisbane is in many ways blazing a trail for other airports to follow."

BAC bought the 2700-hectare airport in 1997 to capitalise on prime development land that was less than 15 kilometres from Brisbane's growing central business district. Rooijmans says the airport has a number of other advantages, including being free of curfews, adjacent to a major international seaport, on the primary north-south road corridor and, since a few years ago, rail access.

"Internationally, the modern airport is becoming a destination in its own right: a tourism drawcard, a business centre, a commercial hub and a key player in outward-focused economies," he says. "I have worked at many international airports and I have to say Brisbane is at the leading edge of advanced airport planning."

He says airports are also beginning to play a more pivotal role in driving regional economies, building communities and creating wealth. Brisbane airport accounts for nearly 8 per cent of Queensland's entire gross domestic product.

There are about 130 businesses with nearly 8000 full-time staff at Brisbane airport, which is used by 13 million passengers each year. Those numbers are forecast to exceed 35,000 workers and 35 million passengers within 20 years. This is the equivalent of the workforce of a regional Australian city, but with a focus on "smart and sustainable" jobs in sunrise industries such as aviation and aerospace support, as well as the more traditional industries of tourism, transport and freight.

"Around successful airports you see cities growing and this is in part an organic and a managed process," Rooijmans says. "Airports provide economic benefits for the growth of the cities that surround them and create a critical mass of economic activity, generating jobs and opportunity."

He adds that as other major airports feel the pressure of urban development, curfews, costs and congestion, Brisbane is already becoming a realistic alternative for both freight and passenger aviation into and out of Australia. ■

Total concept

Brisbane Airport Corporation (BAC) is different from other Australian airport operators because it offers a complete package with its plan for future development.

Some of the inspiration came from Amsterdam Airport Schiphol, which has a 15 per cent stake in BAC and operates a successful model of a global airport city. This model provided the basis for Brisbane's complete package of shopping, corporate offices, hotels and leisure activities.

"You see elements of it at other airports but you don't see the total concept," BAC chief executive officer Koen Rooijmans explains. "We have identified seven different precincts and we focus every precinct in a different way. I believe that total package concept is unique here in Australia."

By Tim Treadgold

Reinventing the wheel

→ An engineering company that nearly sank has created opportunities from a fresh approach to innovation.

The technologies developed by Perth-based Orbital Engine Corporation have found their way into more than 30 different products worldwide and are instrumental in assisting the environment in the areas of reduced fuel consumption. Orbital first floated on the promise of radical new automotive engine technologies but the expected opportunities did not materialise. The business is now fighting back with a far simpler plan designed to deliver financial stability before a new era of growth.

Engineering services, technology income from licence fees and royalties, and a 50 per cent stake in Synerject, a joint venture with Siemens VDO, form the three legs of the restructured organisation. Orbital now calls itself an intellectual property company. It has developed world-leading direct fuel injection, combustion and control system technologies collectively termed the Orbital Combustion Process.

"It's been a difficult process but we are now on a much firmer footing," says Orbital chief executive, Peter Cook. "The 2003 financial year was our turning point. We have right-sized the company, cut costs and been able to stem the heavy losses and cash burn which had drained our resources."

Two years ago, Orbital recognised the need to find a new direction. Major changes were made at the board level to management and to its workforce. With a background in product development, Cook was recruited and given the job of reshaping Orbital or, in his words, "right-sizing the business" to spur innovation.

"Emerging now is a business with multiple income streams which will provide the immediate cash we need as a business, while the technologies we have developed are progressively adopted by the world's leading engine technology firms," he says.

Orbital has been a household name in Australia since 1972 when it was founded by Perth inventor, Ralph Sarich. His "Orbital" engine was never developed to a commercially viable level and he has long since retired from the business. However, the company has continued to develop a number of spin-off technologies, most notably in the area of direct fuel injection systems. These spin-off technologies are today found in products such as Mercury outboards and Vespa scooters.

Reversal of fortune

The new-look Orbital, which has retained its technical heart and soul, is still located in the Perth industrial suburb of Balcatta, but has shrunk from a peak workforce of 380 to 100.

A loss of $A26.8 million in the 2002 financial year has been reversed, with the second half of the 2003 year expecting Orbital to switch from red to black ink with a modest, but significant, $1 million profit.

"Fee-for-service engineering has been the major contributor in this phase of our revival," explains Cook. "We have secured a series of contracts with some of the world's leading original equipment manufacturers which recognise Orbital's world-class people with a proven track record of innovation that is supported by some of the world's best facilities and testing equipment."

Orbital technology reduces exhaust emissions by 70 per cent in this watercraft.

While the business has been "right-sized" to handle the contract engineering research and development work, the original core of Orbital – its own technologies represented by hundreds of patents, has not been forgotten.

"We are enjoying a steady income stream from royalties and licence fees," says Cook. "This will grow as our technologies gain greater market penetration, a process which always takes time but will deliver handsome rewards as it gains pace. The future, given the solid financial base we have created, and the tremendous technical skills we have developed, looks brighter than it has for many years." ■

As the births of living creatures, at first, are ill-shapen, so are all innovations, which are the births of time.

FRANCIS BACON

Contact information

3M Australia
950 Pacific Highway
Pymble NSW 2073
Ph: +61 2 9498 9333
Fax: +61 2 9498 9666
Website: www.3m.com.au

Aon
Head Office
201 Kent Street
Sydney NSW 2000
Ph: +61 2 9253 7000
Fax: +61 2 9253 7001
Email: enquiry@aon.com.au
Website: www.aon.com.au

ADI Limited
Level 2 Building 51
Garden Island NSW 2011
Ph: +61 2 9562 3333
Fax: +61 2 9562 2390
Website: www.adi-limited.com

ARRB Transport Research Ltd
500 Burwood Highway
Vermont South VIC 3133
Ph: +61 3 9881 1555
Fax: +61 3 9887 8104
Email: info@arrb.com.au
Website: www.arrb.com.au

Amcor Australasia
971-973 Burke Road
Camberwell VIC 3124
Private Bag 34
Camberwell VIC 3124
Ph: +61 3 9811 7111
Fax: +61 3 9811 7171
Email: david.brookes@amcor.com.au
Website: www.amcor.com

ATP Innovations Pty Ltd
Suite 145
National Innovation Centre
Australian Technology Park
Eveleigh NSW 1430
Ph: +61 2 9209 4444
Fax: +61 2 9319 3874
Email: info@atp-innovations.com.au
Website: www.atp-innovations.com.au
www.biznetclub.com.au
www.bizcapital.com.au

ANSTO (Australian Nuclear Science and Technology Organisation)
New Illawarra Road
Lucas Heights NSW 2234
PMB 1
Menai NSW 2234
Ph: +61 2 9717 3626
Fax: +61 9717 9272
Email: commercial@ansto.gov.au
Website: www.ansto.gov.au

Australia Post
321 Exhibition Street
Melbourne VIC 3000
Ph: +61 3 9204 7171
Fax: +61 3 9663 1160
Website: www.auspost.com.au

Australian Academy of Science
GPO Box 783
Canberra ACT 2601
Ph: +61 2 6247 5777
Fax: +61 2 6257 4620
Email: eb@science.org.au
Website: www.science.org.au

Australian Industry InnovationXchange Network
Level 1
51 Walker Street
North Sydney NSW 2060
Ph: +61 1800 332 889
Fax: +61 2 9466 5599
Email: info@ixc.com.au
Website: www.innovationxchange.com.au

Australian Business Limited
Level 15
140 Arthur Street
North Sydney NSW 2060
PO Box 938
North Sydney NSW 2059
Ph: 02 9458 7500
Fax: 02 9954 5007
Email: patricia.birrane@australianbusiness.com.au
Website: www.australianbusiness.com.au

Australian Institute for Commercialisation
Brisbane Technology Park
1 Clunies Ross Court
Eight Mile Plains QLD 4113
Ph: +61 7 3853 5225
Fax: +61 7 3853 5226
Email: info@ausicom.com
Website: www.ausicom.com

Australian Bureau of Statistics
ABS House
45 Benjamin Way
Belconnen ACT 2617
Locked Bag 10
Belconnen ACT 2616

Australian Institute of Marine Science
PMB No 3
Townsville MC QLD 4810
Ph: +61 7 4753 4444
Fax: +61 7 4772 5852
Email: reception@aims.gov.au
Website: www.aims.gov.au

Australian Centre for International Agricultural Research
GPO Box 1571
Canberra ACT 2601
Ph: +61 2 6217 0500
Fax: +61 2 6217 0501
Email: aciar@aciar.gov.au
Website: www.aciar.gov.au

The Australian National University
Canberra ACT 0200
Ph: +61 2 6125 5111
Website: www.anu.edu.au

Australian Genome Research Facility Ltd
c/- Gehrmann Laboratories
Research Road
The University of Queensland
Brisbane QLD 4072
Ph: +61 7 3365 4448
Fax: +61 7 3365 1823
Email: agrf@agrf.org.au
Website: www.agrf.org.au

Australian Proteome Analysis Facility Ltd
Building F7B
Macquarie University NSW 2109
Ph: +61 2 9850 6201
Fax: +61 2 9850 6200

Australian Research Council
Cnr Jerrabomberra Avenue
and Hindmarsh Drive
Symonston ACT 2609
GPO Box 2702
Canberra ACT 2601
Ph: +61 2 6284 6600
Fax: +61 2 6284 6601
Email: info@arc.gov.au
Website: www.arc.gov.au

Bligh Voller Nield
Level 6
11-31 York Street
Sydney NSW 2000
PO Box N646
Grosvenor Place NSW 1220
Ph: +61 2 8297 7200
Fax: +61 2 8297 7299
Email: maria_ferrara@bvn.com.au
Website: www.blighvollernield.com

Australian Trade Commission (Austrade)
GPO Box 5301
Sydney NSW 2001
Ph: +61 13 28 78
Fax: +61 2 9390 2800
Email: info@austrade.gov.au
Website: www.austrade.gov.au

BlueScope Steel
Level 11
120 Collins Street
Melbourne VIC 3000
Ph:+61 3 9666 4000
Fax: +61 3 9666 4111
Email: BlueScopeSteelDirect@BlueScopeSteel.com
Website: www.bluescopesteel.com

Bayer CropScience
391-393 Tooronga Road
East Hawthorn VIC 3123
Ph: +61 3 9248 6888
Fax: +61 3 9248 6800
Email: enquiries.australia@bayercropscience.com
Website: www.bayercropscience.com.au

Boral Limited
Level 39 AMP Centre
50 Bridge Street
Sydney NSW 2000
Ph: +61 2 9220 6300
Fax: +61 2 9233 6605
Email: info@boral.com.au
Website: www.boral.com.au

BioMelbourne Network
Milton House
25 Flinders Lane
Melbourne VIC 3000
Ph: +61 3 9650 8800
Fax: +61 3 9650 6066
Email: info@biomelbourne.org
Website: www.biomelbourne.org

Bovis Lend Lease
PO Box H186
Australia Square NSW 1215
Ph: +61 1800 850 922
Fax: +61 2 9277 8086

Bishop Technology Group Limited
10 Waterloo Rd
North Ryde NSW 2113
Ph: +61 2 94916000
Fax: +61 2 98786373
E-mail: bruce.grey@aebishop.com
Website: www.aebishop.com

Brisbane Airport Corporation Limited
Banksia Place
Brisbane Airport QLD 4007
Ph: +61 7 3406 3000
Fax: +61 7 3406 3111
Email: info@bne.com.au
Website: www.brisbaneairport.com.au

Canon Australia Pty Ltd
1 Thomas Holt Drive
North Ryde NSW 2113
Ph: +61 2 9805 2000
Fax: +61 2 9888 3650
Website: www.canon.com.au

Cochlear Limited
14 Mars Road
Lane Cove NSW 2066
Ph: +61 2 9428 6555
Fax: +61 2 9428 6352
Website: www.cochlear.com

Commonwealth Bank
Level 7
48 Martin Place
Sydney NSW 1155
Ph: +61 2 9378 2663
Fax: +61 2 9378 3146
Email: fitzgeb@cba.com.au
Website: www.commbank.com.au

The Como Melbourne
630 Chapel Street
South Yarra VIC 3141
Ph: +61 3 9825 2222
Fax: +61 3 9824 1263
Email: reservations@como.mirvac.com.au
Website: www.mirvachotels.com.au

CSIRO
The Gatehouse
Bayview Avenue
Bag 10
Clayton South VIC 3169
Ph: 1300 363 400
Fax: +61 3 9545 2175
Email: enquiries@csiro.au
Website: www.csiro.au

Curtin University of Technology
Office of Research & Development
GPO BOX U1987
PERTH WA 6845
Ph: +61 8 9266 3045
Fax: +61 8 9266 3048
Email: b.glover@curtin.edu.au
Website: www.curtin.edu.au

Defence Science and Technology Organisation (DSTO)
Level 6, Building 1
Russell Offices (R1-6-A055)
Department of Defence
Canberra ACT 2600
Ph: +61 2 6265 7905
Fax: +61 2 6265 2741
Email: assier@cbr.defence.gov.au
Website: www.dsto.defence.gov.au

Department of the Premier & Cabinet
Office of Science and Innovation
Government of Western Australia
2nd Floor
197 St George's Terrace
Perth WA 6000
Ph: +61 8 9222 8880
Fax: +61 8 9222 8888
Email: bhobbs@dpc.wa.gov.au
Websites: www.sciencecouncil.dpc.wa.gov.au
www.dpc.wa.gov.au
www.scitech.org.au

Department of Primary Industries, Government of Victoria
PO Box 500
East Melbourne VIC 3002
Ph: +61 3 9637 8000
Fax: +61 3 9637 8119
Email: george.riffkin@dpi.vic.gov.au
Website: www.dpi.vic.gov.au

Drake International Pty Ltd
Level 31
55 Collins Street
Melbourne VIC 3000
Ph: +61 3 9245 0245
Fax: +61 3 9245 0231
Email: marketing@au.drakeintl.com
Website: www.drakeintl.com

DuPont (Australia) Ltd
Level 16
168 Walker Street
North Sydney NSW 2060
Ph: +61 2 9923 6111
Fax: +61 2 9923 6011
Email: enquiries@aus.dupont.com
Website: www.dupont.com.au

Feelike Pty Ltd
32 View Street
Annandale NSW 2038
Ph: +61 2 9566 1283
Fax: +61 2 9566 1593
Email: hans.wijgh@feelike.com
Website: www.feelike.com

Edison Mission Energy Holdings Pty Ltd
Level 20 HWT Tower
40 City Road
South Melbourne VIC 3205
Ph: +61 3 9696 6477
Fax: +61 3 9696 8420
Email: rdriscoll@edisonmission.com
Website: www.edison.com

Financial Planning Association of Australia
PO Box 109
Collins Street West
Melbourne VIC 8007
Level 8
570 Bourke Street
Melbourne VIC 3000
Ph: +61 3 9627 5200
Fax: +61 3 9627 5280
Email: fpa@fpa.asn.au
Website: www.fpa.asn.au

Electrolux Home Products Pty Ltd
175 Bonds Road
Riverwood NSW 2210
Ph: +61 2 9717 2111
Fax: +61 2 9717 2222
Website: www.electrolux.com.au

Gartner Australasia Pty Ltd
Level 7
40 Miller Street
North Sydney NSW 2060
Ph: +61 2 9459 4600
Fax: +61 2 9459 4601
Email: ap.gartnerweb@gartner.com
Website: www.gartner.com

EMC Corporation
Level 6
60 Miller Street
North Sydney NSW 2060
Ph: +61 2 9925 7800
Fax: +61 2 9922 4287
Email: emc_infoanz@emc.com
Website: www.anz.emc.com

General Property Trust
The Bond
30 Hickson Road
Sydney NSW 2000
Ph: +61 2 9236 6999
Fax: +61 2 9236 6505
Email: GPT@Lendlease.com.au
Website: www.gpt.com.au

Ergon Energy Pty Ltd
61 Mary Street
Brisbane QLD 4000
Ph: +61 7 3228 8222
Fax: +61 7 3228 7578
Email: business@ergon.com.au
Website: www.ergon.com.au

Grains Research and Development Corporation
PO Box 5367
Kingston ACT 2604
Ph: +61 2 6272 5525
Fax: +61 2 6271 6430
Email: grdc@grdc.com.au
Website: www.grdc.com.au

Griffith Hack
Level 3
509 St Kilda Road
Melbourne VIC 3004
Ph: +61 3 92438300
Fax: +61 3 92438333
Email: ghmelb@griffithhack.com.au
Website: www.griffithhack.com.au

The Institute of Chartered Accountants in Australia
National Office
Level 14
37 York Street
Sydney NSW 2000
Ph: +61 1300 137 322
Email: support@icaa.org.au
Website: www.icaa.org.au

Horticulture Australia Limited
Level 1
50 Carrington Street
Sydney NSW 2000
Ph: +61 2 8295 2300
Fax: +61 2 8295 2399
Email: info@horticulture.com.au
Website: www.horticulture.com.au

The Institute of Patent and Trade Mark Attorneys of Australia
711 High Street
East Kew VIC 3102
Ph: +61 3 9857 0311
Fax: +61 3 9857 0411
Email: mail@ipta.com.au
Website: www.ipta.com.au

Hydro Tasmania
4 Elizabeth Street
Hobart TAS 7000
Ph: 1300 360 441
Fax: +61 3 6230 5823
Email: webmaster@hydro.com.au
Website: www.hydro.com.au

Invetech Pty Ltd
495 Blackburn Road
Mt Waverley VIC 3149
Ph: +61 3 9211 7700
Fax: +61 3 9211 7703
Email: kmc@invetech.com.au
Website: www.invetech.com.au

IDP Education Australia
Level 4
210 Clarence Street
Sydney NSW 2000
Ph: +61 2 8251 2700
Fax: +61 2 8251 2728
Email: info@sydney.idp.com
Website: www.idp.com

John Curtin School of Medical Research
The Australian National University
54 Mills Road
Canberra ACT 0200
Ph: +61 2 6125 2550
Fax: +61 2 6125 2337
Email: director@jcsmr.anu.edu.au
Website: www.jcsmr.anu.edu.au

IP Australia
PO Box 200
Woden ACT 2606
Ph: +61 1300 65 1010
Fax: +61 2 6282 5810
Email: assist@ipaustralia.gov.au
Website: www.ipaustralia.gov.au

Johnson & Johnson Family of Companies
1-5 Khartoum Road
North Ryde NSW 2113
Ph: +61 1800 226 334
Fax: +61 2 8875 3300
Website: www.jnj.com

Lend Lease
PO Box H186
Australia Square NSW 1215
Ph: +61 2 9236 6111
Fax: +61 2 9277 8086
Website: www.Lendlease.com.au

National Health and Medical Research Council
32 Corinna Street
Woden ACT 2606
Ph: +61 2 6289 9184
Fax: +61 2 6289 9197
Email: exec.sec@nhmrc.gov.au
Website: www.nhmrc.gov.au

The Macfarlane Burnet Institute for Medical Research and Public Health (Burnet Institute)
Commercial Road
Melbourne VIC 3004
Ph: +61 3 9282 2111
Fax: +61 3 9282 2100
Email: burnet@burnet.edu.au
Website: www.burnet.edu.au

National ICT Australia
Bay 15 Australian Technology Park
Garden Street
Eveleigh NSW 1430
Ph: +61 2 9209 4750
Fax: +61 2 9209 4748
Email: enquiries@nicta.com.au
Website: www.nicta.com.au

Macquarie Bank Limited
No 1 Martin Place
Sydney NSW 2000
Ph: +61 2 8232 3333
Fax: +61 2 8232 7780
Website: www.macquarie.com

Orbital Engine Corporation
4 Whipple Street
Balcatta WA 6021
Ph: +61 8 9441 2311
Fax: +61 8 9441 2111
Email: info@orbeng.com
Website: www.orbeng.com

Mayne Health Diagnostic Imaging
60 Waterloo Road
North Ryde NSW 2113
Ph: +61 2 9005 7810
Fax: +61 2 9005 7706
Email: jane.gregory@maynegroup.com
Website: www.maynegroup.com

Orica Limited
Level 9
1 Nicholson Street
East Melbourne VIC 3002
Ph: +61 3 9665 7111
Fax: +61 3 9665 7937
Email: companyinfo@orica.com
Website: www.orica.com

National Australia Bank
National @ Docklands
800 Bourke Street
Docklands VIC 3008
Ph: + 61 3 8641 3500, Peter Affleck
+61 2 9957 8000, Rosemary Kirkby
Email: peter_affleck@national.com.au
rosemary_kirkby@mlc.com.au
Website: www.national.com.au

Pacifica Group Technologies Pty Ltd
264 East Boundary Road
East Bentleigh VIC 3165
Ph: +61 3 9575 2022
Fax: +61 3 9575 2555
Email: pgt_enquiries@pgt.com.au
Website: www.pgt.com.au

The Pharmacy Guild of Australia
PO Box 7036
Canberra Business Centre ACT 2610
Level 2
15 National Circuit
Barton ACT 2600
Ph: +61 2 6270 1888
Fax: +61 2 6270 1800
Email: guild.nat@guild.org.au
Website: www.guild.org.au

RMIT University
PO Box 71
Bundoora VIC 3083
Ph: +61 3 9925 7233
Fax: +61 3 9466 9980
Email: sophie.james@rmit.edu.au
Website: www.rmit.edu.au

PricewaterhouseCoopers
Website: www.pwc.com/au

Science, Technology and Innovation Directorate
Department of Further Education, Employment, Science and Technology
Government of South Australia
Level 11
1 King William Street
Adelaide SA 5000
Ph: +61 8 8303 2133
Fax: +61 8 8303 2135
Website: www.sa.gov.au

Queensland Institute of Medical Research
The Bancroft Centre
300 Herston Road
Herston QLD 4006
Ph: +61 7 3362 0222
Fax: +61 7 3362 0111
Email: info@qimr.edu.au
Website: www.qimr.edu.au

SC Johnson & Son Pty Ltd
160 Epping Road
Lane Cove NSW 2066
Ph: +61 2 9428 9111
Fax: +61 2 9428 9264
Email: amiedle@scj.com
Website: www.scjohnson.com.au

RABiT (Research Centre for Advanced By-Wire Technologies)
264 East Boundary Road
East Bentleigh VIC 3165
Ph: +61 3 9575 2014
Fax: +61 3 9575 2432
Email: info@rabit.com.au
Website: www.rabit.com.au

Spotless Group Limited
Level 3
350 Queen Street
Melbourne VIC 3000
Ph: +61 3 9269 7600
Fax: +61 3 9269 7603
Email: enquiries@spotless.com.au
Website: www.spotless.com

ResMed Ltd
97 Waterloo Road
North Ryde NSW 2113
Ph: +61 2 9886 5000 or +61 1800 658 189
Fax: +61 2 9878 0120
Email: reception@resmed.com.au
ResMed Corp
14040 Danielson Street
Poway CA 92064-6857 USA
Ph: +1 858 746 2400 or 1 800 424 0737
Fax: +1 858 746 2900
Email: reception@resmed.com
Website: www.resmed.com

Starpharma Pooled Development Limited
Level 6 Baker Heart Research building
Commercial Road
Melbourne VIC 3004
PO Box 6535
St Kilda Road
Central VIC 8008
Ph: +61 3 8532 2700
Fax: +61 3 9510 5955
Email: info@starpharma.com
Website: www.starpharma.com

Standards Australia
286 Sussex Street
Sydney NSW 2000
Ph: +61 2 8206 6000
Fax: +61 2 8206 6001
Email: mail@standards.org.au
Website: www.standards.org.au

Titan Resources NL
Level 1
24 Outram Street
West Perth WA 6005
Ph: +61 8 9481 6040
Fax: +61 8 9481 6035
Email: titan@titanresources.com.au
Website: www.titanresources.com.au

Steritech Pty Ltd
160 South Gippsland Highway
Dandenong VIC 3175
Tel: +61 3 9793 5566
Fax: +61 3 9701 3158
Email: gwest@steritech.com.au
Website: www.steritech.com.au

University of Ballarat
University Drive
Mt Helen VIC 3353
PO Box 663
Ballarat VIC 3353
Ph: +61 3 5327 9100
Fax: +61 3 5327 9544
Email: ub@ballarat.edu.au
Website: www.ballarat.edu.au

Swinburne University of Technology
Hawthorn Campus
John Street
Hawthorn VIC 3122
Ph: +61 3 9214 8063
Fax: +61 3 9214 5483
Email: fpr@swin.edu.au
Website: www.swin.edu.au/corporate/fpr

The University of Queensland
Brisbane QLD 4072
Ph: +61 7 3365 3367
Fax: +61 7 3365 1488
Email: marketing.communications@uq.edu.au
Website: www.uq.edu.au

Telstra Corporation Limited
242 Exhibition Street
Melbourne VIC 3000
Website: www.telstra.com

The University of Sydney
NSW 2006
Ph: +61 2 9351 2222
Fax: +61 2 9351 4596
Email: cbrown@mail.usyd.edu.au
Website: www.usyd.edu.au

Thales Underwater Systems
274 Victoria Road
Rydalmere NSW 2116
Ph: + 61 2 9848 3500
Fax: + 61 2 9848 3720
Email: enquiries@au.thalesgroup.com
Website: www.thales-underwater.com

University of Southern Queensland
Toowoomba QLD 4350
Ph: +61 1800 811 380
Fax: +61 7 4631 2893
Website: www.usq.edu.au

University of Technology, Sydney
No. 1 Broadway
Sydney NSW 2007
PO Box 123 Broadway
Sydney NSW 2007
Ph: +61 2 9514 2000
Fax: +61 2 9514 1551
Email: info.office@uts.edu.au
Website: www.uts.edu.au

University of Western Sydney
Locked Bag 1797
Penrith South DC NSW 1797
Ph: +61 2 9685 9800
Fax: +61 2 9685 9850
E-mail: obd@uws.edu.au
Web: www.uws.edu.au/obd

The Walter and Eliza Hall Institute of Medical Research
1G Royal Parade
Parkville VIC 3050
Ph: +61 3 9345 2555
Fax: +61 3 9347 0852
Website: www.wehi.edu.au

WMC Resources Ltd
Level 16 IBM Centre
60 City Road
Southbank VIC 3006
Ph: +61 3 9685 6000
Fax: +61 3 9686 3569
Email: public.affairs@wmc.com
Website: www.wmc.com

Any sufficiently advanced technology is indistinguishable from **magic**.

ARTHUR C CLARKE

I made discoveries by always thinking about them.

I keep the subject constantly before me

and wait till the first dawnings

open little by little into the full light.

ISAAC NEWTON

William Henry Bragg

William Lawrence Bragg

Howard Walter Florey

Frank Macfarlane Burnet

John Carew Eccles

Aleksandr Mikhailovich Prokhorov

Bernard Katz

Patrick White

John Warcup Cornforth

Peter Charles Doherty